Organic Bioelectronics for Life Science and Healthcare

Edited by

Akio Yasuda, Technology Consultant
Wolfgang Knoll, Austrian Institute of Technology

Published by **Materials Research Forum LLC**
Millersville, PA 17551, USA

Published as part of the book series
Materials Research Foundations
Volume 56 (2019)
ISSN 2471-8890 (Print)
ISSN 2471-8904 (Online)

Print ISBN 978-1-64490-036-9
eBook ISBN 978-1-64490-037-6

Distributed worldwide by

Materials Research Forum LLC
105 Springdale Lane
Millersville, PA 17551
USA
http://www.mrforum.com

Manufactured in the United States of America
10 9 8 7 6 5 4 3 2 1

Table of Contents

"For Yuri and Christine"

Preface

1. Purpose and background

Looking deeply into the organic molecule-based fusion domain of bio and electronics in its history of evolution and practical uses, the new horizon of this growing field is reviewed and clearly described by the distinguished experts in the field of today.

Organic molecule-based active electronic devices are in the market already built up represented by liquid crystal displays(LCDs) for TVs, monitors, smartphones and watch applications. The LCDs practical applications started in calculators and wrist watches by Sharp Corporation and Suwa Seiko Corporation, Japan in 1973, respectively. The quality of the display was not so good at first actually but the low power consumption, thin device structure and battery operation were the point of interest for the users. Furthermore, thanks to the development of a-Si, TV application was realized 1980`s and gradually overcoming CRT(cathode ray tube) TVs. Today we are not able to imagine the display without LCDs. The success of LCDs can be analyzed that there are applications for LCDs which cannot be realized by conventional displays in the terms of power consumption, thickness and cost although there were always some excuse in the display quality. The history of LCDs gives us an important message that a novel device needs novel applications. This message can be adapted to the organic bioelectronic devices.

One of the important purpose of this book is to share the state of the art materials and device technologies with the application of the novel bioelectronic device in the field of life science and healthcare, where there are lots of sensing targets still remaining.

Life science and healthcare applications are shown and explained with the use of concrete examples as well as needs in those fields. Especially from the sensing point of view, the detection methods and biomarker molecules, together with sensor devices with the working principles, to get biological information such as vital signs, cell behavior, protein and DNA molecule concentration are described in detail.

The advantages of organic materials are also shown by examples and explained in molecular base in terms of not only high efficiency, high sensitivity and flexibility but also biocompatibility and novel fabrication method as printing and coating at low cost.

2. Book chapter structure

There are three sections in this book.

Section1: Organic Field Effect Transistors based sensors

Section2: Carbon based biosensors

Section3: Applications of biosensing technologies

Each section consists of materials, device structure, device working mechanism and biosensing applications except Section 3. Section 3 consists of fabrication method specific for organic devices, their performance and point of care applications for practical uses. Each section is listed with key topics, chapter title and corresponding author.

(1) Organic Field Effect Transistors (OFETs) basics and biosensing applications

Chapter 1: FBI-OFET (Functional Bio Interlayer OFET) (Organic Bioelectronic Transistors: from fundamental investigation of bio-interfaces to highly performing biosensors. Luisa Torsi, et.al.)

Chapter 2: EGOFET (Electrolyte-gated OFET) (Biosensing with electrolyte Gated Organic Field Effect Transistors. Carlo Augusto Bortolotti, et.al.)

Chapter 3: OCMFET (Organic Charge-Modulated FET) The Organic Charge-Modulated Field-Effect Transistor: a flexible platform for application in biomedical analyses. Annalisa Bonfiglio, et.al.)

(2) Carbon based biosensors

Chapter 4: Graphene basics and FET (Graphene based field effect transistors for biosensing: importance of surface modification. Sabine Szunerits, et.al)

Chapter 5: Graphene based FET and MEA biosensing (Graphene as an Organic and Bioelectronic Materials. Andreas Offenhaeusser, et.al.)

Chapter 6: Graphene based Materials and sensor device applications (Graphene based Materials for Bioelectronics and Healthcare. Vivek Pachauri, et.al.)

(3) Applications of biosensing technologies

Chapter 7: Inkjet printing technology for biosensing (Inkjet Printing for Biosensors and Bioelectronics. Bernhard Wolfrum et.al.)

Chapter 8: Point of Care for Stroke, comparison of different systems (Rapid Point-of-Care-Tests for Stroke Monitoring. Alfred I.Y.Tok, et.al.)

(4) Summary and Future Outlook

Finally, the future outlook which is going to be opened up by organic bioelectronics will be discussed based on up-to-date science and technology achievements in this world.

Guidance for the readers

The reader who wants to know

1) the organic transistor materials and working principles of organic transistors to get comprehensive understanding : Chapter 1 covers organic transistor review in detail.

2) more specific organic FET based sensing : Chapter 2 and 3

3) carbon based(carbon nanotube, graphene variation) sensing devices and material modifications relevant to sensing capability. Starting from Chapter 4 to 6 may give you the cutting edge situation of carbon based device sensing.

4) the fabrication method using printing: Chapter 7

5) healthcare application especially for the point of care application ; Chapter 8

We hope this book collecting the state of the art of organic bioelectronics for life science and healthcare materials, devices and application can give a trigger to open up the new horizon of the exciting field as LCDs has been doing.

Acknowledgement

The editors, Akio Yasuda and Wolfgang Knoll, would like express their sincere thanks to all the authors contributing to the book. We are grateful to Mr. Thomas Wohlbier, the publisher, for his patience and giving us the chance to publish this book.

This book would not have been completed without the support from Yuri and Christine.

September 2019,
Akio Yasuda and Wolfgang Knoll - Editors

Organic Bioelectronics for Life Science and Healthcare
Materials Research Foundations **56** (2019) 1-70

Materials Research Forum LLC
doi: https://doi.org/10.21741/9781644900376-1

Chapter 1

Organic Bioelectronic Transistors: From Fundamental Investigation of Bio-Interfaces to Highly Performing Biosensors

Eleonora Macchia[1], Rosaria A. Picca[1], Angelo Tricase[1], Cinzia Di Franco[2], Antonia Mallardi[3], Nicola Cioffi[1], Gaetano Scamarcio[4], Gerardo Palazzo[1,5] and Luisa Torsi[1]*

[1]Dipartimento di Chimica –Università degli Studi di Bari– Bari (I)

[2] CNR - Istituto di Fotonica e Nanotecnologie, Sede di Bari (I)

[3] CNR-IPCF, Istituto per i Processi Chimico-Fisici - Via Orabona, 4 70126 Bari, Italy

[4] Dipartimento InterAteneo di Fisica "M. Merlin" – Università degli Studi di Bari - Bari (I)

[5] CSGI (Centre for Colloid and Surface Science) – Bari (I)

* luisa.torsi@uniba.it

Abstract

Interfacing biomaterials to electronic devices is one of most challenging research field that has relevance for both fundamental studies and for the development of highly performing biosensors. Important aspects connected to electronic biosensors are discussed. In this chapter different electronic biosensors based on field effect transistors, including the Functional Bio-Interlayer sensing platform, are presented. Particular attention is paid to the biosensors operation mechanism and to its reflect on the device analytical figure of merits. The main features of thermodynamic of ligand–bioreceptor interaction confined at solid–liquid interfaces along with the strategies for biomolecule deposition on the biosensors transducer surface are presented.

Keywords

Organic Transistors, Biosensors, Surface Confined Proteins, Functional Bio-Interlayers, Analytical Bioassay, Electronic Bio-Detection

Organic Bioelectronics for Life Science and Healthcare   Materials Research Forum LLC
Materials Research Foundations **56** (2019) 1-70   doi: https://doi.org/10.21741/9781644900376-1

Contents

1.   Introduction

Organic bioelectronics is a highly interdisciplinary research field, whose main focus is on the development of key technologies to gain insight into different biological interfaces as well as to face novel challenges in life sciences [1a]. It currently deals with the development of biosensing devices [2] and of novel tools to study neural interfaces [3] as well as other applications as detailed in recent reviews [1b]. The study of mechanisms involved in processes such as those governing the biochemical interactions between a recognition element (R) integrated into an electronic device, and its affinity ligand (L) plays a key role in bioelectronic devices. The study of R-L binding of biological species confined on a surface is critically important, though very seldom investigated. In fact, proteins confined on a surface are routinely being used in several key relevant biological measurements in laboratory and clinical assays. Just to mention a few of them, the cases of enzyme-linked immunosorbent assay (ELISA) [4] or of surface-plasmon resonance (SPR) [5] methods con be taken as examples of gold standards in bioassays. Indeed, they rely on biomolecules deposited on solid surfaces, such as silica for ELISA and gold for SPR. From a more fundamental point of view, these techniques show that surface-segregated biological species can retain their biofunctional activity. More interestingly, besides the abovementioned artificially produced bio-systems, Nature offers us many examples of surface confined proteins, performing fundamental biological activities. Indeed, cells and organelles are enclosed by lipid bilayers, which separate them from their external environments. Embedded in these membranes there are specialized proteins that span the full width of the bilayer and facilitate communication between the two sides. These membrane proteins carry out a multitude of tasks, including transportation of small molecules, catalysis of enzymatic reactions and signal transduction [6]. In this perspective, bioelectronics may allow to achieve important pieces of information on biofunctional interfaces. The acquired knowledge can also be critically important to realize ultrasensitive biological and chemical sensors [7]. As far as the sensors are concerned, the digital quality of the analytical electronic response and the ultralow detection limits foreseen will offer the possibility of performing reliable quantitative assay of proteins and biomarkers in general in biological fluids such as saliva, blood, urine or tears. Eventually, a label-free, non-invasive strip test can be conceived, based on bioelectronics sensors printed at low cost. Therefore, bioelectronics would pave the way to the so called point-of-care (POC) testing. Such a novel approach aims at the development of clinical tests directly performed at or near the site of patient care, namely at the medical doctor's office or even the patient's house, allowing for timely beginning of an appropriate therapy and/or eventually facilitating the linkages to intensive care units. POC tests should in fact be simple enough to be used for screening in primary care clinics and in remote settings where no laboratory infrastructures are present. POC therefore has the potential to improve the management of diseases as well as of regular medical check-up testing. To this end, organic electronic could be the privileged choice to fabricate such devices. Several research groups have contributed to the advancement in the field, proposing novel platforms that integrate a layer of biological recognition or functional elements, directly coupled to one of the electronic interfaces of an organic thin-film transistor (OTFT). In general, OTFT sensors use π-conjugated organic semiconductors (OSCs) as electronic materials and are endowed with biological recognition capabilities by proper functionalization or integration of bio-systems such as DNA strains, antibodies, enzymes and capturing proteins. Typical materials for OSCs include polymers such as poly-(3-hexylthiophene) (P3HT) and alkyl-substituted triphenylamine polymers (PTAA) but also oligomers such as pentacene and its soluble derivatives as well as many other organic materials capable of providing OTFT devices with a mobility as high as 1 cm^2 V^{-1} s^{-1} [8]. A beneficial exercise is to compare OSC properties and silicon, the material par excellence of electronics. Silicon is held together by a network of covalent bonds, where each atom shares its valence electrons with four neighbors. In contrast, the organic material consists of (macro-)molecular blocks within which atoms are covalently bonded to each other. However, these blocks are held together by means of weak *van der Waals* interactions and, in the case of doped materials, electrostatic interactions as well. The prevalence of *van der Waals* interactions in the "soft" organic material defines the key difference with the "hard" silicon. Firstly, organic semiconductors offer facile chemical modification. The toolbox of organic chemistry can be used to modify the structure of the molecular blocks with near-infinite possibilities. In fact, functionality can be engineered by altering the π-conjugated backbone or the side groups, thus changing opto-electronic, mechanical or biologically relevant properties. Organics offer the possibility of low temperature processing as well. The weak interactions that hold the molecular blocks together can be easily overcome by mild heating (mostly in small molecules) and a film can be deposited from vapor or solution on a wide variety of substrates. Organics also allow for oxide free interfaces with aqueous electrolytes. In fact, if a silicon crystal is cleaved and exposed to the atmosphere, the growth of an oxide layer will be observed. In contrast, the surface of the organic contains no broken covalent bonds and is oxide-free. Organics can therefore be in direct contact with the biological *milieu*, which offers important opportunities in bioelectronics, as it will be discussed in the following.

2.   Organic Thin Film Transistors as highly performing bioelectronic sensors

OTFTs have been proved to be very powerful tools in studying processes occurring at biological interfaces [1,9,10]. One of the main advantages of this approach is the possibility to address key information by measuring the device multiple figures of merits. The idea of monitoring the variation on more than one parameter simultaneously during sensing can be dated back to 2000 by Torsi *et al.* [11] In this visionary paper bulk and two-dimensional conductivity, as well as field effect mobility and threshold voltage have been demonstrated for the first time to be parameters that can change when an OTFT is exposed to a changing gas atmosphere. We will see in the following that the concept of multi-parametric response can be successfully applied to OTFT based biosensing platforms, as well. Recently, it has been shown that OTFTs allow the accurate determination of the free-energy balances associated with a binding process too. In turn, these energetic components (including the electrostatic one), as well as the dynamics involved in conformational changes, have been quantified [7,12]. In the last years, different TFT configurations have been proposed for biosensing applications. However, before entering into the details of OTFT based biosensors, an introduction to the basic functioning principles of an organic TFT is necessary.

2.1   OTFT operating principles

First introduced in the 1980s [13] OTFTs have now reached performance levels comparable to that of their polycrystalline inorganic homologue and a number of p-type and n-type organic semiconducting materials with different chemical and physical properties exhibiting field-effect mobilities higher than 1 cm^2 V s^{-1} can be found [14].A typical OTFT structure is shown in Fig. 1. In its simplest form, an OTFT comprises a gate contact that is defined on the substrate which could be rigid (a silicon wafer) or flexible (plastic or even paper) [15,16].

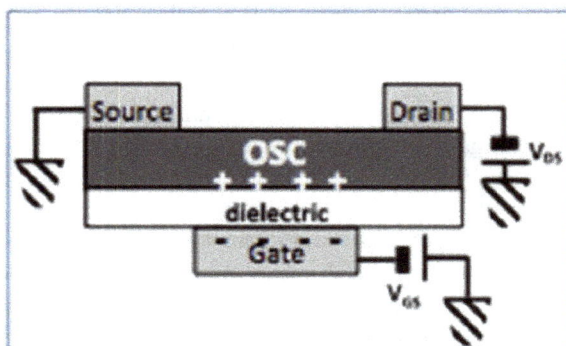

Figure 1.   Schematic representation of a typical OTFT device structure. Figure adapted from reference 9 and reproduced with permission.

The gate electrode is contact with a dielectric that is interfaced to the organic semiconductor (OSC) film. The gate dielectric should have a high capacitance either because it holds a high dielectric constant k, or because it is a thin-film, as will be discussed later on [17]. The OSC can be made of oligomers or polymers that are deposited as films (a few tens of nanometers thick at most) by solution casting, spin coating or sublimation. Other deposition methods are also used, including printing techniques [18,19]. Solution processed OSCs are generally polycrystalline films composed of contiguous grains with linear dimensions of a few hundred nanometers. Source and drain contacts to the OSC can be easily defined either by thermal evaporation through a screen-mask, by photolitography or again by printing. Gold is the most convenient contact metal because its work function is the closest to that of most p-type organic materials. For p-type OSCs the device is operated by independently negatively biasing the drain (D) and the gate (G) contacts applying the V_{DS} and V_G potentials with respect to the grounded source (S). This is called *common source configuration*. Eventually a channel of positive charges, whose geometrical length (L) is the distance between the source and the drain pads, is formed between these contacts. The channel width (W)is the geometrical width of the pad. The metallic source and drain contacts are electrically connected to an OSC (p-type in this case) and are meant to inject and collect positive charges, as depicted in Fig. 1. OTFTs operate in the *accumulation mode* and a highly resistive OSC is necessary to obtain a low I_{DS} current in the *off-state* ($V_G = 0$). The I_{DS} current flowing in the *on-state* ($V_G < 0$) must be, instead, as high as possible. The switching between the two transport regimes is achieved as the gate-contact and the OSC channel are capacitively coupled through the dielectric layer, allowing positive charges to be accumulated and confined in the OSC at its interface with the dielectric layer. In that, V_G controls the accumulation of charges at this interface while, under an imposed bias V_{DS}, the I_{DS} current flows between the source and drain electrodes. In other words, the field generated by the negative V_G bias applied across the dielectric layer leads to a band-bending in the OSC as reported in Fig. 2. In Fig. 2(a) the band structures of the metal gate, the insulator (gate dielectric) and the OCS are depicted when no gate bias is imposed. The few positive charges in the OSC are due to the presence of p-type dopants whose control, at the ultra-low trace level, is very difficult, particularly in not fully ordered materials. Fig. 2(b) shows what happens at the interface when a negative V_G is imposed. This generates a potential well that allows positive charges to be confined and accumulated at the OSC/dielectric interface, forming a conductive channel (between source and drain contacts) running through an ideal path perpendicular to the drawing plane. Due to this confinement the field-induced transport is two-dimensional (2D), that is to say independent of the OSC thickness, provided that a continuous OSC thin layer is deposited.

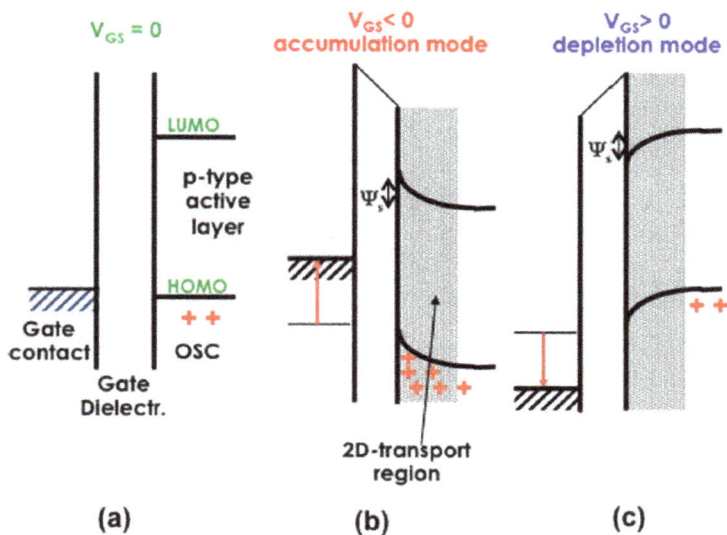

Figure 2.   Band diagram of metal–insulator–(p-type) semiconductor (MIS) structure at (a) zero gate, (b) accumulation and (c) depletion mode. Figure adapted from reference 9 and reproduced with permission.

Conversely the transport occurring at $V_G = 0$ is three-dimensional (3D) as it involves the charges present in the whole p-type OSC film. The presence of these charges (that must be orders of magnitude lower than those induced by the gate field) is due to doping processes connected with the presence of impurities or structural defects that are difficult to control at the trace level. The field generated by the V_{DS} bias allows the charges present in the OSC potential well (OTFT channel) to migrate in a direction perpendicular to the Fig. 2 plane. Indeed, the larger the negative gate bias, the larger the accumulated charge density, and the more intense the I_{DS} on-current is. This is the reason why this is addressed as accumulation mode operation. For an n-type OSC, negative charges are accumulated at the interface by applying positive V_G bias and current flows for positive V_D. To make sure that the transport in the channel is 2D, it is necessary that the field generated by V_G (normal to the channel plane) is always much larger than the one generated by the V_{DS} bias along the channel (*gradual channel approximation*). Again for a p-type OSC, if a positive gate voltage is applied, the field causes the semiconductor band edges (or more correctly the molecular HOMO and LUMO levels) to bend upwards, causing the positive charges to be accelerated towards the bulk of the OSC, see Fig. 2(c). The OTFT channel region is then depleted and the I_{DS} current flow reduced (*depletion mode*). It is worth mentioning that as V_G is applied the conditions for charge accumulation are set but the I_{DS} on-current flow does not start until a threshold gate voltage (V_T) is reached. In order to fully understand the physical meaning of V_T, let's first consider a Metal-Insulator-Semiconductor (MIS) structure. For charge neutrality of the system, it is required that the charges per unit area on the metal (Q_M) are counterbalanced by the total charges per unit area in the semiconductor (Q_S). Therefore, the following equation holds

$$Q_M = -(Q_n + qN_AW_D) = -Q_S \tag{1}$$

where Q_n is the electrons per unit area near the surface of the inversion region and qN_AW_D is the ionized acceptors per unit area in the space-charge region with depletion width W_D. A typical variation of the space-charge density Q_S, as a function of the surface potential ψ_S is shown in Fig. 2.4, for a p-type silicon with $N_A = 4 \bullet 10^{15}\text{cm}^{-3}$ at room temperature. It is worth mentioning that for negative ψ_S, Q_S is positive, corresponding to the accumulation regime. When $\psi_S = 0$ the flat-band condition is settled and $Q_S = 0$. When $0 < \psi_S < 2 \psi_B$, with ψ_B being the Fermi level referred to the midgap, Q_S is negative and the depletion and then weak-inversion is achieved. For $\psi_S > 2 \psi_B$ the strong inversion is established. In the ideal case of absence of any work-function difference, the applied voltage will partly drop across the insulator and partly across the semiconductor. Therefore, the following equation can be written

$$V = V_i + \psi_S \tag{2}$$

where Vi is the potential across the insulator and is given by

$$V_i = \frac{|Q_S|d}{\varepsilon_i} = \frac{|Q_S|}{c_i} \tag{3}$$

As shown in Fig. 3 strong inversion begins at $\psi_S \sim 2\psi_B$.

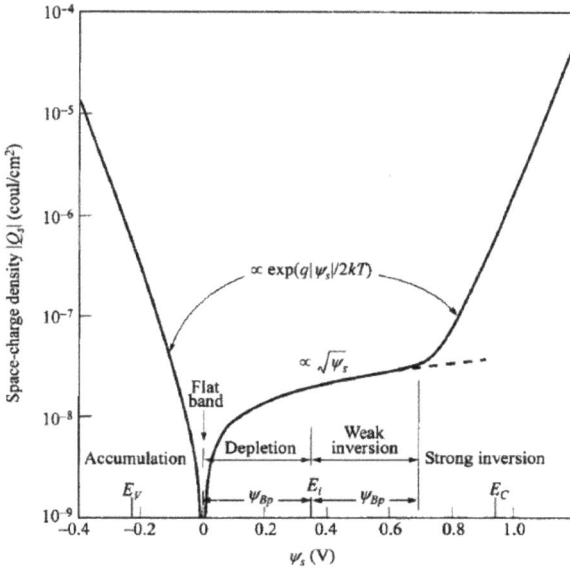

Figure 3.   *Variation of space-charge density in the semiconductor as a function of the surface potential ψ_S for a p-type silicon with $N_A = 4 \times 10^{15}$ cm^{-3} at room temperature. Figure adapted from reference 20 and reproduced with permission.*

Once strong inversion is accomplished, the depletion-layer width reaches a maximum. When the bands are bent down far enough that $\psi_S = 2\psi_B$, the semiconductor is effectively shielded from further penetration of the electric field by the inversion layer and even a very small increase in band bending (corresponding to a very small increase in the depletion-layer width) results in a very large increase in the charge density within the inversion layer. Traditionally, for an ideal MIS-structure, we can define the threshold voltage V_T as the turn-on voltage at which strong inversion occurs, according to the following equation.

$$V_T = 2\psi_B + \frac{Q_S}{C_i} = 2\psi_B + \frac{\sqrt{2\varepsilon_s q N_A (2\psi_B)}}{C_i} \tag{4}$$

Clearly, in a real device the presence of oxide charges, other than that of the interface traps, causes an additional shift of the threshold voltage, whose effect on the voltage shift

Organic Bioelectronics for Life Science and Healthcare   Materials Research Forum LLC
Materials Research Foundations **56** (2019) 1-70   doi: https://doi.org/10.21741/9781644900376-1

is weighted according to the location of the charge. The closer to the oxide-semiconductor interface, the more shift it will cause. As reported in Fig. 4, positive charge is equivalent to an added positive gate bias for the semiconductor so it requires a more negative gate bias to achieve the same original semiconductor band bending.

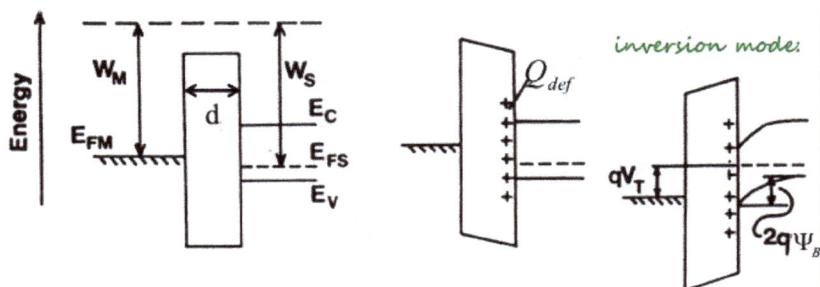

Figure 4.   *Band Diagram of a real MIS-structure, with a work function difference,*
positive oxide charges and under strong inversion.

Importantly, for the preceding discussions on ideal MIS structure, it has been assumed that the work-function difference between the p-type semiconductor and the metal is zero. Actually this condition is never fulfilled for real MIS structure. Consequently, the experimental V_T value will be further shifted from the theoretical value calculated from Eq. 4. So the V_T becomes the sum of the following three contributions

$$V_T = \frac{W_M - W_S}{q} - \frac{d}{\varepsilon_i} Q_{def} + 2\psi_B + \frac{\sqrt{2\varepsilon_S q N_A (2\psi_B)}}{C_i} \tag{5}$$

being the V_T shifts due to the mismatch between the work functions of the metal and the semiconductor and the oxide charges and the turn on voltage to achieve the strong inversion regime. In case of a thin film transistor, which works in the accumulation region, the V_T should be ideally zero, as depicted in Fig. 3. However, even in this case the contributions of the work function difference and the oxide charges need to be taken into account, as reported in Fig. 5. It is worth mentioning that, what is traditionally called threshold voltage in a TFT structure by the organic electronic community, strictly speaking is indeed the flat band potential. From a practical, although less physically rigorous, point of view the V_T value for a non-ideal TFT represents the transistor turn-on voltage, thus maintaining the name of threshold voltage.

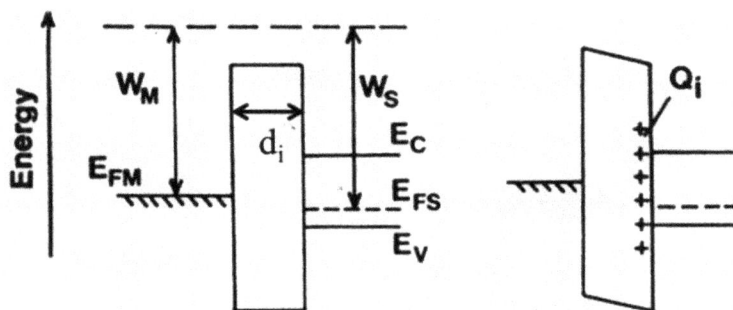

Figure 5.   *Contributions of the work function difference and positive oxide charges to the band diagram of a real TFT-structure.*

In a trap-free device, V_T is in fact the bias needed to offset the built-in potential generated by the mismatch existing between the metal-gate and the OSC electrons electrochemical potentials or Fermi levels. As V_G equals V_T, the energy barrier is flattened and, when the V_G bias exceeds V_T, further injected charges can drift through the channel with a given field effect mobility (μ_{FET}) [20]. In case of interface trap, historically also called interface states, an extra applied gate voltage is required to accomplish the same band bending. Therefore, V_T is increased by a factor of Q_{def}/Ci where Q_{def} is the density of charges that, once injected in the channel, are trapped, while C_i is the dielectric capacitance per unit area. Once all traps are filled the further injected charges can migrate along the delocalized molecular orbital with a given mobility, through the channel under the imposed V_{DS} bias. The V_T value for a non-ideal TFT device can be thus expressed according to the following equation

$$V_T = \frac{W_M - W_S}{q} - \frac{d}{\varepsilon_i} Q_{def} \tag{6}$$

Typical current–voltage, I–V, curves for a P3HT p-channel OTFT are shown in Fig. 6(a).

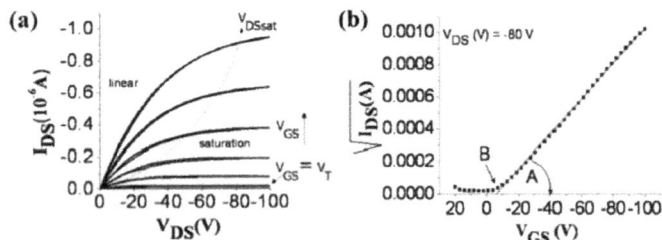

Figure 6. *(a) Current–voltage (I_{DS}–V_{DS}) characteristic curve of a P3HT p-channel OTFT device at different V_G gate voltages. (b) Square root of the current–voltage (I_{DS}–V_G) transfer characteristic measured at V_{DS} = -80 V. Figure adapted from reference 9 and reproduced with permission.*

As for all TFT-based devices the set of I_{DS} currents is measured as a function of V_{DS} and each individual curve at a different, fixed V_G bias. The curves are characterized by a linear region at $V_{DS} \ll (V_G - V_T)$ and a saturation region at $V_{DS} > (V_G - V_T)$. At low drain-source voltages the I_{DS} current follows Ohm's law being proportional to V_{DS} at a fixed V_G (linear regime). In this regime the *gradual channel approximation* holds as the field generated by the gate potential is much larger the one generated by V_{DS}. As the drain-source voltage becomes more negative, a point is reached where the positive charges accumulated in the channel region are depleted at the drain contact region. As this ends, the field along the channel (generated by V_{DS}) becomes too high compared to the gate one and the two-dimensional charge confinement is lost. The presence of this charge-depleted region generates a *pinch-off* of the channel and the current flow is limited to a constant value $I_{DS}{}^{sat}$ (saturation regime). These features are well reproduced by the MOSFETs analytical expressions, generally used also to describe the OTFT I–V curves. The equations are reported in the following:

$$I_{DS} = \frac{W}{L} C_i \mu_{FET}(V_G - V_T)V_{DS} \qquad for \ V_{DS} \ll (V_G - V_T) \qquad (7)$$

Here W and L are the already introduced channel width and length, respectively; μ_{FET} is the field-effect mobility (cm^2 V s^{-1}), measuring how fast charges migrate under the imposed electric field. In the saturated region the following equation holds:

$$I_{DS}^{sat} = \frac{W}{2L} C_i \mu_{FET} (V_G - V_T)^2 \qquad for\ V_{DS} > (V_G - V_T) \tag{8}$$

Fig. 6(b) shows the square root of the I_{DS}^{sat} *vs.* V_G transfer characteristic at constant V_{DS} taken in the saturation region. The field-effect mobility and the threshold voltage can be estimated from this curve. Operatively, Eq. 8 can in fact be written as follows:

$$\sqrt{I_{DS}^{sat}} = \sqrt{\frac{W}{2L} C_i \mu_{FET}} V_G - \sqrt{\frac{W}{2L} C_i \mu_{FET}} V_T = AV_G - B \tag{9}$$

where the value of μ_{FET} (from the saturated region) and V_T can be graphically extracted from the linear fit to Eq. 9, reported as an example in Fig. 6(b) with A and B being the slope and the x-axis intercept:

$$\mu_{FET} = \frac{2L}{WC_i} A^2$$
$$V_T = -\frac{B}{A} \tag{10}$$

Another figure of merit in an OFET is the *on/off ratio*, defined as the ratio between of the I_{DS} current values in the on and off states. This is indicative of switching performance of the device between two distinct conduction regimes, *i.e.* 2D and 3D, taking place in an OTFT device. As it is clear already from the basic description of the device electronic transport properties provided, OTFTs are interfacial devices and the interplay between the dielectric and the OSC surfaces is complex and not yet completely understood. Nonetheless it is clearly received that the dielectric layer interfacial properties influence carrier transport and mobility in different ways. Specifically, the chemical and surface properties affect the morphology of the OSC and the orientation of small molecules or polymer segments. These impact on the transport properties that are strongly related to the molecules' orientation, as higher mobility hopping conduction is determined by the length of the π-delocalization, as discussed in the following. Moreover, the semiconductor/dielectric interface roughness can modulate the mobility of charge carriers. This is the interface that will act as the key relevant one in the OTFT sensing processes.

2.2 OTFT analytical biosensors with different configurations

Several research groups have contributed to the advancement in the field of organic bioelectronics, proposing novel structures that integrate a layer of biological recognition or functional elements directly coupled to one of the electronic interfaces of the TFT. These systems are based on organic bioelectronic transistor structures (mostly TFT, but also field effect transistors, FETs) with diverse device configurations and operating principles. These structures include electrochemical transistors [21], electrolyte gated (EG) TFTs [22], organic charge modulated FETs [23] as well as structures involving back gates or double gates. The devices listed above can be classified into two main categories, depending on the dielectric medium. Specifically, one class of devices is constituted by transistors that involve an ionic conductor and the other by those using an insulator as dielectric medium. In Fig. 7 an overview of the different bioelectronic device structures involving a bottom gate or a top gate is reported. The OTFT sensing process relies on the interaction of biological species with biological receptor molecules suitably integrated in the OTFT structure.

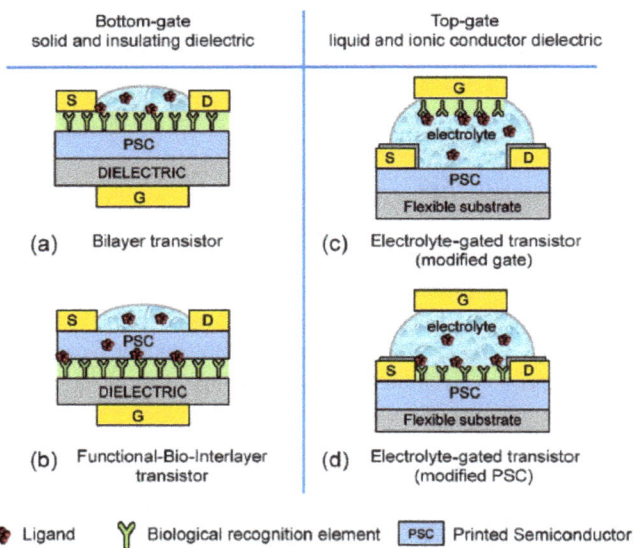

Figure 7.   *Overview of different bioelectronic TFT structures. Figure adapted from reference [24] and reproduced with permission.*

Upon interaction, the electronic properties of the OSC and/or the dielectric / OSC interface are modified, affecting the TFT figures of merit [24]. Indeed, the electrical and sensing performance of an OTFT can be dramatically influenced by the nature of the functionalized active region of the transistor. Moreover, OTFTs offer different active regions suitable for the biorecognition element / transducer coupling. One of the first attempts at this kind of TFT based biosensors is represented by the long-known class of ion-sensitive FETs (ISFETs) [25]. In an ISFET any charge variation on the gate side is able to induce, by capacitive coupling across the gate dielectric, a current variation in the channel. Thus, an ISFET is in fact a charge sensor and any bio-reaction which implies or induces a charge variation on the gate could be in principle detected with this device. As depicted in Fig. 8, the metal gate is replaced by a so-called electrolytic gate, i.e. a metal contact, connected to an electrolytic solution, in direct contact with the gate dielectric, and a reference electrode. In an ISFET based biosensing platform the bio-recognition elements, such as enzymes but also antibodies/antigens, DNA and whole cells, can be either deposited on the metallic gate (as sketched in Fig. 8) or interfaced directly to the dielectric. The analyte interacts with the recognition molecules layer giving rise to electrochemical reactions whose charged products can act as extra gating to the ISFET, leading to a measurable V_T shift. These devices have been extensively used for measuring a variety of bio-related effects provided that the effect to be measured induces, directly or indirectly, a charge variation on the gate side of the gate dielectric. However, the diffusion of ISFETs has been affected by an important limitation: to operate these devices, a reference electrode, normally an Ag/AgCl ideal electrode, is needed. This is due to the fact that the applied gate voltage is dropped across several interfaces; thus, in order to give a reliable and measurable effect, only the interface between the solution and the dielectric layer must be sensitive to the charge variation.

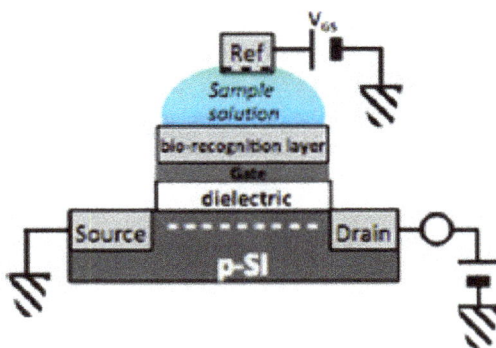

Figure 8.   Scheme of an ISFET biosensor having the bio-recognition layer deposited on the gate metal contact surface. Figure adapted from reference 9 and reproduced with permission.

Unfortunately, reference electrodes cannot be miniaturized and this has severely limited the practical applicability of this kind of device. Another major drawback of this strategy is represented by the fact that the detection is generally limited to electro-active or charged species. Another interesting ISFET-like sensing approach, that overcome the use of a reference electrode, is the so called Charge Modulated (CM) OTFT. In this extended gate OTFT, the gate sensing area and the channel region are physically separated and DNA label-free detection has been reported in the sub nM range.26 However, in this case the bio-recognition is also limited to charged species and there is, by definition, no direct coupling of the bio-recognition and electronic channel. This is however a good experimental tool to measure the sole capacitive effects in a given bio-organic interface. So far other active region of the TFT have been exploited to integrate the biorecognition element into a biolectronic device. In fact, bioelectronic back-gate structures might integrate the bio-recognition element either on top of the OSC, as in the bilayer (BL) transistor depicted in Fig. 7(a) [27], or underneath it, as in the functional biointerlayer (FBI) transistor of Fig. 7(b) [28]. The former structure represents one of the first attempt to realize a biosensor based on a back gate TFT device. One of the first papers dealing with this structure involved the deposition of amino acid or glucosidic units on the OSC to endow the OTFT with chiral-recognition capability. Indeed, electronic chiral differential detection was achieved for citronellol and carvone enantiomers at the ppm concentration level, pushing the limit of solid-state chiral determination down by three orders of magnitude [29]. In the FBI-OTFT geometry, that will be extensively discussed in the following of the *Chapter*, the interaction of the affinity ligand with the integrated recognition elements has been demonstrated to sensitively and selectively impact the characteristics of the transistors. Indeed, these label-free devices are able to reach detection limits down to the pico Molar (pM) level and the reproducibility of the devices is characterized by a relative standard deviation of a few percent over several hundred repeated determinations [30]. As far as concern the device structures outlined in Fig. 7(c) and (d), a top gate is coupled to an OSC through an electrolyte. The biological recognition layer can be deposited on the gate or on the OSC surface. In both cases, an ionic conductor acts as the dielectric.

2.3 Impact of biological recognition on TFT performance features

The response of a bioelectronic TFT can be conveniently evaluated by measuring the device transfer characteristics, namely the I_{DS} versus V_G curve at a constant V_{DS} value in the saturated region. According to Eq. 9, the square root of I_{DS} is linear in (V_G - V_T), as depicted by the plots in Fig. 9 for the different bioelectronic TFT structures.

In Fig. 9 a representation of the square root of I_{DS} vs. V_G before and after the binding event is shown. The red curves refer to the $\sqrt{I_{DS}}$ vs. V_G recorded before the exposure to the analyte. This is addressed as the I_0 current *base-line*. The blue curves are gathered instead after the receptor (R) / affinity ligand (L) complex formation has been accomplished. This is addressed as the I *signal* current. According to Eq. 10, a linear interpolation allows easy extraction of the TFT performance features. Indeed, $C_i\mu_{FET}$ is proportional to the slope of the $\sqrt{I_{DS}}$ vs. V_G curve, while V_T is the intercept with the abscissa (V_G).

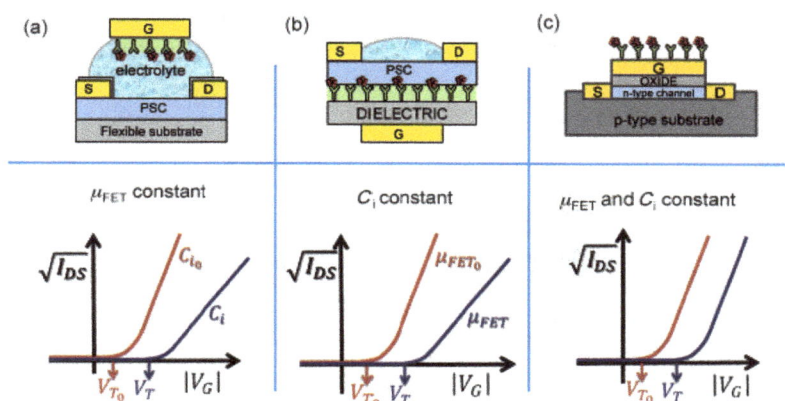

Figure 9.   *Representation of the transfer characteristics of different bioelectronics TFTs before and after a binding event (a) in an EG-TFT, (b) in the FBI transistor and (c) in an ISFET. Figure adapted from reference [10] and reproduced with permission.*

In general, if the semiconductor transport is affected by the formation of the RL complex, μ_{FET} will change. When the complex formation has an impact on the gating capacitance, C_i will be affected. If the impact is on the free electrons electrochemical potential $\overline{\mu_e}$ of the semiconductor of the metal gate, then a V_T shift will be registered. These three occurrences can take place contemporarily, but the cases where these effects can be decoupled are interesting to study, as shown for the device structures reported in Fig. 9. For biolectrionic TFTs, as shown in the equivalent circuits reported in Fig. 10(a) and (b), the gating system capacitance is the result of at least three capacitances in series and therefore it can be written as

$$\frac{1}{C_i} = \frac{1}{C_{OSC}} + \frac{1}{C_{Diel}} + \frac{1}{C_{BIO}} \tag{11}$$

where C_{OSC} and C_{Diel} are the capacitance per unit area associated to the semiconductor and to the dielectric respectively, while C_{BIO} is the capacitance per unit area of the biological layer. If one of the contributing capacitance is much lower than the others, C_i can be approximated to this value. In this respect, the case of the EG-OTFT is a peculiar one. Charge double layers at the gate / electrolyte and electrolyte / OSC interfaces, termed as C_{CDL-1} and C_{CDL-2} respectively in the equivalent circuit shown in Fig. 10(a), eventually form. The presence of a biorecognition layer adds a further contribution to the gating system capacitance, reported as C_{Bio} in the equivalent circuit of Fig. 10(a).

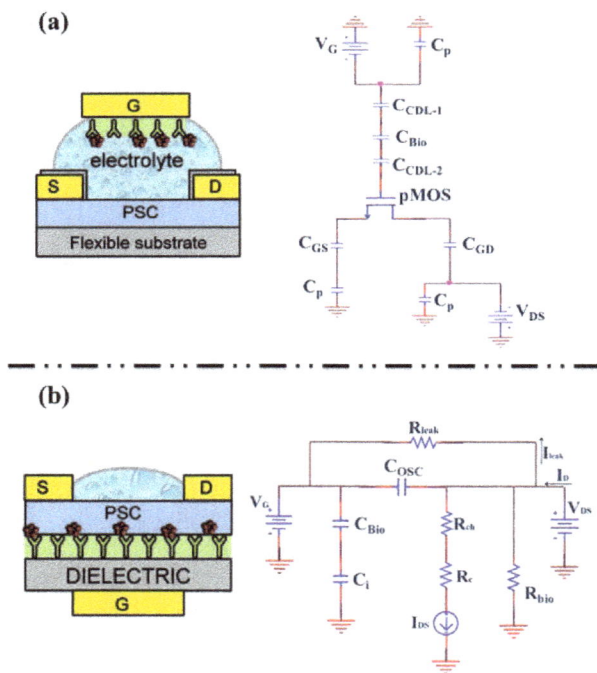

Figure 10.   *Schematic representation along with the corresponding equivalent circuit for (a) the EG-TFT and (b) FBI-OTFT configuration.*

Organic Bioelectronics for Life Science and Healthcare
Materials Research Foundations **56** (2019) 1-70

Materials Research Forum LLC
doi: https://doi.org/10.21741/9781644900376-1

Indeed, the highest capacitance of the series are those of the gate/electrolyte Helmholtz double layer, which can reach tens of $\mu F\ cm^{-2}$, and the electrolyte/OSC double layer capacitance a few of $\mu F\ cm^{-2}$. C_{Bio} can be modeled as a planar capacitor, $C = \varepsilon_0 \varepsilon_r d^{-1}$, with ε_0 an ε_r being the vacuum and the relative permittivity, while d is the distance between the capacitor plate. Taking $\varepsilon_r = 3$, typical for a protein system, and d as the height of a monolayer of proteins it is possible to estimate the capacitance per unit area of the order of magnitude of $0.1\ \mu F\ cm^{-2}$. Therefore, the lowest capacitive contribution to the TFT gating system comes from the layer of biological molecules. Eventually, it is the C_{Bio} contribution that drives or modulates the output current of the EG-TFT. As the dielectric properties of the R layer will change upon the binding of L, the EG-TFT can be very sensitive to subtle changes connected with the conformational changes, even when no net charges are involved. At the same time, the mobility of the semiconductor will remain constant (particularly for the case where the biolayer is deposited on the gate), while V_T can effectively transduce the electrostatic component of the R-L interaction. Comparison with a bottom gate FBI transistor is interesting. The FBI transistor equivalent circuit, reported in Fig. 10(b), has been recently proposed by Macchia *et al.* along with an analytical modelling of the I-V characteristics [31].It is important to highlight that the hundreds of nm thick SiO_2 insulating dielectric (capacitance in the $nFcm^{-2}$ range) typically present is by far the lowest contribution to the capacitance series and, therefore, in this case it is the insulating dielectric capacitance (which will not change upon binding) that gates the device. Therefore, as a consequence of the proximity between the biolayer and the semiconductor 2D channel in FBI transistors, μ_{FET} is affected by the binding process, while the capacitance stays constant. This configuration is ideal for studying the effect of changes to the biolayer on the semiconductor transport, while V_T, as a consequence of its nature, will transduce the electrostatic component of the interaction. Another quite relevant aspect is that EG-TFTs do not require the use of the reference electrode. Importantly it could be even detrimental as the faradaic current systematically depolarizes the charge double layer, thus cancelling the effect of the device being capacitively driven by the bio-layer. In fact, the EG-TFT gate should be an ideally polarisable electrode. In order to fulfill this assumption, the potential window for V_G is chosen so as to avoid any electrochemical process.

The relative variation of the saturated current I_{DS}^{sat} ($\Delta I/I_0$), given by

$$\frac{\Delta I}{I_0} = \frac{I - I_0}{I_0} \tag{12}$$

Is a robust parameter that can be taken as the bioelectronic TFT response, because it normalizes the device-to-device variation, thus contributing to obtaining a highly reproducible response [32]. $\Delta I/I_0$ also allows the contributions coming from the different performance features to be separated when second-order terms are neglected, by differentiating Eq. 8. Consequently, for the FBI device, assuming C_i to be constant, the following holds, up to first order

$$\frac{\Delta I}{I_0} = \frac{\Delta \mu}{\mu_0} - 2\frac{\Delta V_T}{(V_G - V_{T0})} \tag{13}$$

with V_{T0} and μ_0 being the threshold voltage and the field-effect mobility values extracted from the base-line current curve. The differentiation of Eq. 8 for an EG-TFT device leads to the following expression for $\Delta I/I_0$

$$\frac{\Delta I}{I_0} = \frac{\Delta C_i}{C_{i0}} - 2\frac{\Delta V_T}{(V_G - V_{T0})} \tag{14}$$

with C_i being dominated by the capacitance of the biolayer. If $V_T << V_G$ and $\Delta V_T << V_G$, Eq. 14 can be approximated as $\Delta I/I_0 \approx \Delta C_i/C_{i0}$. This shows that an EG-TFT can actually be operated as a capacity-modulated transistor, which is particularly suited to the study of interactions occurring between biological species that do not bear a net charge. Moreover, the output current of the EG-TFT can be modulated by the change in the capacitance of the biolayer and, strikingly, this would work better when the changes in capacitance are small. Therefore, in a TFT modulated by the biolayer capacity, all the capacitances that are in series with the biolayer one, become less and less effective as the biolayer capacitance (and its changes) becomes smaller and smaller. This holds true, provided parallel parasitic capacitances are kept low. This can be viewed as a "reverse" concept in the detection of biological species, and can be ideal for sensing ultralow concentrations.

3. Integration of recognition elements

3.1 Strategies for recognition elements immobilization

One of the crucial step in the fabrication of a biochip is represented by the biomolecules immobilization, such as enzymes, antibodies, DNA or whole cells, tissue, or organs, on an active surface of the device. Immobilization can be defined as the attachment of molecules to a surface resulting in reduction or loss of mobility [33]. In some cases, immobilization may lead to partial or complete loss of protein activity, due to random orientation or structural deformation. In order to fully retain biological activity, protein should be attached onto surfaces without affecting conformation and functions. Besides, the final performance of the biochip strongly depends on the parameter related to the

Organic Bioelectronics for Life Science and Healthcare   Materials Research Forum LLC
Materials Research Foundations **56** (2019) 1-70   doi: https://doi.org/10.21741/9781644900376-1

immobilization process itself. The fundamental parameters, that should be taken into account, are summarized in the following [34].

- The chemical and physical properties of the surface, as they influence both specific and non-specific binding of target and non-target molecules;

- The distance between the immobilized biomolecules and the surface;

- The orientation of the immobilized biomolecules, which might hinder binding, especially to large analytes;

- The density of the proteins on the surface, which determines the device sensitivity and limit of detection.

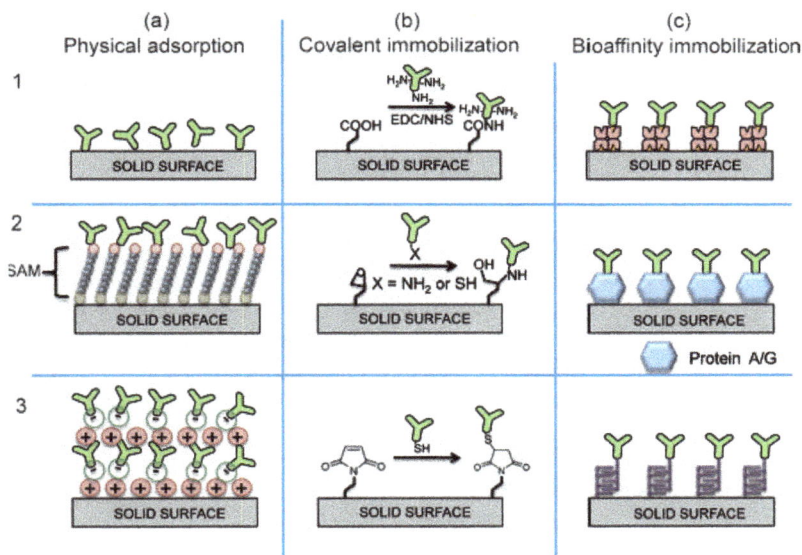

Figure 11.   Methods for the immobilization of biomolecules on solid surfaces. a) Physical immobilization techniques, b) Covalent immobilization of biomolecules through their functional residues and c) Bioaffinity immobilization reactions used for biomolecules orientations on a solid surface. Figure adapted from reference [10] and reproduced with permission.

Organic Bioelectronics for Life Science and Healthcare   Materials Research Forum LLC
Materials Research Foundations **56** (2019) 1-70   doi: https://doi.org/10.21741/9781644900376-1

The selection of the solid surface employed for the fabrication of the biochip depends on the specific application. For instance, gold surfaces are often used for the development of biosensors with electrical [10] and SPR read-out [35], because of their outstanding electrical conductivity. In this case, a convenient functionalization is based on the thiol chemisorption. On the other hand, glass or silicon is typically preferred for optical sensors because of their transparency, in the case of the glass, and low intrinsic fluorescence. Many immobilization techniques have been developed in the past years, essentially based on the following three mechanisms: physical, covalent and bioaffinity immobilization [10].The most used immobilization procedures for the integration of biomolecules are depicted in Fig.11.

3.1.1 Physical Immobilization

Biomolecules can adsorb on surface via intermolecular forces, mainly hydrophobic, ionic and/or Van der Waals interactions. Which of these contributions is the dominant one depends on the biomolecules and surface involved. Physical immobilization is particularly suited to deposit proteins on a variety of surfaces including carbon-based materials, noble metals and metal oxides, as shown in Fig. 11(a1). This method has been also exploited for DNA and protein immobilization on the surface of organic semiconductors such as pentacene, poly(3-hexylthiophene-2,5-diyl) (P3HT) and poly(3,4-ethylenedioxythiophene) (PEDOT) in thin film transistor (TFT) structures [36]. The adsorption capacity of the surface is limited by the geometric size of the immobilized proteins. High-density packing may sterically hamper the active sites of the proteins, interfering with functional proteins. Surface loading can be increased by constructing Self-Assembled Monolayers (SAMs) that interact with proteins through hydrophobic or electrostatic interactions, see Fig. 11(a2). Layer by Layer (LbL) adsorption, depicted in Fig. 11(a3), through stepwise deposition of biological species having opposite charge is an other way to control the biomolecules immobilization on the surface. LbL deposition is achieved by immersing a substrate to positively and negatively charged polymer solutions or particles dispersions. However, the resulting biolayer is likely to be not homogeneous and randomly oriented [33]. Indeed, each molecule can form many contacts in different orientation in order to minimize repulsive interactions with the substrate and previously adsorbed molecules. An other major drawback of the adsorption mechanism is the weak attachment, since proteins might be removed by some buffers or detergents when performing the assays. Both these issues impact on the device long term stability.

Organic Bioelectronics for Life Science and Healthcare   Materials Research Forum LLC
Materials Research Foundations **56** (2019) 1-70   doi: https://doi.org/10.21741/9781644900376-1

3.1.2 Covalent Immobilization

As an alternative, biomolecules can be stably anchored on a solid surface by covalent immobilization through the formation of chemical bonds between complementary functional groups present on the biomolecules and on the solid surface. Covalent bonds are mainly formed between exposed side-chain functional groups of the biomolecules with suitably modified surfaces, as sketched in Fig 11(b). Covalent immobilization results in an irreversible binding and produces a high surface coverage. The functional groups on the surface can be generated by chemical treatment based on the use of polymeric films or SAMs. Indeed, SAMs can have head groups such as thiols and disulfides and a variety of functional tail groups (-COOH, -NH_2, -CH_3) separated by alkyl chain of different length. Several pretreated surfaces are commercially available as well [33].Chemical binding is commonly used to anchor proteins to a surface via accessible functional groups of exposed amino acids. It is worth mentioning that the protein's attachment may occur simultaneously through many residues, resulting in a not oriented immobilization. Lysine residues are the most commonly used anchoring points, being present on the exterior of proteins. Consequently, their abundance might create a multipoint attachment on the surface, increasing heterogeneity and unpredictable orientation. N-Hydroxysuccinimide (NHS) is the most used agent for coupling the complementary functional groups present on the protein and on the solid surface, forming stable amide bonds [37,38]. Briefly, the surface, after activation through NHS, is exposed to the protein dissolved in a buffer. NHS esters react with nucleophilic groups of the proteins, creating a strong amide bond. In other words, immobilization occurs by displacement of the NHS group by lysine residues of the protein. Successful formation of the ester intermediate is reliant on the accessibility of the terminal carboxylate groups; steric packing of these acid groups can limit the rate of intermediate formation, with full conversion of accessible acid groups occurring only after several repeated reaction cycles. This method is experimentally more difficult and often exposes the protein to a harsher environment. However, the resultant irreversible binding which can be produced with high levels of surface coverage makes this approach more popular, although in some cases covalent binding can alter the conformational structure and active center of the protein, causing a reduction in activity. The use of NHS chemistry on SAM has been reported for the first time, to the best of our knowledge, by Patel *et al.* [37], who assessed the effect of accessibility of terminal carboxylated groups of SAM on the reaction with NHS and the immobilization of proteins. As shown in Fig 1.2, they have generated SAMs from short- and long-chain carboxylic acid-terminating alkanethiols to assess the effect of chain length on the reactivity of SAMs toward protein immobilization.

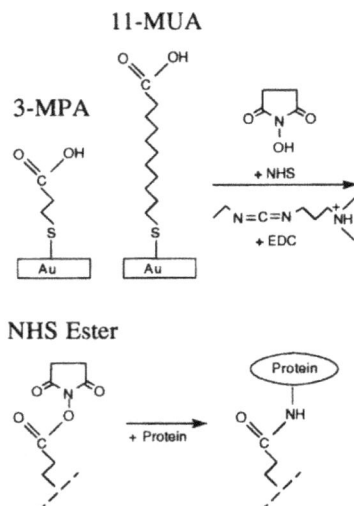

Figure 12.   Strategy for immobilizing protein onto carboxylate-terminated SAMs on gold. Addition of NHS and the water-soluble carbodiimide EDC to the SAMs results in the formation of an NHS ester. Reaction of protein side-chain lysine residues with the ester results in the formation of an amide bond. Figure adapted from reference [37] and reproduced with permission.

In this study SAMs formed on gold from solutions of 3-mercaptopropanoic acid (3-MPA), 11-Mercaptoundecanoic acid (11-MUA) and a mixture of 3-MPA/11-MUA (10:1 ratio) were prepared. The immobilization of protein onto gold involves the formation of an NHS ester with the carboxylate-terminated SAMs using the water-soluble carbodiimide EDC (Fig. 12). Side-chain amino groups of lysine residues on the protein surface displace the terminal NHS groups, resulting in covalent immobilization of the protein. The authors proved that the reactivity of the SAMs toward protein immobilization decreases in the order mixed > 11-MUA > 3-MPA and concluded that the mixed SAMs allow improving accessibility for protein binding due to reduced steric hindrance. Further studies by Lee *et al.* [38] investigate the immobilization efficiency as a function of the composition of a mixture of the thiols 11-MUA and 3-MPA. For this purpose, both the amount of protein adsorption and the kinetic parameters of protein binding for mixed SAMs solution consisting of 20:1, 10:1, 1:1 ratio of 3-MPA to 11-MUA and homogeneous 11-MUA were compared. The authors demonstrate that a molar

Materials Research Forum LLC
doi: https://doi.org/10.21741/9781644900376-1

ratio of 10:1 (3-MPA: 11-MUA) is the best combination for mixing the two alkanethiols. After protein immobilization, the bio-functionalized slide might eventually be immersed in bovine serum albumin (BSA) – containing buffer in order to prevent nonspecific binding [39]. Thiol groups are also important in biomolecules conjugation. Thiols are the terminal group of the side-chain of cysteine (Cys). Thiol groups in proteins are often involved in disulfide bonds (–S–S–). Conjugation via thiol groups is not affecting the protein structure or the binding sites. In addition, immobilization using cysteine residues is considered to be more selective and precise. The number of -SH groups that are present in proteins is less than amines, thus reducing the possibility of multisite attachment [40]. Two thiol compounds can be covalently coupled through alkylation forming a thioether bond or disulfide exchange forming a disulfide bond [41]. Maleimide reagents are used for conjugating to a thiol group. The maleimide group reacts specifically with -SH at pH ranging from 6.5 to 7.5 forming a stable thioether linkage [42]. In more alkaline conditions, maleimide can react with –NH_2 groups or even undergo hydrolysis forming non-reactive maleamic acid. Homo-bifunctional and heterobifunctional maleimide crosslinkers exist so as to effectively conjugate two different proteins, such as bis(maleimidohexane) (BMH) and NHS-ester/maleimide compounds respectively. Pyridyl disulfides also react with thiols over a broad pH range to form disulfide bonds. During the reaction, a disulfide exchange occurs between the biomolecule's –SH group and the reagent's 2-pyridyldithiol group while pyridine-2-thione is released. The disulfide exchange can be performed at physiologic pH, although the reaction rate is slower compared to acidic environment. However, these reagents are insoluble in aqueous solution. The reaction is carried out in aqueous/organic mixed solutions. Vinyl sulfone is another reagent that reacts with thiol groups by Michael addition forming disulfides. The reaction is effective at pH range 7- 9.5 [43]. Combining thiol and amine reactive groups broads the application of covalently binding biomolecules to each other or to a solid support. Eventually epoxy chemistry seems to be an appropriate system for developing easy protocols, due to its stability at neutral pH values, wet conditions and reactivity with several nucleophilic groups to from strong bonds with minimal chemical modification of the protein. Specifically, epoxy groups can be generated on the surface through agents like glycidoxypropyltrimethoxysilane (GOPS) that could be coupled to amino groups present in the protein [44].

3.1.3 Bioaffinity Immobilization

Bioaffinity binding approaches, shown in Fig. 11(c), provide advantages over other immobilization techniques, because biomolecules are precisely oriented on the surface and can be attached with high density. The basis of bioaffinity technique is the biospecific complex formation of biologically active compounds [45] A biospecific complex formation for oriented immobilization of biologically active proteins was used for the first time in 1978 with covalently attached Protein A for oriented immobilization of immunoglobulin G (IgG) [46]. The mostly used oriented immobilization of biomolecules is the biotin - avidin or streptavidin technique [47]. Specifically, the binding of water-soluble vitamin biotin to the egg white protein, avidin, or to its bacterial counterpart, streptavidin, is accompanied by a higher decrease in free energy compared to that observed for other non-covalent interactions. The extraordinarily high affinity constant of biotin-(strept) avidin complex (10^{15} M^{-1}) makes this system an efficient tool to bind biotinylated molecules on (strept)avidin modified surfaces or vice versa. The basic concept is that biotin, coupled to either low or high molecular weight molecules, is recognized by (strept)avidin. Methods for the biotinylation of membranes, nucleic acids, antibodies and other proteins have been developed in many laboratories. Biotin-binding proteins have been widely investigated so far [48]. Both proteins share the same tetrameric structure characterized by four identical monomeric subunits each endowed with a β-barrel consisting of a series of eight juxtaposed β-structure connected by turns to host a biotin molecule. Besides both proteins have a molecular weight of about 67 kDa and show an overall homology of 40%. Despite the fact that streptavidin is more expensive than avidin, its use is sometimes justified since immobilized streptavidin exhibits less nonspecific binding. Avidin is highly positively charged at neutral pH, being its isoelectric point (pI) above 10. Consequently, it binds negatively charged molecules such as nucleic acids, acid proteins or phospholipids in a nonspecific manner. Avidin is also a glycoprotein and therefore interacts with other biological molecules such as lectins or other sugar-binding materials via the carbohydrate moiety. The advantage of streptavidin lies in the fact that it is a neutral (at neutral pH), non glycosylated protein (pI of about 6). Very recently, genetically engineered chimeric avidin molecules containing C-terminal cysteine groups have been immobilized directly onto the gold surface [49]. The former bilayer offer higher binding capacity of biotinylated molecules compared to wild-type avidin anchored on the same surface.

3.2   Bulk and Surface Equilibrium Constants of Ligand–Receptor

As discussed so far, advancing knowledge of ligand–receptor interactions confined at solid–liquid interfaces have fundamental implications for physical chemistry and it is a key issue in order to develop solid-phase bioassays, as microarrays and biosensors. In the simplest and most widely applied model a ligand–receptor surface binding is assimilated to the analogous bimolecular reaction occurring in the bulk ideal solution (Fig. 13) [50].

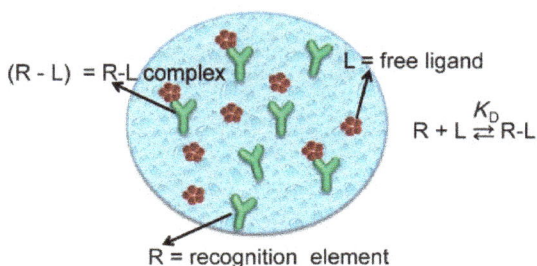

*Figure 13.   Schematic representation of binding process in solution. Figure adapted
from reference [10] and reproduced with permission.*

This model provides a simple way of thinking of a complex phenomenon and in turn to easily evaluate it. However, it is oversimplified for many solid–liquid biological interfaces [51] and its application to these cases implies heavy misinterpretation of the experiments. The limits of validity of the model is still a subject of lively and interdisciplinary discussion. Specifically, let's consider the chemical equilibrium of the binding of a single ligand molecule, L, to a single receptor molecule, R, occurring in closed system in a non-ideal (water) solution at isothermal and isobaric conditions:

$$R + L \rightleftarrows RL \tag{1}$$

If the reaction takes place in the solution bulk (Fig. 13), then the law of mass action (van't Hoff isotherm) holds:

$$\Delta G_b^0 = -RT \ln K^b = -RT \ln \frac{a_{RL}}{a_R a_L} \tag{2}$$

where $\Delta G_b{}^0$ is the change of the standard molar Gibbs free energy of the reaction in bulk solution, R is the gas constant, T the temperature and a_R, a_L, a_{RL} the equilibrium thermodynamic activities of R, L, and of the receptor-ligand complexes, RL, respectively. $K^b = a_{RL} \bullet (a_R a_L)^{-1}$ is the bulk thermodynamic equilibrium constant, often referred in molecular biology as association constant or binding affinity. In the case of dilute water solutions, K^b can be numerically expressed by substituting the activities of the reacting species with their equilibrium molar concentrations (indicated as [R], [L], and [RL]). By further manipulations, K^b can be expressed as a function of the binding efficiency θ, i.e. the fraction of occupied binding sites $[RL] \bullet ([R]-[RL])^{-1}$, through the equation:

$$[L]K^b = \frac{\theta}{1-\theta} \qquad (3)$$

that is the Langmuir isotherm of the ligand–receptor binding reaction [52]. The form given in Eq. 3 is often refined by inserting on the left hand side a multiplying function that accounts for the dependency of K^b from the solution ionic strength [53] as the ligand–receptor binding is normally conducted in dilute saline buffer solutions. Being an expression of K^b, the Langmuir isotherm given by Eq. 3 holds for the systems described by the law of mass action and thus matching the underpinning assumptions. Therefore, when it is used to describe a ligand–receptor binding reaction confined at the solid-solution interface, the Langmuir model essentially assumes that the reaction goes on at the interfacial region as it would do in bulk solution. On the other hand, by tradition the Langmuir isotherm is precisely introduced to describe adsorption of a species A onto solid-surfaces and it is derived by thermodynamic as well as kinetic argumentations. Thermodynamics is preferred for adsorption at the solid–liquid interface [54] while kinetics for adsorption at the solid-gas interface [55]. The Langmuir adsorption isotherm describes the surface as an ensemble of fixed and independent adsorption sites, which can be either empty (S) or occupied by adsorbed molecules (Ad), (Fig. 14).

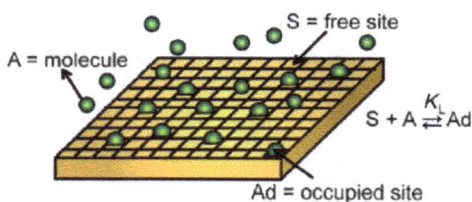

Figure 14.   Schematic representation of adsorption process to a surface. Figure adapted from reference [10] and reproduced with permission.

Similarly to binding isotherm, the adsorption isotherm relates the fraction θ of sites occupied by the adsorbed molecules, $\theta = Ad \bullet (S + Ad)^{-1}$, to the equilibrium molar concentration of A

$$\theta = \frac{[A]}{\frac{1}{K_L}+[A]} \tag{4}$$

where K_L is the Langmuir constant, that is the equilibrium constant for the adsorption process. As opposed to the K^b, the Langmuir constant of the adsorption process involves the activity of species, such as S and Ad, which are confined to a surface. Therefore, even at very low concentration of A (infinite dilution), K_L cannot be approximated using only the equilibrium molar concentrations of [S] and [Ad]. In fact, the surface density should be used for the adsorption sites. Besides, a real ensemble of biological recognition elements can be generally described as a homogeneous system. In fact, the affinity constant among the R molecules are usually negligible. Thus, Eq. 3 typically gives an accurate description of the binding process in case of single and independent binding sites. Conversely, the adsorption sites of a real solid surface are generally characterized by different adsorption energies and consequently different Langmuir constants. The overall adsorption isotherm is obtained by integrating Eq. 4 for all the K_L values [56]. When multiple binding sites on the same recognition element are available, another phenomenon connected with inhomogeneity can occur. The ligand affinity for a binding site, characterized by a certain K^b value, may depend on the occupancy of the other binding site of the same biorecognition element. This is the case of cooperative binding. The cooperativity is said to be positive, if the presence of occupied binding sites facilitates further ligand binding. On the contrary, in case of negative cooperativity, the binding to a site hinders the binding to the other binding sites of the receptor. The case of strong cooperative ligand binding to biological macromolecules can be described by the Hill binding isotherm. In case of adsorption process, lateral interaction may also have a non-negligible impact on solid surfaces, meaning positive or negative influence of adsorbed molecules on the adsorption of further molecules. This might give rise to cooperative adsorption on solid surfaces [57]. To the best of our knowledge, it is not common to find thermodynamic studies correlating the properties of binding process taking place in the bulk of a solution and at a surface. However, this approach can be a powerful tool to gather fundamental pieces of information on a given binding process, occurring, for example, at an active surface of a bioelectronics device. The comparison between the binding equilibria involving species confined at a solid-liquid interface and in the bulk of a solution can be performed considering the thermodynamic path, shown in Fig. 15 [10].

The overall process starts with state 1, in which a ligand L, dissolved in solution, binds to a recognition element R, segregate to the surface, to form an immobilized complex (RL) (state 4). The corresponding change of the standard molar Gibbs free energy ΔG_σ^0 can be

Organic Bioelectronics for Life Science and Healthcare   Materials Research Forum LLC
Materials Research Foundations **56** (2019) 1-70   doi: https://doi.org/10.21741/9781644900376-1

described in terms of an apparent thermodynamic equilibrium constant K^{σ} as $\Delta G_{\sigma}^{0} = RTln(K^{\sigma})$.

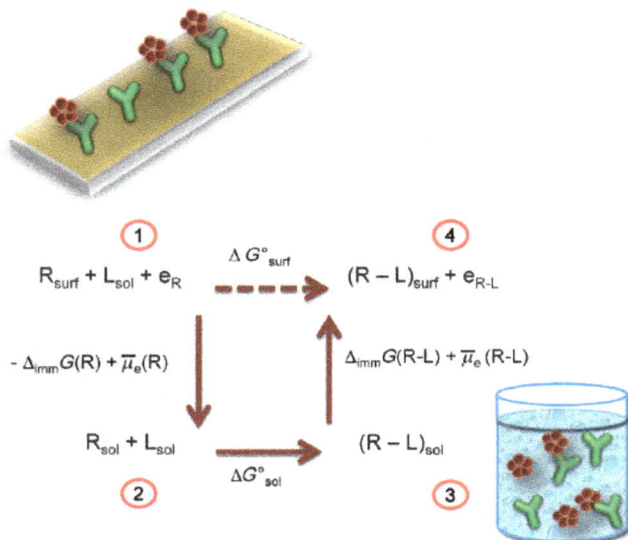

Figure 15.   Thermodynamic path of the binding equilibria involving species confined at solid-liquid interface and in solution. Figure adapted from reference [10] and reproduced with permission.

The process $1 \rightarrow 4$ might be conceptualized considering the three steps sketched in Fig. 15, each one characterized by a change in the standard molar Gibbs energy. In the first step $(1 \rightarrow 2)$ the recognition element R is discharged from the surface into the solution. The change in the free energy associated with step $1 \rightarrow 2$ cannot be computed as the sole work needed to release R, which is opposite to the surface work $\Delta_{imm}G(R)$ required to immobilize it. The latter will depend on the strategy used to immobilized the R species, which will determine the aspect such as the surface density and the orientation of R. However, since in the following we will focus on recognition species anchored on a metallic gate or on the surface of organic semiconductors (OSC), the change in the electrochemical potential $\overline{\mu_e}(R)$ of the gate or of the OSC electrons has to be taken into account as well. The second step $(2 \rightarrow 3)$ is the binding of L to R in the bulk of the solution, with a free energy change of $\Delta G_b^{0} = RTln(K^b)$, that is a function of the equilibrium constant in solution. The final step $(3 \to 4)$ is the immobilization of the RL complex on the surface with a change in the free energy, including the surface work spent on immobilizing the complex $\Delta_{imm}G(RL)$ and the change in the metal-gate potential or in the OSC electron electrochemical potential $\overline{\mu_e}(RL)$. Therefore, the standard molar Gibbs energy for the surface binding of the biological recognition elements segregated to a surface can be written according to Eq. 5.

$$\Delta G_\sigma^0 = \Delta G_b^0 + \overline{W} = \Delta G_b^0 + \Delta E_F + W \tag{5}$$

The binding surface work $\overline{W}$ is composed of a non-electrostatic term W given by the following equation

$$W = \Delta_{imm}G(RL) - \Delta_{imm}G(R) \tag{6}$$

and an electrostatic term, being equal to

$$\Delta E_F = \overline{\mu_e}(RL) - \overline{\mu_e}(R) \tag{7}$$

The latter is the change in the electrochemical potential of the gate or of the OSC electrons as the surface changes from being coated with the bare R layer to the RL complex. Therefore, according to Eq. 5, the surface binding energy (ΔG_σ^0) can be split into a term that takes into account the molecular interaction energy (ΔG_b^0) and one accounting for the surface work contribution $(\overline{W})$. Importantly, the molecular interaction term is the RL recognition energy in solution. The surface work contribution describes the work carried out to accommodate the ligand into the recognition element bound to a surface. This term includes the conformational rearrangements as well as the change in the electrochemical energy of the gate or of the OSC. For justifying this statement, it is convenient to analyze the relation between K^b and K^σ that comes by equating Eq. 2 and 5. It reads

$$\frac{K^\sigma}{K^b} = e^{\frac{\overline{W}}{RT}} \tag{8}$$

Eq. 8 directly links the parameters that describe the equilibria of a ligand–receptor binding reaction in bulk solution and confined at a solid-solution interface. The solution equilibrium is defined by a single constant parameter, K^b, referred to standard conditions, while the interface equilibrium is defined by a couple of parameters, K^σ and $\overline{W}$. Now, for a given value of K^b there exist infinite couples of values of K^σ and $\overline{W}$ that satisfy Eq. 8. Consequently, it follows that a given ligand–receptor binding reaction, that is characterized in bulk solution by a given (constant) value of K^b, at the solid-solution interface can be described by several (potentially infinite) values of K^σ, depending on the values of $\overline{W}$, that are determined by the physicochemical characteristics of the interface. In conclusion, we must expect K^b to be numerically different from K^σ, and the same ligand–receptor binding reaction conducted in different interfacial environments to be described by K^σ of different values. Finally, $K^b = K^\sigma$ only if $\overline{W} = 0$, that is when the work associated with the immobilization of the free recognition element and of the RL complex has the same value. In the more general case of non-negligible net surface work, $\overline{W}$ will exponentially affect the ligand affinity for the immobilized biological recognition element. This model has been successfully applied to describe data of DNA surface hybridization obtained from different solid-solution interfaces by different solid-phase assays [58]. In particular, it was found that W can range between -90 kJ mol^{-1} and +75 kJ mol^{-1}. Upon substitution of these values into Eq. 8, neglecting the electrostatic contribution that cannot be evaluated by means of a microcantilever set-up, a K^σ/K^b ratio ranging from 10^{-16} to 10^{13} can be achieved. This ratio spans over 29 orders of magnitude, consistently with the literature experimental observations.

4.   OTFT devices with a biolayer between the gate dielectric and the organic semiconductor

A biosensor comprising an OTFT device where the bio-recognition elements are deposited underneath the OSC, resulting in the Functional Biological Interlayer (FBI)-OFET, has been conceived for the first time by Torsi *et al.* with the aim of creating a direct interface between the OSC and the capturing molecules [28]. Clearly in such a device an intimate contact is created between the region where the bio-recognition takes place and the OCS FET channel. This unconventional approach involves a FET electronic channel working on top of a biological layer. Indeed, a successful FET transport was evaluated as highly unlikely, as the FET transport critically depends on the dielectric/OSC interface quality. Consequently, such an approach required a wide assessment, involving the characterization of devices embedding three different biosystems as prototype for membranes, membrane proteins and hydrophilic proteins as detailed in the following.

Materials Research Forum LLC
doi: https://doi.org/10.21741/9781644900376-1

4.1   Interfacial electronic effects in functional bio-layers integrated into OTFT

The very first proof of concept on FBI-OTFT has been provided by Torsi *et al.* [28] using phospholipid (PL) bilayer, purple membrane (PM) film and a streptavidin (SA) protein layer as FBI. The phosphatidylcholine PLs are one of the major components in all cell membranes and are permeable to volatile anesthetics. The PM, taken from the bacterium *Halobacteriumsalinarum*, is constituted of a PL bilayer including the sole bacteriorhodopsin (bR) protein and it shows exceptional stability against thermal and chemical degradation. The PM explicates its bio-activity when exposed to light or to other external stimuli such as general anesthetics capable to change the bR local pK_a values. Streptavidin is a hydrophilic tetrameric protein with an extraordinarily high affinity for biotin (dissociation constant on the order of fM) thus representing a prototype system for biosensor technology bench-test. The details of the FBI-OTFT layers morphological and structural characterizations are reported in the following section. For the moment, it is important to observe that the bio-interlayers retain their bio-activity after integration in the OTFT. In fact, it is not straightforward, considering that the bio-layers have to withstand to many processing steps that could cause a loss of functionality. PL layers are important constituent in many bio-systems. However, they do not explicate a specific bio-activity. On the other hand, the bio-activity of bR and SA interlayers have been demonstrated after integration into the OTFT. One way to assess the bR bio-activity is to investigate the changes occurring in the PM absorption spectrum upon exposure to yellow light ($\lambda > 500$ nm). The bR is a light-driven ion-pump, whose biological activity is explicated by protons driven from the cytoplasmic (CP) to the extracellular (EC) side as photons are absorbed. In fact, the spectrum measured in the dark is dominated by the absoption peak at ~ 570 nm ascribed to the bR's all-*trans* retinal bound to the Lys-212 by a protonated Schiff base. Upon exposure to yellow light the retinal chromophore switches to the 13-*cis*-configuration triggering the proton traslocation cycle. Indeed, the PM integrated into the OTFT shows an absorbance spectrum under yellow light, which exhibits a pronounced bleaching at 570 nm with the appearance of a band at 410 nm. This evidence proves that the bR molecules in the PM bio-interlayer system are still capable to generate a proton flux, meaning that they retain their bio-activity. On the contrary, the bare P3HT film spectrum is not affected by exposure to the same yellow light. Also, the SA layer bioactivity is not affected when subjected to the relevant FBI-OTFT device fabrication processing. Specifically, it has been proven that the SA capability to bind biotin molecules remains unaffected by the organic semiconductor deposition. Besides, not only the FBI layers retain their bioactivity while integrated in the OTFT structure, but also FBI-OTFTs show a successful FET transport. In fact, the FBI-OTFT output current-voltage characteristics reported in Fig. 16 furnish clear evidence that the devices are capable of good electronic performances.

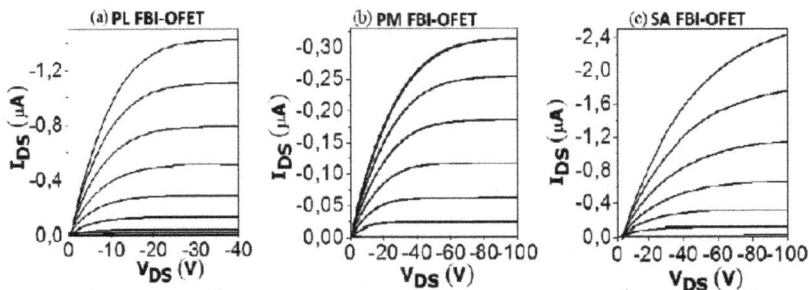

Figure 16. Typical I-V characteristics curves for (a) PL FBI-OTFT, (b) PM FBI-OTFT, (c) SA FBI-OTFT devices. Figure adapted from reference [28] and reproduced with permission.

Recently, the output current-voltage characteristics of a SA embedding FBI-OTFT have been modeled by means of the analytical equations of an enhancement mode p-channel OTFT modified according to the ad hoc designed equivalent circuit reported in Fig. 10 (b) [31]. This allows to reliably extracting the OTFT figures of merit taking into due account leakage currents and contact resistance. V_{DS} and V_G negative voltages generate the electric fields along the channel and orthogonal to it, respectively. The field generated by V_G involves a very thin OSC region (*ca.* 10-20 nm) at the interface with the protein layer where the FET two-dimensional transport occurs [59]. The field generated by V_{DS} develops across the channel ($L \geq 200\mu m$). Indeed, it is commonly assumed that the V_G generated field is orders of magnitudes more intense than that along the channel, as required by the *gradual channel approximation*. This occurrence is required in order to confine the field - effect transport at the interface with the dielectric. In the ideal case the total resistance, R_T, is the sum of the channel resistance (R_{ch}) and the source and drain contact resistance R_c. The R_{leak} element in Fig. 10 takes into account the gate leakage current I_{leak}, coming both from the source and the drain contacts, since no patterning of the OSC is performed. C_{GD} hold the usual meaning and is not computed as the measurements are all performed in dc [60]. The streptavidin layer is modeled with a capacitance (C_{SA}) in series with the gate dielectric capacitance C_i and a resistance (R_{SA}) in parallel to R_T. It can be assumed that the value of C_{SA} is buffered by C_i, as a biological material typically holds a capacitance of the order of $\mu F/cm^2$ [61]. The R_{SA} symbol stands both for a R_{SA-w} or a R_{SA-b} resistances alternatively introduced in the circuit to account for the resistance of the protein layer after exposure to water or to biotin. In the equivalent circuit R_{SA} is grounded as during the gold metal deposition on the granular P3HT the source can contact also the SA layer, being slightly conductive itself. The simulation of the I-V curves with PSPICE allows estimating the resistance values, used as a cross-check for the analytical modeling. The output I-V curves have been modeled starting from the Poisson equation [62] for an enhancement mode p-channel MOSFET implemented with gate bias dependent leakage current (I_{leak}), field-effect mobility (μ_{FET}) and threshold voltage (V_T). Accordingly, the well-known second degree polynomial equation in V_{DS} becomes:

$$I_D \cong \frac{W\mu_{FET}C_i}{L}\left[(V_G - V_T)V_{DS} - \frac{V_{DS}^2}{2}\right] + I_{leak}(V_G) = I_{DS} + I_{leak} \quad for \ |V_{DS}| \le |V_{DS}^{sat}| \quad (9)$$

$$a = \frac{WC_i\mu_{FET}}{L}(V_G - V_T)$$
$$b = -\frac{WC_i\mu_{FET}}{2L} \quad (10)$$
$$c = I_{leak}$$

with a, b and c being the fitting parameters from which μ_{FET}, V_T and I_{leak} can be computed at different V_G. The gate voltage dependence of μ_{FET} in polycrystalline (disordered) OSCs is expected, as the mobility is a function of the charge density, which is controlled by V_G. This dependence is theoretically predicted by charge transport models based on hopping of charged species in an exponential energetic distribution of traps. V_T dependence from V_G can be explained by the same token. Concerning the saturation region, the following expression has been used in agreement with the proposed equivalent circuit:

$$I_D^{sat} \cong \frac{W\mu_{FET}C_i}{2L}(V_G - V_T)^2 + \frac{V_{DS}}{R_{SA}} = I_{DS}^{sat} + \frac{V_{DS}}{R_{SA}} \quad for \ |V_{DS}| > |V_{DS}^{sat}| \quad (11)$$

The SA FBI-OTFT electrical characteristics (I_D vs. V_{DS} for different V_G) for a L=200 μm upon exposure to water and to 1 ppb of biotin are shown in Fig. 17 (a) and (b), respectively. The dots (along with the relevant error bars over three replicates) are the experimental data, while the solid lines are the result of the analytical modeling. It is worth mentioning that very well shaped I-V curves can be measured on the SA FBI - OTFTs also after exposure to water and to the analyte. This is not trivial as the FET channel is set to work over a biological layer before and after the SA-biotin complexes have been formed. A second comment concerns the excellent agreement achieved between the experimental data and the modeling obtained exploiting Eq. (9) for the linear region and Eq. (11) for the current flowing in the saturation regime ($|V_{DS}| > |V_{DS}^{sat}|$). According to Eq. (11) I_D^{sat} sees a contribution accounting for the resistive behavior of the SA layer, exposed to water and biotin. The resistance values extracted from the analytical model are in agreement with the values evaluated from the PSPICE simulation within few percentages, as it can be seen from the data reported in Table I.

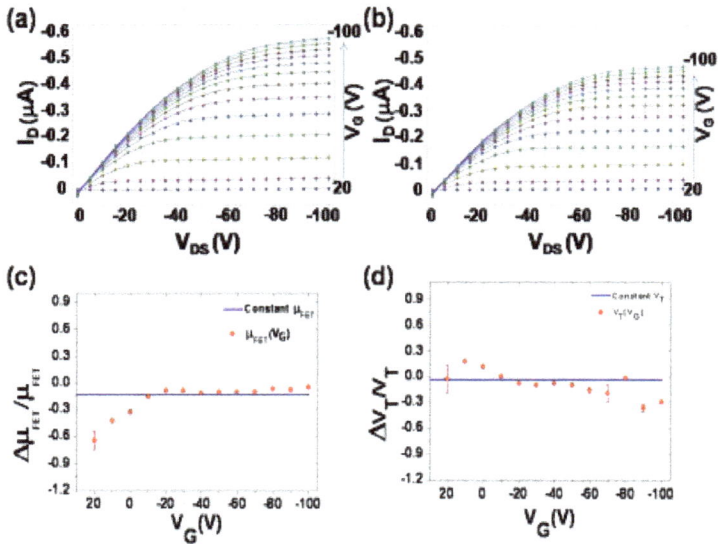

Figure 17.   *Output characteristics (I_D vs. V_{DS} at different V_G) for a SA FBI-OTFT after exposure to pure water (a) and to 1 ppb (5nM) of biotin (b). The dots are the experimental data, while the solid lines are the result of the model. Error bars are the relative standard deviations (RSDs) over three replicates measured on the same FBI-OTFT. In panel (c) and (d) μ_{FET} and V_T relative variation upon exposure to 1 ppb of biotin are reported as a function of the V_G. The values are extracted from the fitting of the I-V characteristics. The solid lines are the constant mobility and threshold relative variations graphically extracted from the transfer curves. Figure adapted from reference [31] and reproduced with permission.*

Table I. Comparison between the resistance values extracted from the analytical model and from the PSPICE simulation for the device with channel length L = 200 μm. The results are reported for the relevant V_G values.

V_G (V)	R_c (MΩ)	R_{ch} (MΩ) L = 200 μm	R_{SA-w} (GΩ)		R_{SA-b} (GΩ)		R_{leak} (GΩ)	
	Analytical	Analytical	Analytical	PSPICE	Analytical	PSPICE	Analytical	PSPICE
−20	158.75	168.75	4.05	3.95	8.13	8.05	3.19	3.19
−40	121.46	136.04	2.56	2.40	6.22	5.90	3.46	3.46
−60	96.44	114.31	2.05	1.97	3.24	3.11	4.00	4.02
−80	80.58	98.42	1.22	1.21	1.9	1.85	4.20	4.21
−100	69.29	86.21	0.94	0.93	1.79	1.76	4.50	4.50

The relative variations of μ_{FET} and V_T are also reported in Fig. 17 (c) and (d), respectively along with their values derived from the analytical model, see Table II. These data are compared to the relative variations of the constant μ_{FET} and V_T values, graphically extracted from the square root of I_{DS}^{sat} plotted as a function of V_G at V_{DS}= -100 V. To this aim Eq. (11) is used (neglecting the second term) with μ_{FET} taken from the slope, while V_T taken from the x-axis intercept. It is readily apparent that the graphical extraction of μ_{FET} and V_T returns relative variation values that are in very good agreement with the analytical model data in the -10 ÷ -80 V, V_G range.

Table II. Comparison between the figures of merit of the device after exposure to 1 ppb of biotin extracted from the analytical model and from the graphical extrapolation. The results are reported for the relevant V_G values.

V_G (V)	μ_{FET} (cm²/Vs) @ 1 ppb biotin		V_T (V) @ 1 ppb biotin		$\Delta\mu_{FET}//\mu_{FET}$		$\Delta V_T/V_T$	
	Analytical	Constant	Analytical	Constant	Analytical	Constant	Analytical	Constant
−20	9.16×10^{-4}	8.70×10^{-4}	32.10	32.92	−0.10	−0.15	−0.08	−0.04
−40	8.03×10^{-4}		26.56		−0.12		−0.08	
−60	7.54×10^{-4}		15.42		−0.11		−0.16	
−80	7.80×10^{-4}		9.57		−0.08		−0.02	
−100	7.74×10^{-4}		18.13		−0.06		−0.30	

Indeed the comparison does not hold in the sub-threshold region while for V_G beyond -80 V a mobility saturation is usually seen. The evaluation of the R_{ch} and R_c was also performed by means of the *transfer line method* (TLM) [63], extrapolating R_C as R_T at L = 0. To perform this extrapolation, I-V curves at different channel lengths (L = 200 μm, 600 μm, 1000 μm) were measured and the relevant R_T values were extracted from the linear region of the I-V characteristic. To this aim, the linear regime characteristics should be used, in order to satisfy the gradual channel approximation for the whole channel length. From the equivalent circuit and the reported I_D equation (neglecting the second order dependence from V_{DS}) the relation between R_T and R_c becomes the following:

$$R_T W = R_c W + \frac{L}{C_i \mu_{FET}(V_G - V_T)} \tag{12}$$

In Eq. (12) the R_{leak} contribution has been neglected, since its value (evaluated analytically as V_G/I_{leak} at $V_{DS} = 0$ V, as well as from a PSPICE simulation) is at least two orders of magnitude larger than R_T. $R_T W$ is then plotted *vs.* the channel length, L, and R_c is extrapolated as the resistance at $L = 0$ at different V_G. The R_T contribution has been computed as the I_D first derivative with respect to V_{DS} at $V_{DS} = 0$ from Eq. (9). Indeed, the R_T values are relevant to V_G dependent μ_{FET} and V_T values. For the sake of comparison, the same R_T data were computed also for the graphically extracted constant μ_{FET} and V_T. The application of the TLM method allowed the extraction of R_{ch} for the three channel lengths along with R_c values. The comparison of the R_c values as a function of V_G for the constant and gate bias dependent $\mu_{FET}(V_G)$ and $V_T(V_G)$, is reported in Fig. 18 (a). All data refers to the SA FBI-OTFT upon exposure to 1 ppb of biotin. It is readily apparent that the R_c values are not significantly different, this being particularly important as it justify the use of the very simple and straightforward parameter graphical extraction form the square root of I_D. The channel resistance R_{ch}, depending on channel lengths, and the contact resistance R_c extracted for the model involving fixed μ_{FET} and V_T are shown in Fig. 18 (a). R_c is shown to be lower than R_T for all the channel lengths although the two resistances hold similar values for the $L = 200$ μm channel. Similar features were already demonstrated also in the case of OFET chemical sensors [64]. Moreover the resistance data obtained with $\mu_{FET}(V_G)$ and $V_T(V_G)$ show the same trends of those derived for fixed μ_{FET} and V_T values, while quantitatively the data differ for few percentage.

As a cross-check for the validity of the approach, the model has been applied also to a bare P3HT OTFT. Also in this case, within the error bars, the resistances values estimated with the two procedures are in very good agreement. This is quite a relevant result as it validates the use of the transfer curves that are much easier to measure and handle, in order to extract reliable electronic sensor parameters. The differential changes of R_C and R_{Ch} after exposure to 1ppb of biotin, with respect to water, at different V_G and for different channel lengths are shown in Fig. 18 (c). In this case the relative variation of the contact resistance is sizably lower than that of all the other channel resistances. Moreover, the variation of the contact and of the channel resistances are in opposite direction, $\Delta R_c/R_c$ being negative in value, while $\Delta R_{ch}/R_{ch}$ values are positive. Also these data are in agreement to what previously reported [24, 64].

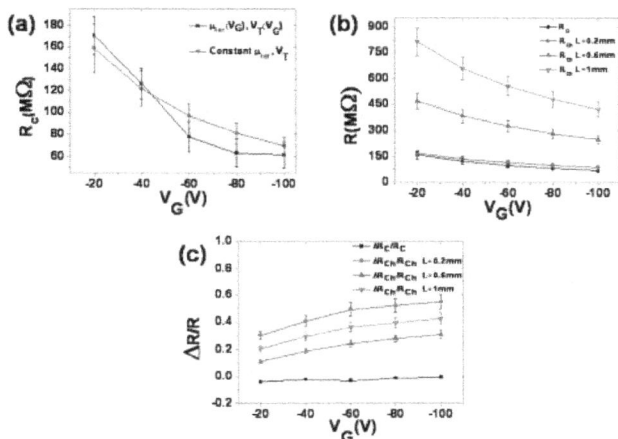

Figure 18.   In panel (a) the comparison of the contact resistance extracted with $\mu_{FET}(V_G)$ and $V_T(V_G)$ (solid squares) and with constant μ_{FET} and V_T (solid rumble) is reported. Panel (b) shows the SA FBI-OTFT contact and channel resistances extracted from the model with constant μ_{FET} and V_T, for L=200 μm, 600 μm, 1000 μm after exposure to 1ppb of biotin. Panel (c) reports the differential changes of R_c and R_{ch} after exposure to 1ppb of biotin at different V_G and for different channel lengths. Figure adapted from reference 31 and reproduced with permission.

4.2   Structural and morphological characterization of functional biointelayers

Assessing the quality of the interface generated by the biological layer and the OSC is of great relevance to fabricate OTFT devices with high electronic performances particularly as to the field-effect mobility is concerned. In this respect, it is critical that the properties and the quality of the dielectric / OSC interface are controlled as they also strongly affect the device performance and stability. Torsi *et al.* has investigated the morphology of the SA FBI layers by combining Scanning Electron Microscopy (SEM) and X-ray specular reflectivity (XSR) analyses and X-ray photoelectron spectroscopy (XPS), for characterizations at micro and nano scales [65,66]. SEM pictures of a spin-coated SA film are reported in Fig. 19 (a) and (b) proving that, at the μm length scale, SA molecules aggregate into clusters of different size (> 1 μm), showing an island-like structure, with inhomogeneous grains in size and distributions.

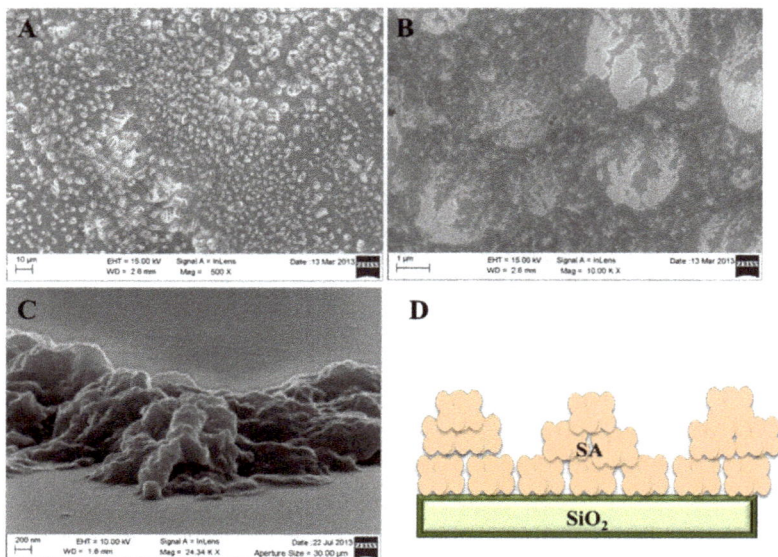

Figure 19.   *SEM images of the streptavidin layer obtained by spin-coating a 10 μg/mL protein solution on the SiO$_2$ dielectric surface. The images were acquired by using the In-lens detector at different magnification (a, b) and by tilting the sample of 75° (c). (d) Schematic of the presumed SA adsorption mechanism in the spin-coating deposition. Figure adapted from reference [65] and reproduced with permission.*

A single agglomerate of proteins has been studied, at higher magnification, in Fig. 18 (c), showing an irregular structure, presenting regions with uneven height ranging from 1 to 2 μm. The SA proteins assemble into clusters under spin-coating may be due to the very low rotational velocity (200 rpm) of the deposition procedure. However, the presence of a thin streptavidin layer, in direct contact with the SiO$_2$ surface, acting as "ad-layer" or as a continuous system of "seeds" that promotes the growth of the larger agglomerates cannot be excluded. Such a structure could act as a continuous ionic conduction path between larger SA clusters and this explains why a consistently higher off-current is measured in FBI-OTFTs in the presence of the protein layer. In Fig. 20 (a) and (b) the micro-morphology of a spin-deposited SA, covered by a spin coated P3HT film, is shown. Agglomerates, up to few μm in size, very similar to those present in the spin-deposited SA films, are seen in Fig. 20 (a). These structures, better visible at higher magnification in Fig. 20 (b), reveal features easily ascribable to SA agglomerate smoothly covered by a much thinner P3HT layer. Interestingly, the P3HT layer covering the SA cluster presents hollows and voids and indeed, inspecting within a hole, the structural features characteristic of a protein cluster can be seen. In Fig. 20 (c) the SEM image of a bare P3HT film reveals that the hollows are present also when the film is deposited directly on the SiO_2 surface. The pores in the P3HT film can be originated by the occurrence of de-wetting processes that happening when a hydrophobic organic semiconductor such as P3HT is deposited on hydrophilic surfaces such a not-silanized SiO_2 and a protein layer. It is in fact known that the silanizing process improves the organic semiconductor adhesion on the SiO_2 surface. The presence of holes in the P3HT film having dimensions of *ca.* 10 – 100 nm was also confirmed by Atomic Force Microscopy (AFM), as reported in the previous section.

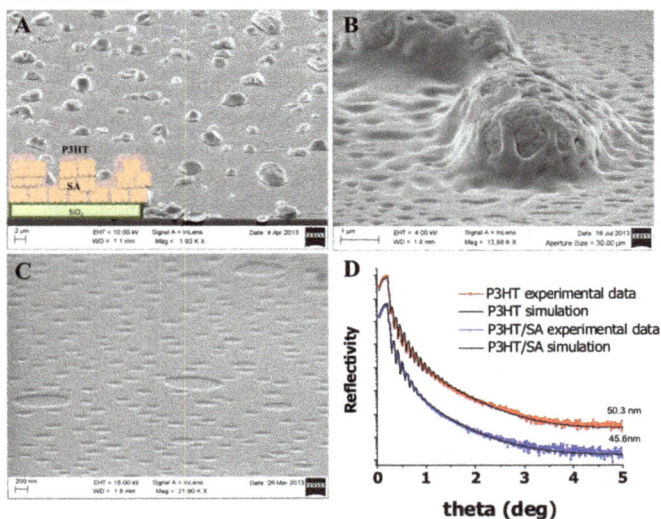

Figure 20.   SEM images of a spin-coated SA layer covered by the P3HT organic semiconductor film. The sample was prepared by spin-coating a 10 μg/mL SA solution followed by P3HT at 2.6 μg/mL in chloroform. The images were acquired by tilting the sample of 60° (a) and 75° (b) at different magnifications; (c) SEM image of a P3HT film deposited by spin-coating on a bare Si/SiO₂ substrate, the image was acquired by tilting the sample of 75°; (d) XSR experimental (symbols) and simulated (lines) curves from a bare P3HT film (red curve) and a SAV/P3HT bi-layer obtained by spin-coating a 10 μg/mL SAV solution and P3HT at 2.5 μg/mL in chloroform (blue curve). The actual thickness values as derived from the fit are reported next to each curve. Figure adapted from reference [65] and reproduced with permission.

However, the peculiar morphological features occurring in the P3HT film deposited on the SA layer works in favors of the FBI-OFET biosensors operation, as it allows also rather large molecules, such as insulin or streptavidin, to percolate through the organic semiconductor film. The P3HT layer structure evidences also that a continuous semiconductor layer spanning all the surface can be easily deposited. Such a two-dimensional percolation of the P3HT domains sustains electronic paths explaining why the OTFT electronic performances are only weakly affected in the SA FBI-OTFT device. In the light of these data it is also worth mentioning that the field-effect mobility of the SA FBI-OTFT is slightly lower than that of the P3HT. Indeed, the SA film surface roughness is much larger (> 1 μm) than the average P3HT grain size (~ 0.1 μm), this allowing the charge carriers mobility (or delocalization length) of the P3HT not to be substantially affected as the film lies on the SA clusters. The morphology of the SA / P3HT bilayer was also investigated by means of XSR. The reflectivity curves, collected on bare P3HT (red circles) and on the SA/P3HT bilayer (blue triangles) are shown in Fig. 20 (d) along with the best fittings curves (solid lines). Both curves presented show very similar trends. The structure of the P3HT layer appears not affected by the presence of SA layer, and both the SA and SA / P3HT structures can be fitted with a single layer whose thickness is invariably ~ 50 nm. Consequently, it is possible to infer that a P3HT film with an average thickness of 50 nm fills the space between the large aggregates, as evidenced by SEM. In such a space, only SA clusters smaller than 50 nm, embedded in the P3HT layer, are possibly allowed. A schematic of the bilayer structure is reported in the inset of Fig. 20 (a). In Fig. 20 (d) the difference between model and experiment, around 2.65° (corresponding to 1.6 nm d-spacing), in the upper curve, is due to the appearance of the (100) diffraction peak related to molecular order (crystallinity) in P3HT, not taken into account in the reflectivity calculations. Other important pieces of information on non-destructive depth profiling of SA-P3HT bilayer have been gathered by means of parallel angle-resolved XPS (PAR-XPS) [66]. PAR-XPS data suggested that a thick and irregular SA film is obtained by spin coating deposition. Protein is rather aggregated on the surface, and its organization is dictated by interaction forces with the substrate. When the organic semiconductor layer is added, it perfectly resemble the feature of SA deposited underneath. A large interpenetration of the SA and the OSC layer has been proven, as the PAR-XPS protein-related signals show the same trend observed in bare SA spin-coated films.

4.3 FBI-OTFT based biosensing platforms

The aforementioned spectroscopic and morphological studies confirmed the wide applicability of the FBI-OTFT sensing platform to detect large target molecules. FBI-OTFT integrating a PL layer has been proposed to detect archetypal general volatile anesthetics, such as diethyl ether and halothane by directly interfacing the electronic channel to the PL layer [67], as shown in the inset of Fig. 21. Typical transfer characteristics for a PL FBI-OTFT are reported in Fig. 21, as measured in N_2 and in a controlled atmosphere of diethyl ether at a concentration of 3.0 wt% in N_2 atmosphere.

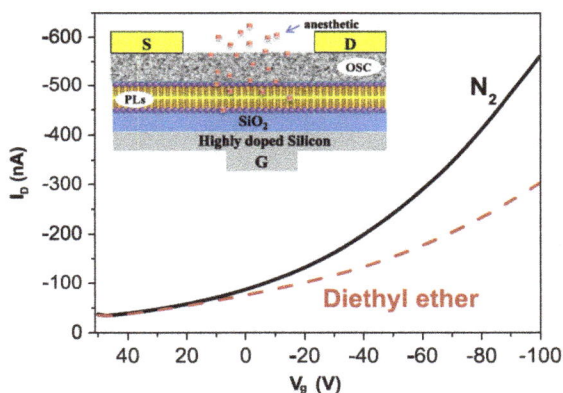

Figure 21.   Transfer characteristics for a PL FBI-OTFT measured in N2 and in a controlled atmosphere of diethyl ether at a concentration of 3 wt% in N2 atmosphere. Inset: Schematic structure of a PF FBI-OTFT. Figure adapted from reference [67] and reproduced with permission.

To study the effect of volatile anesthetic on the PL FBI-OTFT sensors, the device has been exposed to diethyl ether vapors in a concentration range of 0.6 – 3 wt%. As a negative control experiment the device was exposed to acetone, a non-anesthetic ketone with a vapor tension similar to the one of diethyl ether. Both the anesthetic and the acetone were tested also on a bare P3HT OTFT. The relevant calibration curves are reported in Fig. 22.

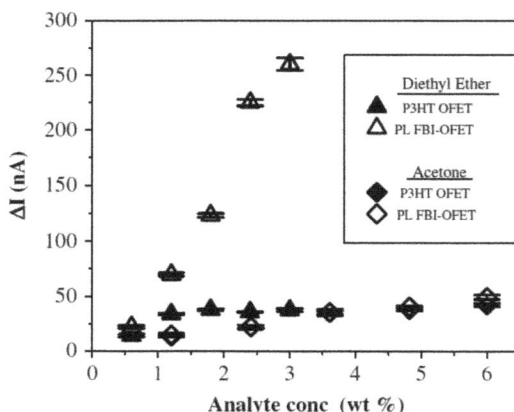

Figure 22.   Calibration curve of the PL FBI-OTFT and bare P3HT OTFT exposed to diethyl ether and acetone vapors. Each data point has been calculated as the average between three replicate measurements, while the error bars as the standard deviation. Figure adapted from reference [67] and reproduced with permission.

A significant current variation has been demonstrated only when the device is exposed to the volatile anesthetics, while very low response is reported to acetone. Comparable low responses have been demonstrated on the bare P3HT OTFT to both diethyl ether and acetone. This report shows a sizable current decrease when exposing the PL FBI-OTFT device to anesthetics at physiological concentrations. In fact the interaction of PL layer with volatile anesthetics causes a change in the PL film domain structure. This domain changes occurs at the interface between the PL layer and the OSC where the field-effect two-dimensional electronic transport occurs. The disorder induced by this interaction into the PL bilayer is responsible for affecting the two-dimensional electronic transport at this interface. Another interesting example of FBI-OTFT based sensing platform has been proposed by Torsi et al. in 2014 [68]. Specifically, a PM FBI-OTFT embedding bR nanolamellae has been proposed so far to detect proton translocation processes occurring when the device is exposed either to green-yellow light or to anesthetic vapors.

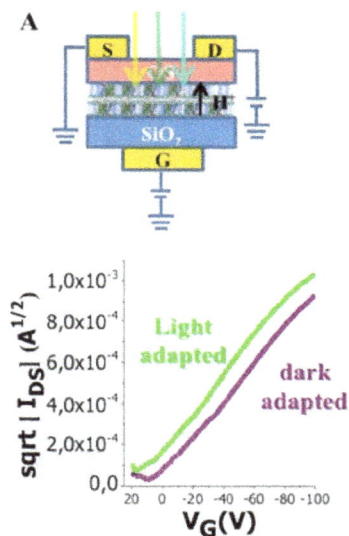

Figure 23.   *PM FBI-OTFT structure under illumination with green-yellow light, along with the square root of* $|I_{DS}|$ *vs.* V_G *curves measured in the dark (purple curve) and under green-yellow light (green curve). Figure adapted from reference [68] and reproduced with permission.*

The electronic detection of the bR photoinduced conformational change has been investigated by exposing the PM FBI-OTFT to green-yellow light while registering the transfer characteristics. Fig. 23 shows the square root of IDS vs. VDS measured in the dark and under green-yellow light. The current flowing between source and drain increases its intensity upon illumination. Such an effect has not been observed on a bare P3HT OTFT device. The observed light-induced increase of I_{DS} has been attributed to the presence of the negative gate bias, hindering the H+ flux in the direction opposite to the one of the applied field. This leads to a net proton injection into the OSC. Besides, the proton translocation process in bR is known to be influenced by the presense of exogenous chemical compounds, such as volatile general anesthetics. The electronic detection of these effects has been demonstrated in this report for the first time. The PM FBI-OTFT response to halothane vapors in N_2 atmosphere has been investigated and the electronic responses are reported in Fig. 24 (a).

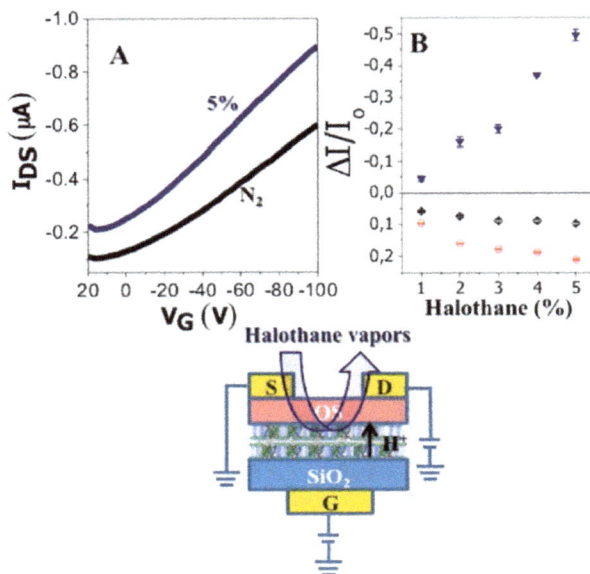

Figure 24.   (a) Transfer characteristics measured in inert N2 flux (black curve) and in a 5 % halothane atmosphere (blue curve) along with the schematic representation of PM FBI-OTFT. (b) Responses of PM FBI-OTFT (blue triangles), bare P3HT OTFT (black diamonds) and PL FBI-OTFT (red circles) to clinically relevant halothane concentrations. Figure adapted from reference [68] and reproduced with permission.

In Fig. 24 (b) the responses upon exposure to halothane in the 1 – 5 % concentration range are reported for the PM FBI-OTFT compared to that of a bare P3HT OTFT. The effect of halothane interaction with PM FBI-OTFT is opposite of that observed in the case of bare P3HT OTFT, where a very weak current decrease has been registered. Interestingly a current decrease has been demonstrated also upon exposure to halothane of a PL FBI-OTFT. Therefore, the effect of the PM interaction with halothane has been ascribed to the bR rather than to the phospholipids of the PM or to the P3HT. Another interesting case of study is represented by an FBI-OTFT embedding a streptavidin capturing layer, capable of performing label-free selective electronic detection of biotin at 3 part per trillion (mass fraction) or 15 pM [69]. The FBI-OTFT using the streptavidin-biotin receptor-analyte system have been studied to provide for the first time picomolar detection limit with an electronic bioanalytical assay. In fact, the SA−biotin complex has been proposed as a prototype system, since the dissociation constant is extremely low (K_D 10^{-15} M). Due to the high affinity, SA−biotin chemistry has been widely exploited in previous works either to detect biotin (by using SA as receptor) or to quantify SA (by using biotin as receptor). However, the detection sensitivity of the two approaches is quite different, as the binding of a very large protein (SA weights *ca.* 60 kDa) to a small molecule (just 244 Da for biotin) is likely to perturb the transducing system much more than in the opposite way around. Accordingly, label-free electronic quantification has been limited to SA detection, so far. Specifically, biotinylated lipid bilayers supported on carbon nanotube FETs have been proven to provide SA detections in the 5 µM or 312 part per million (mass fraction, ppm) range [70], while silicon nanowires transistor delivered 2.5 µM (156ppm) [71]. Better performances were achieved with a biotinylated OSC blend, capable of detecting 0.8 nM (0.05 ppm) of SA [72]. The best performance level reached was 10 pM (0.6 part per billion, ppb) SA detection obtained with biotinylated single silicon nanowire transistor. None of these studies involved the assessment of analytical figures of merit such as limit of detection, limit of quantification, analytical sensitivity, repeatability, or reproducibility. As anticipated, the quantification of small molecules is arather demanding task, and, in the case of biotin, nM concentration levels were detected at most, even with extremely sensitive non electronic determinations such as an electrochemical and a label needing fluorescent assay, both carried out in solution. In the label-free assay proposed by Torsi *et al.* biotin is detected at the pM level, by using an SA FBI as receptor layer. Such a device configuration allows, for the first time to the best of our knowledge, the electronic (not electrochemical) probing of biotin. The FBI-OTFTs have been measured in the *common-source* configuration, and electronic parameters (μ_{FET}, on/off ratio and V_T) were extracted from the I_D-V_G transfer-characteristics, keeping V_D fixed at −80 V. The measurements were performed in the dark and under a continuous nitrogen purging. The *base-line* I_D-V_G transfer curve measurement has been taken by depositing a droplet (2 µL) of water HPLC grade, directly on the P3HT surface in the OTFT channel, i.e. between two contiguous source and drain pads. The device was subsequently incubated for 15 min and dried under a nitrogen flow. The I_D-V_G transfer characteristic was then measured, and the I_0 value was taken as I_D at $V_G = -100$V. The response to biotin was measured afterward, by placing a 2 µL droplet, at the lowest biotin concentration, on the same OTFT channel and incubating the device for 15 min. Subsequently, the excess solution was rinsed with water and after drying the OTFT in nitrogen, the I_D current was measured, the latter being the *signal*; the I_D (analyte) value was taken also at $V_G = -100$ V. The biosensor response $\Delta I/I_0$ was finally evaluated according to Equation (2.14). Five increasingly higher biotin concentrations were evaluated too on the very same OTFT channel, following the aforementioned procedure. Error bars assessing the responses reproducibility are evaluated as the relative standard deviations (RSDs) on the same set of responses taken from three different devices. In Fig. 24 the I–V characteristic curves are measured on a FBI-OTFT embedding streptavidin, showing that FET transport occurs with figures of merit that, though affected by the integration of the recognition element, are still quite good (μ_{FET} from 4 to 2 • 10^{-3} cm^2 V s^{-1}, V_T from 5 to 20 V, on/off ratio from 10^3 to 10^2). Applied biases would have been lower than 5 V if a high-k thinner dielectric had been used. An improvement of these figures is foreseen by using a higher mobility OSC (tips pentacene for instance) and/or by using different deposition procedures (Langmuir–Blodgett, spray-coatings).

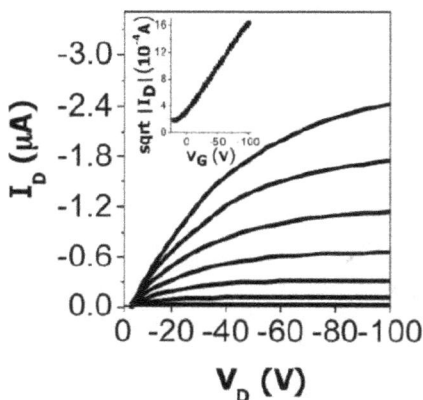

Figure 25. I_{DS} *vs.* V_{DS} *output characteristics for a SA FBI-OTFT;* V_G *ranges from +20 V to −100 V. Inset: sqrt* $|I_{DS}|$ *vs* V_G *at* V_{DS} = −80 V. *Figure adapted from reference [69] and reproduced with permission.*

We extensively discussed about the selective and almost irreversible interaction of SA with biotin, leading to the formation of a very stable complex. In this process, SA undergoes a conformational change which causes the characteristic binding loops to switch to a close on the incorporated biotin. Biotin binding to SA is depicted in Fig. 25 (a) (biotins are the grey triangles, the SA tetrameric protein is sketched by the four black circles) [73].

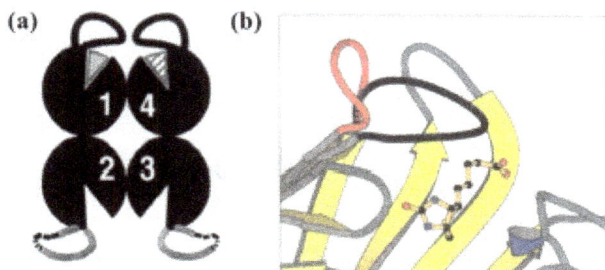

Figure 25.   (a) Schematic structure of streptavidin evidencing the four biotin binding sites and their conformational change after biotin interaction. (b) Detailed structure of a streptavidin recognition site after the biotin binding. Figure adapted from reference [73] and reproduced with permission.

Fig. 25 (b) shows a detail of one of the SA binding pocket, showing how the red loop turns into the black one as the biotin–SA complex is formed. As the FBI-OTFT the complex is formed just underneath the OSC, a very sensitive response is expected. The results of the SA FBI-OTFT exposed to biotin solutions of different concentrations (from pM to nM) are reported in Fig. 26 (a) as the relevant I_{DS} *vs.* V_G characteristics.

Here a systematic and scalable decrease in the current is observed as the device is exposed to different biotin concentrations, showing that the response appears to be directly connected with the complex formation [74]. The relevant dose curve, reporting the relative current decrease is shown in Fig. 26 (b). For both panels, error bars, as standard deviations over three replicates on different SA FBI-OTFT, show how the current variation is in fact significant even at the lowest concentration (50 pM). The other data points present in Fig. 26 (b) are relevant to sets of negative control experiments that provide a zero response assessing the selectivity of the streptavidin–biotin interaction in the FBI-OTFT assay proposed. Also, error bars are within a few percent while the LOD is 50 pM and the LOQ is 100 pM. The quantification of small molecules is a rather demanding task and, in the case of biotin, nM concentration levels have been detected at most, even with extremely sensitive non electronic determinations such as an electrochemical and a label-needing fluorescent assay, both carried out in solution. The label-free FBI-OTFT approach allows the reaching of detection levels that are comparable to those so far achieved by much better performing nanostructured sensors. Also relevant is that the FBI-OTFT determination does not need a reference electrode and can be performed also in the case of neutral species. One drawback is the necessity to operate the device in dry conditions, thus in a non physiological environment for the biorecognition layer.

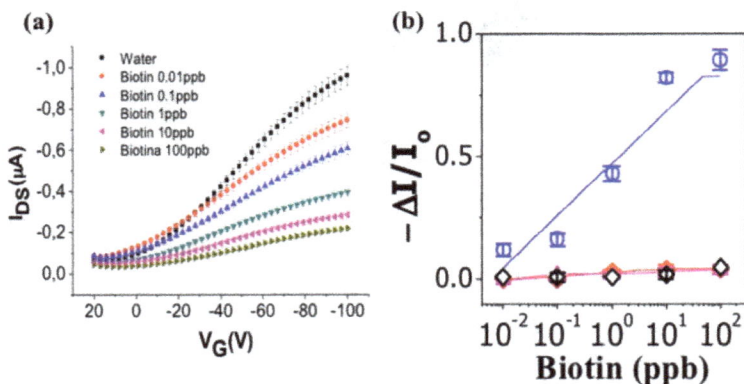

Figure 26.   Typical $I_{DS}-V_G$ curves obtained for a streptavidin FBI-OFET exposed to pure water and biotin solutions at different concentrations. (b) Response of SA FBI-OTFT to different biotin concentrations. Each data point is the $\Delta I/I0$ mean value over three replicates measured on different OTFT devices. Error bars are taken as the relative standard deviations. The response to different biotin concentrations for FBI-OFETs embedding other capturing layers used as negative controls is also reported. Triangles: saturated streptavidin–biotin complexes FBI-OTFT; squares: P3HT-OTFT; diamonds: bovine serum albumin FBI-OTFT. Figure adapted from reference [9] and reproduced with permission.

4.4   Organic bioelectronics probing conformational changes in surface confined proteins

The study of proteins confined on a surface has recently attracted a great deal of attention. The confinement of proteins or peptides on a surface results in the aggregation of structures that may be characterized by different degrees of order. However, it is quite difficult to control the aggregation processes, and to understand how the molecular details impact on and/or determine the macroscopic structure and its relative mechanical and electric properties. A thorough understanding of the self-assembly of highly ordered peptides or biomolecules has, in general, a twofold purpose: *(i)* it can offer insights into the onset of pathological diseases where protein aggregation plays a crucial role (e.g.

neurodegenerative diseases) [75]; *(ii)* it provides valuable information for the bottom-up design and fabrication of nanoscale devices and sensors for biomedical applications [76]. In this respect, organic bioelectronics can be a precious tool to achieve a deeper and more complete understanding of the processes involving protein interfaces, since the energetic and electrostatic contributions play the main role in shaping the response of its devices [77]. Biotin-binding proteins such as avidin (AV), streptavidin (SA) and neutravidin (NAV) have been widely studied as prototype systems to investigate binding-induced conformational changes in biological macromolecules [77]. The crystallographic data available for these proteins suggest that each S(AV) monomer consists of a long chain that loops around the bound biotin. The conformation of these loops is largely dependent on the presence of biotin, with biotin binding reducing the fluctuations of the surrounding residues. Moreover, although biotin binding is essentially non-cooperative [78], a more entangled structure involving S(AV) sub-units has been already postulated based on the experimental evidence that biotin binding in solution increases the protein denaturation temperature in excess of about 30 °C. In addition, the binding process induces a conformational change of the loops surrounding the β-barrel cavity, switching from an *open* to a *closed* state that clamps the bound biotin [79]. Despite the broad range of studies involving biotin complexes, the binding energetics and the dynamics of the process are not fully understood yet. For instance, changes in the electrostatic and polarization properties of these proteins upon binding are still poorly experimentally investigated. On the other hand, computational studies have recently predicted that the electrostatic polarization of biotin-binding proteins provides a substantial contribution to the free energy of its extraordinary strong binding process. This is ascribed to the stabilization of the hydrogen bond between the biotin and the tyrosine moiety while the electrostatic interaction between biotin and the nearby residues appears to be responsible for the enhanced binding affinity. Organic thin-film transistors have been proved to be very powerful in studying processes occurring at biological interfaces. One of the main advantages of this approach is the possibility to address key information by measuring the device multiple figures of merit. Recently, it has been shown that OTFTs allow the accurate determination of the free-energy balances associated with a binding process. In turn, these energetic components (including the electrostatic one), as well as the dynamics involved in conformational changes, have been quantified. In this respect, very sensitive determinations become possible when the direct interfacing of a biological recognition layer with a transistor electronic channel is exploited, as in the FBI-OTFTs. This device configuration has allowed achieving unprecedented performance level in terms of selectivity, accuracy and reproducibility as well. On the other hand, the morphology of the organic semiconductor is of great relevance to obtain high electronic performances, particularly as far as the field-effect mobility and on/off ratio are concerned. It is, thus, crucial to assess the FBI-OTFT fabrication protocol leading to a stable bio-recognition layer / organic semiconductor interface. More importantly, it is critical that the properties and the quality of the dielectric/organic-semiconductor interface are controlled as they also strongly impact on the device performance and stability. Recently, Torsi et al. proposed FBI- and the newly introduced pre-formed complex (PFC)-OTFTs [12] as devices capable to accurately probe biotin-biding events occurring at a protein interface. Both devices embed a layer of SA, NAV or AV that resides right at the dielectric / organic-semiconductor (OSC) interface. The output current change upon binding, taken as the transducing signal, is further decoupled into its component figures of merit. This procedure allows to separately addressing key features of the protein-ligand interaction considering both electrostatic and conformational changes. All-atoms molecular dynamic (MD) simulations were performed on the whole AV tetramer with and without ligands. This provides a molecular rationale for the hypotheses that are needed to interpret the experimental findings. The most relevant steps in the fabrication of an FBI-OTFT are illustrated in Fig. 27 (a), (b) and (c). The FBI layers are deposited by very low speed spin-coating from a water solution, on a SiO_2 surface, resulting in an island-like structure, with features that are inhomogeneous in size and distributions as recently proven by Scanning Electron Microscopy (SEM) characterization. Subsequently, a regioregular poly-3-hexylthiophene (P3HT) p-type OSC is deposited by spin-coating from chloroform on top of the FBI layers, see Fig. 27 (a), smoothly covering the protein clusters. It has been proven that the P3HT forms a layer full of hollows and voids, covering the protein cluster.[65] Finally, a row of source (S) and drain (D) contact pads is patterned on top of the OSC. The device is exposed to a fixed aliquot of 1 part-per-billion (1 ppb, 5 nM) biotin solution, Fig. 27 (b), and, after incubation, the unreacted biotin is removed by washing with water, Fig. 27 (c). The PFC-OTFT configuration fabrication steps are shown in Fig. 27 (e), (f) and (c). Also in this case the biotin-binding proteins are deposited by slow spin-coating on the SiO_2 surface, as shown in Fig. 27 (e). At variance with the FBI-OTFT configuration, exposure to biotin, and eventually complex formation, is achieved before OSC deposition. To this aim, two deposits of protein are placed on two different locations in the same SiO_2 substrate: one is exposed to water and the other to the 1 ppb biotin solution. The same aliquots of solution as for the case of the FBI-OTFTs are used. After incubation, the unreacted biotin is washed away. For the sake of clearness, in Fig. 27 (f) only the region of the protein exposed to biotin is represented. The last step consists in the deposition of P3HT on the layer of pre-formed complexes along with the row of S and D contact pads placed on top of the OSC, as in Fig. 27 (c).Typical I-V output characteristics (source-drain current, I_{DS} *vs.* source-drain voltage, V_{DS} with the gate voltage, V_G ranging from +

Organic Bioelectronics for Life Science and Healthcare   Materials Research Forum LLC
Materials Research Foundations **56** (2019) 1-70   doi: https://doi.org/10.21741/9781644900376-1

20 V to - 100 V in steps of -10 V) measured for the FBI and PFC OTFTs embedding AV-biotin complexes, are reported in Fig. 27 (d) and (g), respectively.

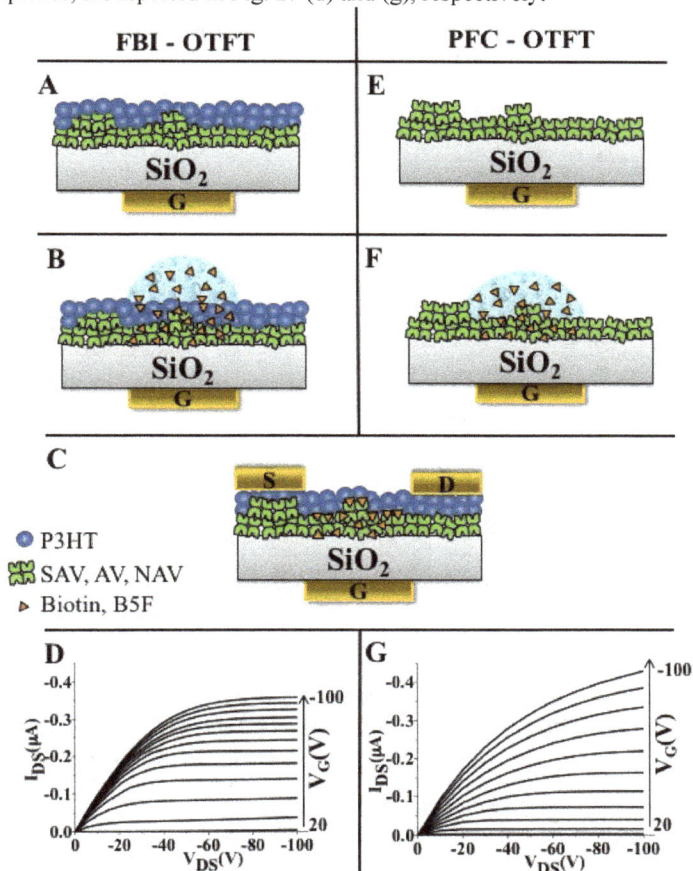

Figure 27. Schematic illustrations of the fabrication steps implemented for the realization of the functional bio-interlayer (FBI) OTFT (a-c) and a pre-formed complex (PFC) OTFT (e, f, c). P3HT is reported in blue, SA, AV or NAV layer is shown in green, while the orange triangles represent biotin or B5F. I-V characteristics (I_{DS} vs. V_{DS} measured for V_G values ranging between + 20 V to - 100 V in steps of - 10 V) for the FBI (d) and PFC (g) structures respectively. The devices integrate an AV layer and have been exposed to 1 ppb of Biotin. Figure adapted from reference [12] and reproduced with permission.

The curves exhibit a good level of current modulation as well as nicely shaped linear and saturated regions along with low leakage currents at low V_{DS}. Typical $\sqrt{I_{DS}}$ vs. V_G curves for AV embedding FBI and the PFC OTFTs measured after exposure to water and 1 ppb of biotin, are reported in Fig. 28 (a) and (b), respectively.

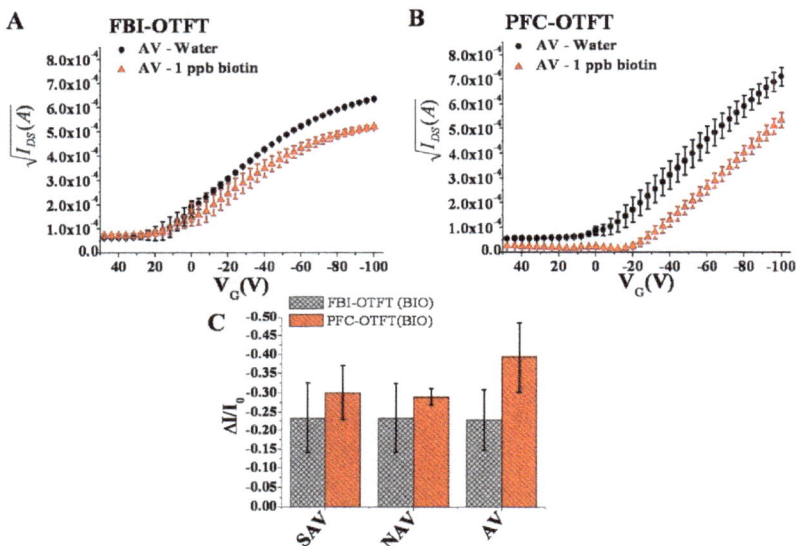

Figure 28.   *Square root of the source-drain channel current vs. V_G for AV embedding OTFTs when exposed to water and to 1 ppb of Biotin. Data relative to (a) FBI-OTFT and (b) PFC-OTFT, respectively. Data are relevant to three repeated measurements and are reported as the average value along with the relevant RSDs. (c) Normalized current response $\Delta I/I_0$ for AV embedding FBI and PFC-OFET upon exposure to 1 ppb of Biotin. Data are relevant to an ensemble of measurements acquired on nine different devices (reproducibility error) and are reported as the average value along with the relevant RSDs. Figure adapted from reference [12] and reproduced with permission.*

Organic Bioelectronics for Life Science and Healthcare   Materials Research Forum LLC
Materials Research Foundations **56** (2019) 1-70   doi: https://doi.org/10.21741/9781644900376-1

Visual inspection of these curves already reveals a change in the slope of the linear region of the curves for the FBI-OTFT devices, while no significant shift along the V_G axis can be observed upon binding. The curves measured for the PFC-OTFT show, on the contrary, a fixed slope and a V_T shift. A linear interpolation of these curves allows quantifying these changes by extracting the devices figures of merit, according to Eq. (9). As discussed in the previous section, in the FBI-OTFT configuration the changes in the I-V curves slope upon binding are ascribable only to changes of the OSC field-effect mobility. The current-voltage transfer curve (I_{DS}, *vs.* V_G at $V_{DS} = -80$ V) is measured before and after the formation of the biotin-protein complex. I_{DS} measured at $V_G = -100$ V after binding is taken as the *signal* value (I). The prior measurement with the same setting but employing a water (biotin-free) solution) provides the *base-line* value (I_0). The normalized current response is evaluated from the measured transfer characteristics, according to Eq. (12). The $\Delta I/I_0$ fractional decreases (averaged over 9 different devices) are shown in Fig. 28 (c), turned out to be very similar for the FBI and the PFC OTFTs. Similar conclusions can be drawn also for the device embedding AV and NAV. This evidence demonstrates that, despite the differences in the FBI- and PFC-OTFT preparation procedures, the two classes of devices yield a comparable number of TFT detectable biotin-binding events. Such an occurrence is expected as, from data gathered in a previous study [80], the percolation of a biotin solution through the OSC in an FBI-OTFT structure is expected to have a yield well in excess of 80 %. Remarkably, while the fractional changes of the current are constant, those of the μ_{FET} and the V_T figures of merit, are significantly different for the FBI- and the PFC-OTFTs. These relative changes can be correlated to the I_{DS} fractional changes. Assuming C_i to be constant, Eq. (13) holds, up to first order. Fig. 29 (a) shows the V_T values measured for the devices embedding the three proteins.

Fig. 29 (b) displays a schematic diagram of the energy levels for an OTFT embedding an AV protein layer at the OSC / dielectric interface. V_T fractional changes for the FBI- and PFC-OTFTs upon exposure to biotin and biotin (5-fluorescein) conjugated (B5F) are reported in Fig. 29 (c). Due to the presence of the carboxyl group, bare biotin carries a net negative charge at the experimental pH of 7.5, while in the B5F this site is bounded to fluorescein thus making B5F neutral. Significant V_T fractional changes are observed for the PFC-OTFTs. Apparently, the $\Delta V_T/(V_G - V_{T0})$ values are all positive. The $(V_G - V_{T0})$ term is always negative, being the V_{T0} *base-line* threshold voltage values consistently less negative than the V_G value at which the fractional changes are evaluated ($V_G = -100$ V). Eventually, it can be stated that all the ΔV_T shifts evaluated are negative. The data in Fig. 29 (c) show also that a negligible V_T fractional change is measured in the PFC-OTFTs for all the protein-B5F complexes. Strikingly, Fig. 29 (c) shows also how biotin binding has a negligible impact on the V_T of the FBI-OTFT devices. Fig. 29 (d) compares the fractional decreases of μ_{FET} for FBI and PFC devices. In the FBI configuration, a decrease of 0.20 ± 0.07 is consistently observed for all three proteins. On the contrary, PFC-OTFT devices produce much more scattered μ_{FET} relative changes showing no correlated values for the three proteins upon biotin binding. The structures of AV and SA were obtained from the Protein Data Bank (PDB) and the protonation state of the residues as well as the charge distributions were accounted for considering the charge associated with the protein residues at the pH of the AV and SAV solutions (7.5). The charge distribution of NAV could not be evaluated since its structure is not available in the PDB. At pH 7.5, arginine and lysine residues carry a positive charge, while glutamate and aspartate ones are negatively charged. To mimic the experimental conditions corresponding to a layer of densely packed dry proteins, MD simulations were performed with an implicit solvent model, assuming a relative permittivity ε_r of 3, typical for a protein system [20]. This was applied on the AV tetramer (taken as a model system for the three proteins) to investigate how the tight interactions with biotin and B5F affect the protein dipole moment and the conformation of the loops surrounding the binding pocket. The L3,4 loop (connecting strands $\beta3$ and $\beta4$ – residues from position 35 to position 45) and the L5,6 (connecting strand $\beta5$ and $\beta6$ – residues from position 70 to position 75) of all the considered systems, are shown in Fig. 30 (a). In Fig. 30 (b) the magnitude of the electric dipole moments computed for AV as well as for its complexes with biotin and B5F, are shown. A single value averaged along the entire MD trajectory is reported for each considered system. Specifically, the bare AV and the AV-B5F complexes exhibit close dipole moments of 154 ± 35 D and 118 ± 29 D, respectively, while the value computed for the AV-biotin complex is 283 ± 33 D. The time-dependence of the computed dipoles supports the evidence that the AV-biotin holds an associated electric dipole moment that approximately doubles the value found for the system in which the biotin binding pocket is empty. This difference is stable for almost the entire time-span of MD simulations. The Root Mean Square Fluctuations (RMSF) values were computed for the L3,4 and the L5,6 loops.The results, reported in Fig. 30 (c), clearly show that the molecular flexibility of both loops is strongly limited upon binding of both biotin and B5F. This is actually consistent with reported experimental and computational data showing how an *open* arrangement of L3,4 is not allowed upon biotin binding in AV, while biotin forces L3,4 in a *closed* conformation [81]. In addition, MD simulations indicate that also the L5,6 has a similar fate. In this case the tendency to adopt a *closed* conformation is even more pronounced. In this respect, the RMSF values and the time-dependent evolutions of the Root Mean Square Deviations (RMSD) values strongly support the hypothesis that the molecular behavior of L3,4 and L5,6 is similarly influenced by biotin and B5F binding.

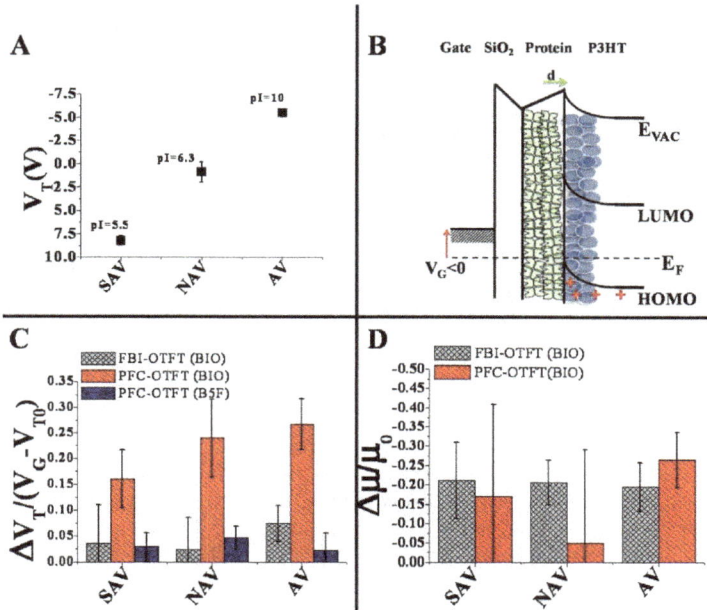

Figure 29. (a) Threshold voltage (V_T) values measured for OTFT devices embedding SA, NAV and AV. Data are relevant to nine measurements acquired on different devices and are reported as the average value along with the relevant RSDs. (b) Band diagram of an OTFT embedding the protein layer underneath the OSC, upon application of a negative gate bias that drives the transistor in the accumulation mode. (c) Fractional changes of the threshold voltage evaluated for the FBI configuration upon exposure to 1 ppb of Biotin and for the PFC configuration upon exposure to 1 ppb of Biotin and B5F. (d) Fractional changes of the field-effect mobility evaluated for the FBI and PFC configurations upon exposure to 1 ppb of Biotin. Figure adapted from reference [12] and reproduced with permission.

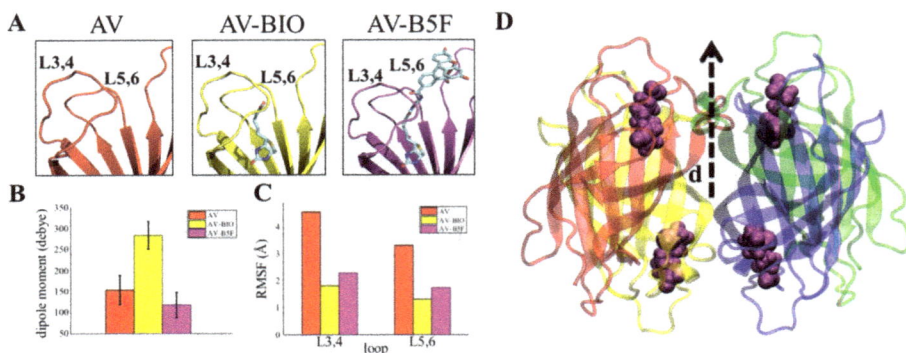

Figure 30. (a) Residues close to the binding pocket of the AV (left), AV-Biotin complex (center) and Avidin-B5F complex (right). (b) Histogram showing the dipole moment (Debye) averaged along the obtained molecular dynamics trajectory for AV (red), AV-Biotin (yellow) and AV-B4F (magenta) systems. A single value is reported for each system averaging statistics from the two monomers. (c) Histogram showing the computed Root Mean Square Fluctuations (RMSF) (Å) values for AV (red), AV-BIO (yellow) and AV-B4F (magenta) systems considering both L3,4 and L5,6 loops. For each system, the averaged RMSF, computed from the two monomers, is reported. (d) Overall tetrameric structure of the simulated AV with bound biotin molecules. Dipole directions are shown (not to scale). Figure adapted from reference [12] and reproduced with permission.

The conformational changes occurring in proteins immobilized underneath the electronic channel of an OTFT can impact on the electronic properties of the device in a two-fold fashion:*(i)* if the gate electrode or the OSC electrochemical potential is affected, the measured shift in V_T is related to a change in the interfacial protein layer electrostatic properties; *(ii)* if the conformational change impacts on the transport properties of the OSC, μ_{FET} is affected by the structural / morphological changes occurring in the protein layer. Interestingly, with the FBI- and the PFC-OTFT configurations, the V_T and μ_{FET} contributions to the measured current fractional changes can be decoupled and separately addressed. Hence, the combined use of these two structures allows gaining a rather complete picture of the electrostatic and conformational changes occurring to a protein while it interacts with its affinity ligand. No other available techniques to study thin-films or self-assembled monolayers of proteins deposited on a solid surface are capable to provide these important pieces of information. Here, we first discuss why V_T shifts allow quantifying the electrostatic changes occurring at the OSC / dielectric interface. The

Organic Bioelectronics for Life Science and Healthcare   Materials Research Forum LLC
Materials Research Foundations **56** (2019) 1-70   doi: https://doi.org/10.21741/9781644900376-1

capacitance coupling between the SiO_2 gate-dielectric and the OSC determines the conditions to generate a two-dimensional (2D) electronic channel of free-charges in the OSC [59]. These charges are injected and eventually drift under the V_{DS} bias. However, although when V_G is applied the conditions for the 2D free-charge accumulation at the dielectric / semiconductor interface are set, the actual I_{DS} current flow in the OSC does not start (under a given applied V_{DS}) until V_G exceeds the transistor V_T. In a trap-free device, V_T is in fact the bias needed to offset the built-in potential generated by the mismatch existing between the metal-gate and the OSC electrons electrochemical potentials or Fermi levels. As V_G equals V_T, the energy barrier is flattened and, when the V_G bias exceeds V_T, further injected charges can drift through the channel with a given μ_{FET}. Of interest to this study is the occurrence that the presence of a protein interlayer between the gate-dielectric and the OSC will determine a further shift of the TFT V_T according to the protein electrostatic properties. Similarly, also the formation of the complex between such an immobilized protein and its affinity ligand will cause a V_T shift that quantifies the electrostatic properties of the protein as the complex is formed. The protein layer electrostatic properties can be modelled by assuming that the positively and negatively charged residues characterizing the SA, NAV and AV tetramers are surrounded by the counter-ions coming from the buffer solution used during the protein purification. The charged residues and the associated counter-ions give rise to a dipole moment whose resultant has been evaluated for the whole AV tetramer, taken as a model for the three proteins investigated, in MD simulations with implicit solvent, considering a relative permittivity $\varepsilon_r = 3$, as reported in Fig. 30 (b). Due to the rather uneven morphology, the overall protein film thickness used in the present study is not uniform, although the estimated protein surface density ($10 \, pM/cm^2$) is compatible with the deposit of a fully packed monolayer. Nonetheless, in the proposed device structures, the uniformity of the protein layer is not critical for the following two main reasons: *(i)* the contribution of its capacitance to the gating system is negligible; *(ii)* only the outermost portion of the protein layer will influence anyhow the electrostatics of the OSC. The latter relies on the assumption that each protein can be represented by a permanent dipole moment that induces a dipole in the OSC (Debye interaction) with the interaction force being a function of d^{-6} [82]. Indeed, the screening length can be computed as the cutoff radius of the Debye interaction force whose estimate for the AV / OSC interface is 1.3 nm. Consequently, only a permanent dipole moment residing in the protein at the interface can affect the OSC interfacial region. In a TFT configuration this is the only portion of the OSC film that matters, as it is where the 2D electronic transport takes place. The interface between the *ca.* 3 nm top protein monomeric subunit and the OSC 2D electronic channel (< 5nm) falls within this range. In other words, the TFT sees only the interfacial monolayer of proteins independently from the presence of multilayer protein clusters. As the symmetric nature of the tetrameric biotin-binding proteins is broken by the presence of the interface with the OSC, the protein layer will induce a net dipole moment in the OSC even if the protein layer is not ordered. This happens because the protein layer induces a net dipole moment as the closer monomeric subunits of the protein will have a much stronger interaction with the OSC than the farther off ones. Clearly, the dipole moment orientation in the OSC interfacial layer will have the same orientation as that of the protein or of the complex under investigation. It is also important to remark that the OSC highly conformable nature allows this layer to follow very nicely the unevenness of the cast protein layer, resulting in a large protein / OSC interface. Data in Fig. 29 (a) show the V_T shifts, measured when the SA, NAV and AV proteins are embedded into the OTFT. The measured V_T values do asses that an overall permanent dipole moment can be associated with the protein interfacial layer that, in the case of the SA, is oriented with its negative pole facing the OSC [83,84] while the opposite direction is observed for the AV as depicted in the schematic model reported in Fig. 29 (b). As expected, a very weak dipole is associated to NAV. MD simulations performed on the AV tetramer confirm the conclusion that this protein is indeed endowed with an intrinsic permanent dipole moment. Fig. 30 (d) displays the AV tetramer and the direction of its dipole moment associated to biotin binding. It is important to compare our results to a recent study performed on an OTFT including a self-assembled monolayer of differently charged quartz-binding polypeptides attached to the SiO_2 dielectric at the interface with a pentacene OSC [85]. In this study, the direction of the V_T shifts are assumed to follow the orientation of the fixed dipoles associated with the polypeptides and the counter-ions distributed as the outermost layer at the interface with the OSC. However, the investigation of the changes in the electrostatic properties of the protein layer upon formation of the protein-biotin complex, to the best of our knowledge, has never been explored before. The data reported in Fig. 28 (c) quantify the electrostatic contribution to the fractional current change for the SA, NAV and AV proteins and their complexes with the biotin or B5F ligands. The positive $\Delta V_T/(V_G-V_{T0})$ implies that the formation of the pre-formed complexes shift V_T towards more negative values. Therefore, upon biotin binding, the induced dipole moment increases in absolute value and is oriented in the direction opposite to the one induced by the polarization field of the gate. This means that the dipole moment of the complex layer points (or has a significant component directed) towards the OSC layer. Moreover, for all proteins, a systematically larger $\Delta V_T/(V_G-V_{T0})$ value is observed for the PFC-OTFTs as compared to the FBI-OTFTs. This occurrence can be explained by considering that the formation of the induced dipole orientation in the OSC can be achieved only in the PFC device structure as the binding-induced dipole moment change occurs before the OSC is deposited. Indeed, during the OSC deposition process, the permanent dipole of the cast proteins can

Organic Bioelectronics for Life Science and Healthcare   Materials Research Forum LLC
Materials Research Foundations **56** (2019) 1-70   doi: https://doi.org/10.21741/9781644900376-1

induce a net polarization only when the OSC solvent molecules are present. This is in line with published theoretical results, proving how the presence of residual solvent molecules increases the free space available to the polymeric chains, thus allowing a rapid chain reorientation [86]. Under such circumstances the OSC can rearrange its structure to minimize the electrostatic interaction energy with the protein layer deposited beneath. Eventually, the dipole moment induced in the OSC at the interface with the cast proteins will be *quenched* when the OSC is dried to form the TFT electronic film. In the FBI-OTFT the OSC is already deposited before the binding process and therefore its electrostatic structure is determined following the empty proteins dipole moment and cannot rearrange due to the lack of solvent molecules. Apparently, a very week V_T shift is observed also when the B5F-binding is detected. This occurrence implies that the charge carried by the ligand critically matters, since only charged species are able to cause a large electrostatic rearrangement in each of the three proteins. The conformational change is therefore related to the charge carried by the biotin ligand and it is this one that generates significant charge redistribution on the protein whole structure upon binding. The increase of the dipole moment associated to the electrostatic changes upon biotin-binding is also proved by the MD simulations whose results are shown in Fig. 3.25(b). A priori, one would expect contribution of this dipole layer to the V_T shift according to the Helmholtz equation [84]:

$$\Delta V_{T,Theor} = \frac{M_{dip}}{\varepsilon_0 \varepsilon_r A} \tag{13}$$

where M_{dip} is the dipole moment computed for the AV-biotin complex, ε_0 and ε_r are the vacuum and the relative permittivity while A is the area of the outermost portion of the protein actually impacting on the electrostatics of the OSC 2D interfacial layer. The V_T shift evaluated for the AV-biotin complex according to Eq. (13) has been found to be of 23.73 V. This value is in excellent agreement with the measured V_T shift of 29.04 ± 5.45 V, within one standard deviation. As shown in Fig. 30 (d), this dipole is also found to be oriented with its positive pole pointing towards the OSC layer in correspondence with the entrance of the AV binding pocket, thus confirming the experimental evidences. Also perfectly in line with the experimental evidences is the finding that the binding to B5F does not significantly affect the dipole moment of the protein, see Fig. 30 (b). In this case the biotin carboxylic group, responsible for the biotin negative charge, is conjugated with the fluorophore 5-fluorescein, which interacts with the lysine and tryptophan functional groups of the AV L7,8 loop, thus neutralizing the positive pole. Another quite relevant aspect of this study involves the quantification of the μ_{FET} changes occurring upon biotin-binding to the three studied proteins. The experimental observations reported in Fig. 28

(d) prove how a significant fractional decrease of the mobility is detected for SA, NAV and AV proteins with both FBI and PFC-OTFT devices. However, while a consistent μ_{FET} decrease for all proteins is seen with the FBI-OTFT, much more scattered data are gathered with the PFC. Such a misalignment between the two approaches can be explained considering that the observed mobility decreases can be ascribed to the disorder induced, locally, by the proteins structural / conformational changes that impact on the OSC transport properties. This effect is clearly much more sensitively detected if the OSC is already deposited on the proteins when binding occurs. This is the case only for the FBI device. A quantitative support to the claim that disorder is induced in the P3HT by the proteins conformational changes, is provided in the following. It is well assessed that the L3,4 and L5,6 loops surrounding the biotin-binding pocket adopt a *closed* conformation upon complex formation. Our MD simulations, carried out on the AV tetramer, confirm that L3,4 and L5,6 loops switch from an *open* to a *closed* conformational state in terms of a decrease in the Root Mean Square Fluctuations (RMSFs) from approximately 4 Å to 2 Å, see Fig. 30 (c). Typically, P3HT self-assembles in lamellae whose spacing determines the transfer integral value and ultimately the charged species hopping mobility [87]. A drop of almost one order of magnitude has been computed for the transfer integral upon increase of the P3HT inter-chain distance of about 1 Å, as shown in Fig. 31. MD simulation quantifies the disordered induced by the fluctuation of the loops in about 2 Å of displacement, thus supporting quantitatively the claim that such a conformational change generates a local disorder on the Å length-scale that may already deeply affect the P3HT delocalization length leading to a drop in the P3HT μ_{FET} of about 20 % as the protein biotin complex formation is accomplished. The scattered mobility changes for the case of the PFC-OTFTs can be accounted for by considering that the induced electrostatic rearrangement can impact also on the mobility as this localized effect can generate a disorder at a length scale that is shorter than the OSC delocalization length.

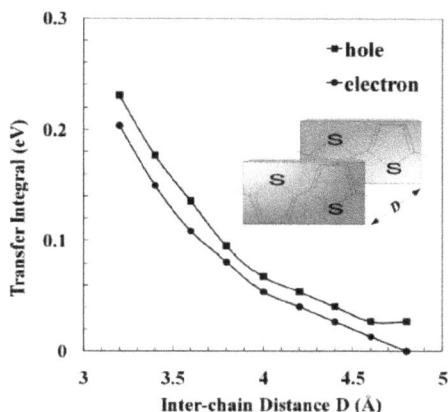

Figure 31.   The transfer integrals of electron and hole along the interchains as a function of the intermolecular separation distance D. Figure adapted from reference [87] and reproduced with permission.

Conclusions

The interfacing of biomaterials to electronic devices is one of most challenging research field that has relevance for both fundamental studies and for the development of highly performing biosensors. In fact, there are many intriguing examples of proteins performing crucial tasks for living systems while segregated on a surface. In this perspective, bioelectronics may allow to achieve important pieces of information on biofunctional interfaces. The acquired knowledge can also be critically important to realize ultrasensitive biological and chemical sensors. Therefore, the thermodynamic tools that can allow comparing biological processes occurring in solution and on a surface have been discussed in details. The most widely used techniques to anchor biomolecules to a surface have been briefly reviewed. In this perspective, the advantages offered by organic thin film transistors as biosensing platforms are discussed along with an overview the operating principles and the possible configurations suitable for biosensing applications. The analytical figures of merit that need to be assessed have been discussed in an operative manner, as well. Particular attention is paid to the Funtional Bio-Interlaye OTFT biosensing platform. The main examples on volatile anesthetic and protein detection with FBI-OTFT have been briefly discussed. Details on the devices fabrication procedure, morphological and electrical characterizations, analytical modelling and molecular dynamic simulations are reported as well.

Specifically, the experiments conducted on the Functional Bio-Interlayer (FBI)-OTFT and on the Pre-Formed Complex (PFC)-OTFT have been presented. These configurations have been proven to separately address the changes occurring in the protein-ligand complex morphology and dipole moment.

References

[1]   J. Rivnay, R.M. Owens, G:G. Malliaras, The Rise of Organic Bioelectronics. Chem. Mater. 26 (2014) 679-685. b) Rinay J., Inal S., Salleo A., Owens R.M., Berggren M., Malliaras G.G., Nat. Mater. Review 3 (2018), 17086. https://doi.org/10.1021/cm4022003

[2]   I. Willner, R. Baron, B. Willner, Integrated Nanoparticle-Biomolecule Systems for Biosensing and Bioelectronics. Biosens Bioelectron. 22 (2007) 1841-1852. https://doi.org/10.1016/j.bios.2006.09.018

[3]   D. Khodagholy, et al. Neurogrid: Recording Action Potentials from the Surface of the Brain. Nat. Neurosci. 18 (2014) 310-315. https://doi.org/10.1038/nn.3905

[4]   R.D. Grange, J.P. Thompson, D.G. Lambert, Radioimmunoassay, Enzyme And Non-Enzyme-Based Immunoassays. Br. J. Anaesth. 112 (2014) 213-216. https://doi.org/10.1093/bja/aet293

[5]   Homola, J. Surface Plasmon Resonance Sensors for Detection of Chemical and Biological Species. Chem. Rev.108(2), 462-493 (2008). https://doi.org/10.1021/cr068107d

[6]   L. Anson, Membrane Protein Biophysics. Nature Insight 459 (2009) 343-385. https://doi.org/10.1038/459343a

[7]   M.Y. Mulla, et al. Capacitance-Modulated Transistor Detects Odorant Binding Protein Chiral Interactions. Nature Commun. 6:6010 (2015) 1-9. https://doi.org/10.1038/ncomms7010

[8]   Wang, C., Dong, H., Hu, W., Liu, Y. & Zhu, D. Semiconducting π-Conjugated Systems in Field-Effect Transistors: A Material Odyssey of Organic Electronics. Chem. Rev.112 (2011) 2208–2267. https://doi.org/10.1021/cr100380z

[9]   L. Torsi, L., M. Magliulo, K. Manoli, G. Palazzo, Organic Field-Effect Transistor Sensors: a Tutorial Review. Chem. Soc. Rev. 42 (2013) 8612-8628. https://doi.org/10.1039/c3cs60127g

[10]   K. Manoli, et al. Printable Bioelectronics to Investigate Functional Biological Interfaces. Angew. Chem. Int. Ed. 54 (2015) 12562-12576. https://doi.org/10.1002/anie.201502615

[11]   L. Torsi, A. Dodabalapur, L. Sabbatini, , P.G. Zambonin, Multi-Parameter Gas Sensors Based On Organic Thin-Film-Transistors. Sens. Actu. B67 (2000) 312–316. https://doi.org/10.1016/S0925-4005(00)00541-4

[12]   E. Macchia, et al. Organic Bioelectronics Probing Conformational Changes In Surface Confined Proteins. Scientific Reports 6 (2016) 1-12. https://doi.org/10.1038/srep28085

[13]   F. Ebisawa, T. Kurokawa, S. Nara, Electrical Properties of Polyacetylene / Polysiloxane Interface. J. Appl. Phys. 54 (1983) 3255–3259. https://doi.org/10.1063/1.332488

[14]   C. Wang, H. Dong, W. Hu, Y. Liu, D. Zhu, Semiconducting π-Conjugated Systems in Field-Effect Transistors: A Material Odyssey of Organic Electronics. Chem. Rev.112 (2011) 2208–2267. https://doi.org/10.1021/cr100380z

[15]   D. Tobjork, R. Osterbacka, Paper Electronics. Adv. Mater. 23 (2011) 1935-1961. https://doi.org/10.1002/adma.201004692

[16]   G. Mattana, P. Cosseddu, B. Fraboni, G.G. Malliaras, J.P. Hinestroza, A. Bonfiglio, Organic electronics on natural cotton fibres. Organic Electronics 12 (2011) 2033-2039. https://doi.org/10.1016/j.orgel.2011.09.001

[17]   L. Basirico, P. Cosseddu, A. Scida, B. Fraboni, G.G. Malliaras, A. Bonfiglio, Electrical Characteristics Of Ink-Jet Printed, All-Polymer Electrochemical Transistors. Organic Electronics 13 (2012) 244-248. https://doi.org/10.1016/j.orgel.2011.11.010

[18]   M. Magliulo, et al. Printable and Flexible Electronics: from TFTs to Bioelectronic Devices. J. Mater. Chem. C.3 (2015) 12347–12363. https://doi.org/10.1039/C5TC02737C

[19]   L. Basirico, P. Cosseddu, B. Fraboni, A. Bonfiglio, Inkjet Printing Of Transparent, Flexible, Organic Transistors. Thin Solid Films 520 (2011) 1291-1294. https://doi.org/10.1016/j.tsf.2011.04.188

[20]   S. M. Sze, K.K .Ng, Physics of semiconductor devices (John Wiley & Sons, ed. 3, 2006). https://doi.org/10.1002/0470068329

[21]   D. Khodagholy, et al. High Transconduttance Organic Electrochemical Transistors. Nat. Commun. 4 (2013) 2133. https://doi.org/10.1038/ncomms3133

[22]   H. Kim, et al. Electrolyte-Gated Transistors for Organic and Printed Electronics. Ad. Mater. 25 (2013) 1822-1846. https://doi.org/10.1002/adma.201202790

[23]   M. Demelas, et al. Charge Sensing by Organic Charge-Modulated Field Effect Transistors: Application to the Detection of Bio-Related Effects. J. Mater. Chem. B1 (2013) 3811-3819. https://doi.org/10.1039/c3tb20237b

[24]   Magliulo, M., Manoli, K., Macchia, E., Palazzo, G. &Torsi, L. Tailoring Functional Interlayers in Organic Field-Effect Transistor Biosensors. Adv. Mater., DOI: 10.1002/adma.201403477 (2014). https://doi.org/10.1002/adma.201403477

[25]   P. Bergveld, Thirty years of ISFETOLOGY. What happened in the past 30 years and what may happenin the next 30 years. Sensors and Actuators B88 (2003) 1–20. https://doi.org/10.1016/S0925-4005(02)00301-5

[26]   S. Lai, M. Demelas, G. Casula, P. Cosseddu, M. Barbaro, A. Bonfiglio, Ultralow Voltage, OTFT-Based Sensor for Label-Free DNA Detection. Adv. Mater.25 (2013) 103–107. https://doi.org/10.1002/adma.201202996

[27]   M.L. Hammock, O. Knopfmacher, B.D. Naab, J.B.H Tok, Z. Bao, Investigation of Protein Detection Parameters using Nano-Functionalized Organic Field-Effect Transistors. ACS Nano 7 (2013) 3970- 3980. https://doi.org/10.1021/nn305903q

[28]   M.D. Angione, et al. Interfacial Electronic Effect in Functional Biolayers Integrated into Organic Field-Effect Transistors, PNAS 109 (2012) 6429-6434. https://doi.org/10.1073/pnas.1200549109

[29]   L. Torsi, et al. A sensitivity-enhanced field-effect chiral sensor. Nat. Mater.7 (2008) 412–417. https://doi.org/10.1038/nmat2167

[30]   B. Crone, et al. Electronic sensing of vapors with organic transistors. Appl. Phys. Lett.78 (2001) 2229-2231. https://doi.org/10.1063/1.1360785

[31]   E. Macchia, et al. An Analytical Model for Bio-Electronic Organic Field-Effect Transistor Sensors. App. Phys. Lett. 103 (2013) 103301. https://doi.org/10.1063/1.4820347

[32]   F.N. Ishikawa, et al. A Calibration Method for Nanowire Biosensors to Suppress Device-to-Device Variation. ACS Nano 3 (2009) 3969-3976. https://doi.org/10.1021/nn9011384

[33]   F. Rusmini, Z. Zhong, J. Faijen, Protein Immobilization Strategies for Protein Biochips. Biomacromolecules 8 (2007) 1775-1789. https://doi.org/10.1021/bm061197b

[34]   P. Jonkheijm, D. Weinrich, H. Schroder, C.M. Niemeyer, H. Waldmann, Chemicals Strategies for Generating Protein Biochips. Angew. Chem. Int. Ed. 47 (2008), 9618-9647. https://doi.org/10.1002/anie.200801711

[35]   H.J. Lee, Y. Yan, G. Marriott, R.M. Corn, Quantitative Functional Analysis Of Protein Complexes On Surfaces. J. Physiol. 563 (2005) 61-71. https://doi.org/10.1113/jphysiol.2004.081117

Materials Research Forum LLC
doi: https://doi.org/10.21741/9781644900376-1

[36]   P. Stoliar, et al. DNA Adsorption Measured With Ultra-Thin Film Organic Field Effect Transistors. Biosens. and Bioelectron.24 (2009)2935–2938. https://doi.org/10.1016/j.bios.2009.02.003

[37]   N. Patel, et al. Immobilization of Protein Molecules onto Homogeneous and Mixed Carboxylate-Terminated Self-Assembled Monolayers. Langmuir 13 (1997) 6485-6490. https://doi.org/10.1021/la970933h

[38]   J.W. Lee, S.J. Sim, S.M. Cho, J. Lee, Characterization of a self-assembled monolayer of thiol on a gold surface and the fabrication of a biosensor chip based on surface Plasmon resonance for detecting anti-GAD antibody. Biosensors and Bioelectronics 20 (2005) 1422–1427. https://doi.org/10.1016/j.bios.2004.04.017

[39]   G. MacBeath, S.L. Schreiber, Printing Proteins as Microarrays for High-Throughput Function Determination. Science 289 (2000) 1760-1763.

[40]   T. Viitala, I. Vikholm, J. Peltonen, Protein Immobilization to a Partially Cross-Linked Organic Monolayer. Langmuir 16 (2000) 4953-4961. https://doi.org/10.1021/la990817+

[41]   M.A. Jongsma, R.H. Litjens, Self-Assembling Protein Arrays on DNA Chips by Auto-Labeling Fusion Proteins With a Single DNA Address. Proteomics 6 (2006) 2650-2655. https://doi.org/10.1002/pmic.200500654

[42]   R. Wacker, H. Schroder, C.M. Niemeyer, Performance of Antibody Microarrays Fabricated by either DNA-Directed Immobilization, Direct Spotting, or Streptavidin-Biotin Attachment: a Comparative Study. Anal. Biochem. 330 (2004) 281-287. https://doi.org/10.1016/j.ab.2004.03.017

[43]   S.C. Rizzi, J.A. Hubbell, Recombinant Protein-co-PEG Networks as Cell-Adhesive and Proteolytically Degradable Hydrogel Matrixes. Part I: Development and Physicochemical Characteristics. Biomacromolecules 6 (2005) 1226-1228. https://doi.org/10.1021/bm049614c

[44]   C. Mateo, et al. Epoxy Sepabeads: a Novel Epoxy Support for Stabilization of Industrial Enzymes Via Very Intense Multipoint Covalent Attachment. Biotechnol. Prog.18 (2002) 629-634. https://doi.org/10.1021/bp010171n

[45]   J. Turková, Oriented Immobilization of Biologically Active Proteins as a Tool for Revealing Protein Interactions and Function. Journal of Chromatography B722 (1999) 11–31. https://doi.org/10.1016/S0378-4347(98)00434-4

[46]   S. Werner, W. Machleidt, Isolation of Precursors of Cytochrome Oxidase From Neurospora Crassa: Application of Subunit-Specific Antibodies and Protein a From Staphylococcus Aureus. Eur. J. Biochem. (1978) 90,99. https://doi.org/10.1111/j.1432-1033.1978.tb12579.x

[47]   M. Wilchek, E.A. Bayer, Introduction to Avidin-Biotin Technology. Methods Enzymol. 184 (1990) 1-746. https://doi.org/10.1016/0076-6879(90)84256-G

[48]   M. Lehnert, et al. Adsorption and Conformation Behavior of Biotinylated Fibronectin on Streptavidin-Modified TiOx Surfaces Studied by SPR and AFM. Langmuir 27 (2011) 7743-7751. https://doi.org/10.1021/la200908h

[49]   I. Vikholm-Lundin, et al. Cysteine-Tagged Chimeric Avidin Forms High Binding Capacity Layers Directly On Gold. Sens. Actuat. B-Chem 171 (2012) 440-448. https://doi.org/10.1016/j.snb.2012.05.008

[50]   J. Homola, Present and Future of Surface Plasmon Resonance. Biosensors. Anal. Bioanal. Chem. 377 (2003) 528–539. https://doi.org/10.1007/s00216-003-2101-0

[51]   E.A.Vogler, Structure and Reactivity of Water at Biomaterial Surfaces. Adv. Colloid. Interfac. 74 (1998) 69–117. https://doi.org/10.1016/S0001-8686(97)00040-7

[52]   I.M. Klotz, Ligand–Receptor Energetics. Wiley: New York or Haynie DT (1997).

[53]   A. Halperin, A. Buhot, E.B. Zhulina, On the Hybridization Isotherms of DNA Microarrays: the Langmuir Model and its Extensions. J. Phys. Condens. Matter18 (2006) S463–S490. https://doi.org/10.1088/0953-8984/18/18/S01

[54]   D.K. Chattoraj, K.S. Birdi, Adsorption and the Gibbs Surface Excess. Plenum Press: New York; 257–298, 339–383 (1984). https://doi.org/10.1007/978-1-4615-8333-2_8

[55]   A. Adamson, A.P. Gast, Physical Chemistry of Surfaces (6th edn). Wiley: New York; 48–100, 90–430, 599–684 (2000).

[56]   R. Sips, On the Structure of a Catalyst Surface. J. Chem. Phys. 16 (1948) 490-495 https://doi.org/10.1063/1.1746922

[57]   S. Cazalbou, G Bertrand, C. Drouet, Tetracycline-Loaded Biomimetic Apatite: An Adsorption Study. J. Phys. Chem B119 (2015) 3014-3024. https://doi.org/10.1021/jp5116756

[58]   G. Oliviero, S. Federici, P. Colombi, P. Bergese, On the Difference of Equilibrium Constants of DNA Hybridization in Bulk Solution and at the Solid-Solution Interface. J. Mol. Recognit. 24 (2011) 182–187. https://doi.org/10.1002/jmr.1019

[59]   A. Dodabalapur, L. Torsi, H. Katz, Organic Transistors: Two-Dimensional Transport And Improved Electrical Characteristics. Science 268 (1995), 270-271. https://doi.org/10.1126/science.268.5208.270

[60]   G. Dacey, I. Ross, The field effect transistor. Bell System Technical Journal 34 (1955), 1149-1189. https://doi.org/10.1002/j.1538-7305.1955.tb03794.x

[61]   M. Riepl, M. Optimization Of Capacitive Affinity Sensors: Drift Suppression And Signal Amplification. Analytical Chimical Acta 392 (1999), 77-84. https://doi.org/10.1016/S0003-2670(99)00195-6

[62]   L.Torsi, A. Dodabalapur, H. Katz, An Analytical Model for Short Channel Organic Thin-Film Transistors. J. Appl. Phys. 78 (1995), 1088-1093. https://doi.org/10.1063/1.360341

[63]   J. Zaumseil, K.W. Baldwin, J.A. Rogers, Contact Resistance in Organic Transistors that Use Source and Drain Electrodes Formed by Soft Contact Lamination. J. Appl. Phys. 93 (2003), 6117-6124. https://doi.org/10.1063/1.1568157

[64]   L. Torsi, et al. Contact Effects in Organic Thin-Film Transistor Sensors. Organ. Electron. 10 (2009), 233-239. https://doi.org/10.1016/j.orgel.2008.11.009

[65]   M. Magliulo, et al. Structural and Morphological Study of a Poly(3-hexylthiophene)/Streptavidin Multilayer Structure Serving as Active Layer in Ultra-Sensitive OFET Biosensors. J. Phys. Chem. C118 (2014), 15853–15862. https://doi.org/10.1021/jp504652u

[66]   M.C. Sportelli, R.A. Picca, K. Manoli, M. Re, E. Pesce, L. Tapfer, C. Di Franco, N. Cioffi, L. Torsi, Applied Surface Science 420 (2017), 313-322. https://doi.org/10.1016/j.apsusc.2017.05.086

[67]   M.D. Angione et al., Biosensors and Bioelectronics 40 (2013), 303-307. https://doi.org/10.1016/j.bios.2012.07.068

[68]   G. Palazzo, M. Magliulo, A. Mallardi, M.D. Angione, D. Gobeljic, G. Scamarcio, E. Fratini, F. Ridi, L. Torsi, ACS Nano 8 (2014), 7834-7845. https://doi.org/10.1021/nn503135y

[69]   Magliulo, M. *et al.*, Anal Chem. 85(2013), 3849-57. https://doi.org/10.1021/ac302702n

[70]   X. Zhou, J.M. Moran-Mirabal, H.G. Craighead, P.L. McEuen, Nat. Nanotechnol. 2 (2007), 185−190. https://doi.org/10.1038/nnano.2007.34

[71]   A. Star, J.C.P. Gabriel, K. Bradley, G. Grüner, NanoLett. 3 (2003), 459−463. https://doi.org/10.1021/nl0340172

[72]   S.C. Lim, et al., ETRI J. 31 (2009), 647−652. https://doi.org/10.1016/j.optmat.2008.07.002

[73]   S. Freitag, I. Le Trong, L. Klumb, P.S. Stayton, R.E. Stenkamp, Protein Sci. 6 (1997), 1157–1166. https://doi.org/10.1002/pro.5560060604

[74]   E. Macchia, M. Magliulo, K. Manoli, F. Giordano, G. Palazzo, L. Torsi, L. Organic Light EmittingMaterials and Devices XVIII, Proc. of SPIE 9183 (2014), 918302.

[75]   F. Chiti, C.M. Dobson, Annu. Rev. Biochem. 75 (2006), 333-366. https://doi.org/10.1146/annurev.biochem.75.101304.123901

[76]   J.I. Peterson, G.G. Vurek, Science 224 (1984), 123-127. https://doi.org/10.1126/science.6422554

[77]   M. Lehnert, et al., Langmuir 27 (2011), 7743–7751. https://doi.org/10.1021/la200908h

[78]   M.L. Jones, G.P. Kurzban, Biochemistry 34 (1995), 11750–11756. https://doi.org/10.1021/bi00037a012

[79]   A. Bykhovski, W. Zhang, J. Jensen, D. Woolard, J. Phys. Chem. B117 (2013), 25–37. https://doi.org/10.1021/jp3075833

[80]   Grubmuller, H., Heymann, B. &Tavan, P. Science 271 (1996), 997-999. https://doi.org/10.1126/science.271.5251.997

[81]   S. Izrailev, S. Stepaniants, M. Balsera, Y. Oono, K. Schulten, Biophys. J. 72 (1997), 1568–1581. https://doi.org/10.1016/S0006-3495(97)78804-0

[82]   F. London, Trans. Faraday Soc.33 (1937), 8-26. https://doi.org/10.1039/tf9373300008b

[83]   K.P. Pernstich et al., J. Appl. Phys. 96 (2004), 6431-6438. https://doi.org/10.1063/1.1810205

[84]   S.K. Possaner, K. Zojer, P. Pacher, E. Zojer, F. Schurrer, Adv. Funct. Mater. 19 (2009) , 958-967. https://doi.org/10.1002/adfm.200801466

[85]   A. Dezieck et al., Appl. Phys. Lett. 97 (2010), 013307. https://doi.org/10.1063/1.3459978

[86]   D. Alberga, G.F. Mangiatordi, L. Torsi, G. Lattanzi, J. Phys. Chem. C118 (2014), 8641-8655. https://doi.org/10.1021/jp410936t

[87]   Y.K. Lan, C.I. Huang, J. Phys. Chem. B112 (2008), 14857–14862. https://doi.org/10.1021/jp806967x

Organic Bioelectronics for Life Science and Healthcare   Materials Research Forum LLC
Materials Research Foundations **56** (2019) 71-96   doi: https://doi.org/10.21741/9781644900376-2

Chapter 2

Biosensing with Electrolyte Gated Organic Field Effect Transistors

Carlo Augusto Bortolotti[1]*, Marcello Berto[1], Matteo Sensi[1], Michele Di Lauro[1], Fabio Biscarini[1]

[1]Department of Life Sciences, University of Modena and Reggio Emilia, Modena, Italy

carloaugusto.bortolotti@unimore.it

Abstract

Electrolyte Gated Organic Field Effect Transistors (EGOFETs) are rapidly emerging as novel players in the field of biosensing: they allow ultra-sensitive, label-free and fast response, and can be employed to sense very diverse analytes, from small molecules to large multimeric proteins. Here, we present the current level of understanding of the working mechanism of EGOFETs, and review some of the most recent and relevant applications as sensors for healthcare and life sciences, discussing advantages and limitations of this technology. EGOFETs appear as a powerful sensing platform that can be readily adapted to the detection of a wide range of biologically relevant species.

Keywords

EGOFET, Biosensor, Immunoassays, Surface Chemistry, Double Layer

Contents

1.   Introduction

It might be difficult to exactly define organic bioelectronics because, as for every rapidly emerging field, its boundaries are rather blurred: one, very effective synthesis of the main scope of bioelectronics is that it serves as a translator between communication within biological systems and that in artificial, man-made electronics, based on organic molecules as active material [1]. It is apparent that one of the applications that might benefit the most from the technological advances related to the rise of organic bioelectronics is biosensing. This is confirmed by the rapidly growing increase of papers describing the use of organic electronics devices as tools to detect and quantify biomolecules in a sample [2]. The panorama is very broad: different architectures have been tested, and each of them has been employed to try to detect biomolecules extremely diverse in terms of size and physico-chemical properties.

Here, we will focus on one class of organic electronics transistors, namely the Electrolyte Gated Organic Field Effect Transistor, commonly abbreviated as EGOFET, and exclusively on its applications in biosensing. Our goal is to provide the reader with an overview of the applications of EGOFETs as novel tools in health and life sciences, grasping their potential, also in comparison with other biosensing platforms.

EGOFETs are three-electrode devices: two electrodes, source (S) and drain (D), are connected to an organic semiconductor (OSC) layer (Fig. 1). The density of charge carriers (electrons or holes, depending on whether an n- or p-type OSC is used) in the channel is controlled by the voltage (V_{GS}) applied at a third electrode, the gate (G), which is separated from the channel by an electrolyte solution (or an ion gel), which acts as a dielectric layer. The gate bias is capacitively coupled to the OSC across the dielectric layer. For p-type OSC and devices working in the so-called accumulation mode, a $V_{GS} < 0$ attracts mobile cations in the electrolyte solution at the gate/electrolyte interface with formation of a first electrical double layer (EDL) [3]. Consequently, anions, being repelled from the gate, tend to accumulate at the electrolyte/OSC interface, causing accumulation and confinements of holes in the interfacial region of the OSC, forming a second EDL. Imposing a negative bias between drain and source (V_{DS}) drives the carriers and leads to the output drain current I_{DS}.

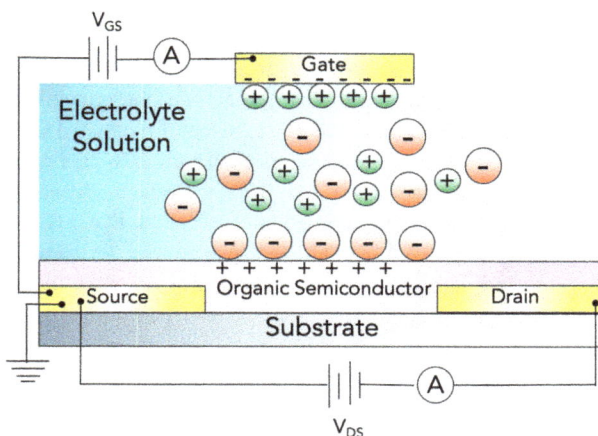

Figure 1.   Schematics of the working principle of an EGOFET.

In a recent review, Rivnay *et al.* [4] effectively described the EGOFET as an extreme case of an Organic Field Effect Transistor OFET, for which reducing the thickness of the dielectric is one largely adopted strategy, together with the use of high-*k* dielectrics, to increase its capacitance and consequently charge carriers density in the channel. The two EDLs in EGOFETs can be considered as nanometer-thick capacitors [5]: the thin separation between the charge layers in EDLs therefore maximizes their capacitance (that can be as large as 10 μF cm^{-2} [3, 6]) and guarantees high drain current in EGOFETs even at applied potentials well below 1V [5].

2.   EGOFETs as sensors for life sciences

In order to exploit an organic electronic device as a biosensor, one obviously needs to generate an electronic signal proportional to a molecular event related to the presence of a certain bioanalyte, or to a specific environmental condition (pH, temperature) affecting structural or functional properties of a given biomolecule. Therefore, two are the main players that define the range of applications of an organic electronics biosensor: one is the physico-chemical process triggered by the presence of bioanalyte, and the second is the signal transduction, both of them tightly connected with the working mechanism of the specific architecture one decides to exploit.

A large fraction of demonstrated organic electronic biosensors bases its sensing capability on biorecognition, typically non-covalent binding between two biological partners. Nevertheless, other physico-chemical processes can be exploited, e.g. electron or proton transfer, conformational changes or phase transitions. When it comes to EGOFETs, the current level of understanding of their working principle indicates that, in order to construct a sensing interface, one has to select a biological process that would impact on one of the following three parameters contributing to the drain current I_{DS}: the threshold voltage V_{th}, the field effect mobility μ or the effective capacitance C_{eff} of the device. The experimental I-V curves recorded to characterize EGOFETs and to construct the dose curves, when they are used as biosensors, can be well described by Eq. 1 and Eq. 2, which are the standard equations for transistors operated in the linear and saturation regime, respectively [7].

$$I_{DS} = \frac{W}{L}\,\mu C_{eff}(V_{GS} - V_{th})V_{DS} \tag{1}$$

$$I_{DS} = \frac{W}{2L}\,\mu C_{eff}(V_{GS} - V_{th})^2 \tag{2}$$

Here I_{DS} is the source-drain current, W and L are channel width and length, C_{eff} is the capacitance per unit area between semiconductor and gate, μ is the charge carrier mobility, V_{GS} is the gate-source voltage, V_{DS} is the drain-source voltage and V_{th} is the threshold voltage.

The EGOFET biosensors described in the last years indicated that a very broad set of biological processes can elicit such changes in one or more of the transistor figures of merit, provided that they cause a rearrangement of the biomolecular structure or of the solvent shell causing redistributions of charges or variations in the capacitance at the interface between the biomolecule and its surroundings.

2.1 Target analytes and biorecognition elements employed to date in EGOFET biosensors

As of today, EGOFET have been demonstrated as biosensors for different molecules, ranging in size from small chemicals to biomacromolecules (see Table 1). Besides the demonstration of an EGOFET as a pH sensor [8], the smallest species, for which a detection scheme has been implemented are neurotransmitters dopamine [9] and carvone

[10], both featuring a molecular weight (MW) of about 150 g/mol, and slightly larger 2,4-dichlorophenoxyacetic acid (2,4-D) [11] and bisphenol A (about 220 g/mol) [12]. Dopamine and 2,4-D are positively and negatively charged, respectively, at physiological pH, at variance with neutral species carvone and bisphenol A: very importantly, EGOFET biosensors can detect even the presence of molecules that bear no net charge at the operational pH. The EGOFET architecture also allows for the quantification of much larger molecules, as is the case of cytokines Interleukin 6 (IL-6, 23.7 kDa) [13] and TNFα (trimeric in solution, with a MW of about 51 kDa) [14] and even C-Reactive Protein (CRP), which is as large as about 100 kDa in its functional form as an omopentamer [15].

The specificity of the EGOFET biosensor response is heavily dependent on the choice of the biorecognition element: therefore, it is apparent that the detection of the diverse group of bioanalytes listed above requires the use of a large variety of biomolecules, to be immobilized on one of the device interface, each ensuring both high affinity and selectivity to the target. Antibodies are by far the most widely employed biorecognition element, a key component in the development of the several described EGOFET biosensors, including those for IL-4 [16], IL-6 [13] and procalcitonin [17]. As it will be described in the next two sections, different surface chemistry strategies have been employed for antibody confinement. While still being the molecule of choice for immunorecognition-based approaches not only in diagnostics but also for therapeutic purposes, antibodies are known to present non-negligible drawbacks [18]: high production costs (typically involving animals), instability, batch-to-batch variability, high molecular weight are only some of the disadvantages related to using antibodies, especially for biosensing purposes. In response to these problems, novel molecules are being widely investigated as an alternative to antibodies: these efforts have led to the development of aptamers, i.e. nucleic acids ligands, selected through a directed *in vitro* evolutionary method [19]. These single-stranded oligonucleotides can be able to bind targets, such as proteins, with affinities comparable to those exhibited by antibodies. In parallel, a conceptually similar approach has led to the demonstration of non-immunoglobulin (non-Ig) binders based on proteins rather than nucleic acids [20]: a protein scaffold is engineered into a protein binder through randomization of a few amino acids (typically, about 10-20) at the binding surface. Affimers, affibodies, DARPins are only some of the many examples of these novel non-Ig-based binders, which are sometimes referred to as "peptide aptamers" [21].

Both classes of aptamers have been tested as core elements of the sensing units of EGOFET biosensors [13, 22]: one of the main advantages of choosing this novel generation of binders is the straightforward functionalization procedure for immobilizing them covalently, typically at the gate/electrolyte interface, exploiting terminal –SH group in DNA/RNA aptamers or single/double engineered solvent exposed cysteines in peptide aptamers. Besides antibodies and aptamers, other proteins with high affinity for a ligand of interest might be employed: within this respect, the most notable example is the streptavidin-biotin couple [23], with the former typically surface-confined to quantify biotin in solution.

Apart from non-covalent interactions, one can exploit chemical reactions between an analyte and the immobilized receptor: this strategy has been exploited to sense dopamine (DA), by immobilizing 4-formylphenyl boronic acid (BA) on the gold gate surface. In this case, an esterification between BA and DA takes place on the surface, resulting in the irreversible formation of a large surface dipole that impacts on the device transconductance g_m.

2.2   Why EGOFET as biosensors?

When compared with other sensing strategies (both commercially available and developed at the laboratory research level), the EGOFETs present important advantages, some of which are intrinsically ascribable to the fact that they rely on organic (bio)electronics, while others are specific features of this class of devices. Organic electronics technology ensures environmental friendly, low cost fabrication procedures and relies on materials that are both biocompatible and able to withstand operation in harsh biological environment [2]. When aiming at wearable sensors, organic electronics also offers the possibility to fabricate devices on flexible or foldable substrates.

A crucial point is that the EGOFET-based detection scheme is label-free, allowing to discard the use of fluorescent or radioactive labels that might impact on the molecular recognition. These devices ensure a direct transduction, i.e. no secondary binding agents are required to detect and/or amplify the device response following the biorecognition event. This feature impacts heavily on their ease-of-use, response time and associated costs.

Perhaps the most striking feature of EGOFETs is their tremendous sensitivity, which can be ascribed to their working mechanism: in fact, the capacitive coupling enables the amplification of subtle electrostatic potential variations at the semiconductor/electrolyte interface by up to three orders of magnitude, so any miniscule potential change is translated into a large change of transistor parameters. The capacitive coupling based detection has allowed for ultra-low limits of detection in EGOFET-based biosensors, down to [17] and even well below [24] the pM range. As anticipated before, the possibility to operate EGOFETs as capacitance modulated transistors [10] allows to reach remarkably high sensitivities even when detecting molecules bearing no net charge, a major advantage over other organic electronic detection schemes, as for example the Organic Charge Modulated Field-Effect Transistor (OCMFET) [25].

Another important characteristic of EGOFET biosensors is their versatility: the examples listed in Table 1 and overviewed in section 2.1 indicate that this detection scheme can be applied to a highly diverse set of analytes, spanning several orders of magnitude in terms of size of detectable molecules. EGOFET biosensors therefore constitute a platform that can be readily adapted to address different problems involving biorecognition events.

Table 1.   Overview of some representative EGOFET-based biosensors reported in literature for detection of Life Sciences- or Healthcare-relevant analytes.

Target molecule	Functionalized interface	Biorecognition element	OSC used	Ref.
DNA	OSC/electrol.	Complementary ssDNA	P3HT-COOH	[26]
Streptavidin	OSC/ electrol	Biotin on supported lipid bilayers	P3HT	[23]
C-reactive protein (CRP)	OSC/electrol	Antibody	P3HT	[15]
Bisphenol A (BPA)	OSC/ electrol	Antibody	pBTTT-C16	[12]
Procalcitonin (PCT)	OSC/ electrol	Antibody	P3HT	[17]
Avidin & Streptavidin	OSC/ electrol	Biotin	P3HT	[27]
Interleukin 4 (IL-4)	Gate/electrolyte	Antibody	Pentacene	[16]
Carvone	Gate/electrolyte	Odorant Binding Protein	P3HT	[10]
Dopamine	Gate/electrolyte	Cysteamine + Boronic Acid derivative	P3HT	[9]
TNFα	Gate/electrolyte	Antibody	Pentacene	[28]
TNFα	Gate/electrolyte	Peptide aptamer	Pentacene	[14]
Interleukin 6 (IL-6)	Gate/electrolyte	Antibody / Peptide aptamer	Pentacene	[13]
DNA	Gate/electrolyte	Complementary ssDNA	P3HT	[29]
Ricin	Gate/electrolyte	Nucleic acid aptamer	P3HT	[22]
2,4-D	Gate/electrolyte	Antibody	DPP-DTT	[11]

2.3   Recognition element immobilization in EGOFETs biosensors

As in the most widely employed bioassays, enzyme-linked immunosorbent assay (ELISA) and surface-plasmon resonance (SPR), one crucial step in the development of an EGOFET biosensor is the surface immobilization of the biorecognition element. From a conceptual point of view, the most straightforward strategy is to implement biorecognition at the OSC/electrolyte interface: binding of the analyte to the capturing molecule tethered to the OSC would impact on the transport within the channel, resulting in a change in at least one of the different figures of merit that contribute to the multi-

Organic Bioelectronics for Life Science and Healthcare   Materials Research Forum LLC
Materials Research Foundations **56** (2019) 71-96   doi: https://doi.org/10.21741/9781644900376-2

parametric response of the EGOFETs. Several EGOFET biosensors with biorecognition element confined at this interface have been demonstrated, despite the inherent difficulties related to the derivatization of OSC building blocks and their functionalization with hydrophilic biomacromolecules.

EGOFET also possess a second addressable interface, namely that between the metal gate and the electrolyte: potentiometric perturbations due to interfacial changes (binding, conformational transitions) are transduced into electrical changes thanks to the capacitive coupling. Gate electrodes are typically made of materials (above all, gold), for which a number of surface derivatization procedures have been optimized in the last decades: this fact has led to several successful examples of EGOFET biosensor with functionalized gate as the core sensing unit.

In the following sections, we will review some of the latest and more relevant examples of biosensors based on EGOFET architecture, classified according to the location of the biorecognition element and to the functionalization strategy adopted to surface confine it.

3.   EGOFET biosensors with functionalized OS/electrolyte interface

Functionalization of the OSC with biomolecules is not trivial, the main reason being that OSC-forming molecules are highly hydrophobic, at variance with the hydrophilic character of soluble biorecognition elements (nucleic acids, proteins, antibodies) [7]. Nevertheless, different strategies have been devised in the recent past to successfully confine biorecognition elements at the OS/electrolyte interface in EGOFET architectures, and we might tentatively classify them into three different groups: i) covalent modification of the OSC forming molecules; ii) surface modification the OSC, eventually further functionalized by tethering a lipid bilayer; iii) physisorption of biomolecules at the OSC/electrolyte interface.

3.1   Immobilization of biorecognition element using OSCs with substituted main conjugated backbone

The first strategy to immobilize biomolecules on the OSC surface is to use building blocks modified with organic chemistry synthetic routes, to endow them with functionalities that can be further used for covalent immobilization of biorecognition elements [30]. This strategy has been employed for OFET with solid state dielectrics, for example by functionalizing phenylenethiophene oligomers with chiral aminoacids or sugars to demonstrate a chiral sensor [31]. Despite the fascinating perspective of endowing the electronic active material itself with biomolecules, this approach has not been employed very often to EGOFET functionalization, most likely because of the worsening of electrical performances that can be induced by the introduction of functional groups on the OSC molecules backbone, reducing charge delocalization [32]. Nevertheless, there are a few notable examples, as is the case of the DNA biosensor described in 2012 by Kergoat et al. [26]. The authors spin coated a –COOH modified semiconductor poly [3-(5-carboxypentyl)thiophene-2,5-diyl], P3PT–COOH, to endow the channel material with the ability to bind biomolecules, namely 28-mer oligonucleotides covalently grafted via an 1-ethyl-3-(3-dimethylaminopropyl)carbodiimide (EDC) - N-hydroxysuccinimide (NHS) click chemistry approach. The EGOFET with ss-DNA modified OSC channel was then used to detect the presence of a complementary DNA strand in solution by monitoring shift of the onset voltage V_{Gmin} and of the off current I_{off}. Changes of the device parameter upon binding of the target, and in particular for I_{off}, are more marked when hybridisation is detected with the EGOFET upon replacement of PBS electrolyte with deionised water, and the authors ascribe the enhanced sensitivity in H_2O to a large increase (more than two orders of magnitude, reaching a value as high as 200 nm) of the Debye length with respect to that calculated for PBS.

Covalent modification of the OSC building blocks can be exploited also to *incorporate the recognition element within the OSC*: this was the case of the P3HT modified by covalently bound biotin by Suspene et al. [27], which resulted in a random copolymer, composed of P3HT, P3HT-COOH and P3HT-biotin moieties, featuring about 8% of biotin groups. Upon operation in deionized water, the EGOFET features a mobility of 7.3 x 10^{-3} cm^2 V^{-1} s^{-1}, V_{th} = -0.1 and I_{on}/I_{off} ratios of about 500: these parameters indicate satisfactory electrical performances of the copolymer, which appears to be stable up to one hour of operation in contact with H_2O. On the contrary, to improve the performances of the EGOFET as a biosensor, non-specific adsorption on the OSC had to be reduced by further incubating the deposited OSC in a 1-octanol solution after deposition, with the purpose of increasing the interface hydrophilicity, as supported by contact angle measurements. The EGOFET featuring the copolymer/1-octanol channel becomes able to selectively detect the presence of both avidin and streptavidin and does not respond to the presence of BSA, which could only bind non-specifically. The binding invariably leads to a decrease in I_{DS}, independently of the charge of the bound analyte: this finding let the authors suggest that, at least in water gated FETs, capacitive effects prevail over charge-related ones.

3.2   Interfacial modification of the OSC surface

Since the recognition events endowing EGOFETs with sensing capability are believed to take place exclusively at the interface with the electrolyte, and due to the potential worsening of the performances of the OSC upon modification of its building blocks, most research efforts were aimed at immobilizing biomolecules by modifying exclusively the OSC surface.

To this end, biomaterials can be modified using non-equilibrium plasma treatments, which allow for controlled surface chemistry modifications without affecting the bulk material properties, while ensuring sterility and industrial scalability with low environmental impact [33]. In an EGOFET, the use of Plasma enhanced chemical vapor deposition (PE-CVD) of a hydrophilic coating from acrylic acid and ethylene was used to modify the surface of a P3HT channel functionalizing it with carboxylic groups [32]. Despite the obvious increase of the hydrophilicity of the surface, satisfactory electrical performances were observed, especially if P3HT was annealed before the plasma deposition.

The carboxylic acid functionality introduced by the PE-CVD process can then be used to covalently bind amine-exposing biomolecules via formation of amide bonds. This approach was demonstrated by Magliulo et al., who grafted phosphatydilethanolamine bilayers on plasma modified P3HT surface [23]. To endow this interface with binding capabilities, the authors prepare phospholipid (PL) vesicles also containing a small fraction of biotynilated lipid: the result is a PL bilayer deposited on the OSC and exposing a small number of streptavidin moieties to the electrolyte solution, that the authors employ as a testbed to demonstrate the EGOFET as biosensor for streptavidin in the 10nM-1µM range with a LOD of 10 nM. They observed increases in the drain current as a consequence of streptavidin binding to the biotynilated PL bilayer, and they ascribed it to the negative charge of streptavidin at the operational pH: bound streptavidin would therefore act as an additional gating agent inducing more holes in the p-type semiconductor upon application of a negative V_{GS}.

A further example of interfacial modification of the OSC to introduce tethering sites for biomolecules was described in 2015 by Mulla et al. [34]: the authors replaced the expensive, though effective, plasma based treatment with the UV-crosslinking of poly(acrylic acid), PAA, previously spin coated on a Poly[2,5-bis(3-hexadecylthiophen-2-yl)thieno[3,2-b]thiophene] (PBTTT-C16) channel. This low cost and straightforward wet chemistry processing led to the presence of solvent-exposed carboxylic groups on the PBTTT-C16 surface, and these served as anchoring sites for covalent attachment of biotynilated PL bilayers as previously described [23]. The EGOFETs fabricated with this channel modification procedure were used again to sense streptavidin, this time down to a 10 pM LOD. The authors describe a decrease of the estimated field effect mobility for the PBTTT/PAA/B-PL EGOFET with respect to untreated PBTTT devices (10^{-3} vs 10^{-2} cm^2 V^{-1} s^{-1}, respectively) and evidences of hysteresis at high V_{GS}, as a consequence of the increased hydrophilicity of the PAA functionalized channel, thus less stable in polar solvents; nevertheless, the overall electrical and sensing performances are better than those of the plasma treated B-PL EGOFETs.

It is apparent that the biotin/streptavidin couple, is often chosen in proof-of-concept demonstration of novel architectures to monitor biorecognition because of its very high affinity constant. Besides sensing of streptavidin, the described approaches can be used for the bottom up construction of multilayered biointerfaces, immobilizing streptavidin on the biotynilated PL bilayer, and then further binding biotynilated biorecognition elements (nucleic acids, proteins) to serve as specific partner toward a bioanlayte of interest.

This strategy was employed by Palazzo et al. [35] to detect C reactive protein (a biomarker for inflammatory processes) with a 20nm thick biosystem composed of a biotynilated PL bilayer, a streptavidin layer and a third layer of biotynilated anti-CRP antibodies with a LOD of about 10 µM.

3.3 Direct physisorption of biorecognition elements on bare OSC

A third approach to confinement of biomolecules at the OSC/electrolyte interface is the non-covalent, direct adsorption on untreated OSC surface, typically by simple drop casting. Considering the above described worsening of the transport properties of the semiconductor as a consequence of covalent modification of the monomers or surface treatments, this latter approach is surely much less detrimental to the electronic performances of the transistor *per se*. Nevertheless, direct adsorption might not be the best strategy for retention of biomolecule activity, especially if aiming at demonstrating a biosensor; this caveat is due, above all, to the well-documented tendency of many proteins to undergo conformational changes and even denaturation upon exposure to a surface, especially in case of a high energy one [36]. Other factors that limit the applicability of this approach are the weakness of the biomolecule/surface interactions, likely leading to detachment of biorecognition elements from the substrate during operation, and the lack of uniform orientation of the biolayer, reducing protein functionality by steric hindrance of the available bioactive sites.

Despite these limitations, there are reports in literature of highly performing EGOFET biosensors with the sensing unit composed by biorecognition elements directly adsorbed on the channel surface. One example is provided by the work of Magliulo et al. [15], who deposited anti-CRP antibodies on P3HT and used this biointerface to sense CRP down to a remarkably low extrapolated LOD of 2 pM, in a wide dynamic range (from 4 pM to 2 µM) and with high reproducibility (the above parameters were averaged over ten devices), especially considering the functionalization strategy with uncontrolled antibody orientation. The drain current decreased for increasing concentration of CRP: the threshold voltage V_{th} was basically unaffected by CRP binding to its Ab, so the I_{DS} variations were attributed almost exclusively to changes in the gating capacitance. The authors also assign an active role in imparting a more uniform Ab orientation to the presence of the hydrophilic N-[tris(hydroxyl-methyl)methyl]acrylamide-lipoic acid conjugate (pTHMMAA), further deposited on the surface and mainly used as a blocking agent to minimize non-specific adsorption at the OSC/electrolyte interface.

A very similar setup was very recently employed by Luisa Torsi's group to demonstrate a biosensor for Procalcitonin C (PCT), a 116aa-protein that serves as a biomarker in sepsis, a systemic inflammatory response that requires very fast diagnostic response to limit its potentially lethal effects [17]. The sensing unit consisted of randomly oriented physisorbed anti-PCT antibodies, and BSA was used as a blocking agent. The biosensor was tested in the 0.8 pM to 4.7 nM range, which can be considered clinically relevant since the PCT levels are increased from the low pM range to the nM range during sepsis; the evaluated LOD was 2.2 pM. As was the case of the CRP EGOFET biosensor by Magliulo et al. [15], a decreased of I_{DS} was observed for increasing [PCT], but, at variance with the previous example, the authors describe threshold voltage changes as a consequence of immunorecognition and minor effects on transconductance. The V_{th} shift was ascribed to the negative charge of PCT at physiological pH (its pI being around 5.1); immobilized PCT is therefore thought to provide additional trap sites to the P3HT hole carriers.

In the previous section, we surveyed papers reporting on a strategy to covalently tether a PL bilayer via amide bond formation on a –COOH derivatized OSC surface. It is also possible to deposit PL layers exploiting non-covalent direct adsorption, as proposed by Cotrone et al. [37]. The authors describe the deposition of stakes of phosphatidylcholine bilayers on a P3HT channel, by simply exposing the bare OSC surface to a suspension of unilamellar vesicles. The functionalization strategy is extremely straightforward, although there is no control on the number of stacked bilayers, which fail to assemble into ordered layered structures along the whole surface. The main claim in the paper concerns the role played by the phospholipid bilayer, which improves the electrical performances rather than serving as an intermediate step for protein immobilization. The EGOFET functionalized with PL bilayers exhibit reduced hysteresis, higher I_{on}/I_{off} ratio and slightly higher field effect mobility when compared to EGOFETs with bare P3HT, but no repercussions on the threshold voltage. These changes in the transistor parameters are ascribed to the minimization of ions migration into the channel due to the presence of (zwitterionic) PL layers. The interpretation is backed up by Electrochemical Impedance Spectroscopy investigations, besides the smaller hysteresis area observed for PL-EGOFETs; according to the authors, the reduction of ions penetration would reduce the "electrochemical doping" of the semiconductor and ensure a purely capacitive gating mechanism.

4. EGOFET biosensors with functionalized gate/electrolyte interface

Despite the diverse strategies for endowing the OSC/electrolyte interface with sensing capability, it is apparent that the metal gate electrode is a much more easily addressable surface for immobilizing the biorecognition element that would turn the EGOFET into a biosensor. In the relatively "young" history of EGOFET-based biosensors, examples of biosensors based on functionalized gate are less numerous than those reporting modified OSC, most likely because of the not fully elucidated yet nature of the coupling mechanism between the molecular recognition at the gate and the transport within a spatially distinct region. Nevertheless, in the last 2-3 years EGOFET biosensors based on binding at the gate/electrolyte interface are rapidly emerging at the solution of choice for many groups working in the field, and this trend is also contributing to the elucidation of the molecular determinants in the signal transduction in EGOFET biosensors.

The first example of an EGOFET biosensor with gate functionalization was reported in 2013 by our group [9]: the work by Casalini et al. described the functionalization of the gold gate surface exploiting thiol based chemistry, which represents the most widely used strategy to immobilize recognition units on the gate. The authors exploited the multistep formation of self-assembled monolayers of cysteamine and boronic acid. By exploiting the chemical affinity of the latter for dopamine (DA), which reacts with the boronic acid in an esterification reaction, the authors demonstrated an EGOFET biosensor that would detect DA down to the pM range thanks to the changes in both transconductance and V_{th}. This example differs from all those described so far, as in this case the biosensing is not based on non-covalent interactions but rather on a chemical reaction, which results in the formation of a large surface dipole that is most likely the main responsible for the [DA] dependent-decrease in drain current.

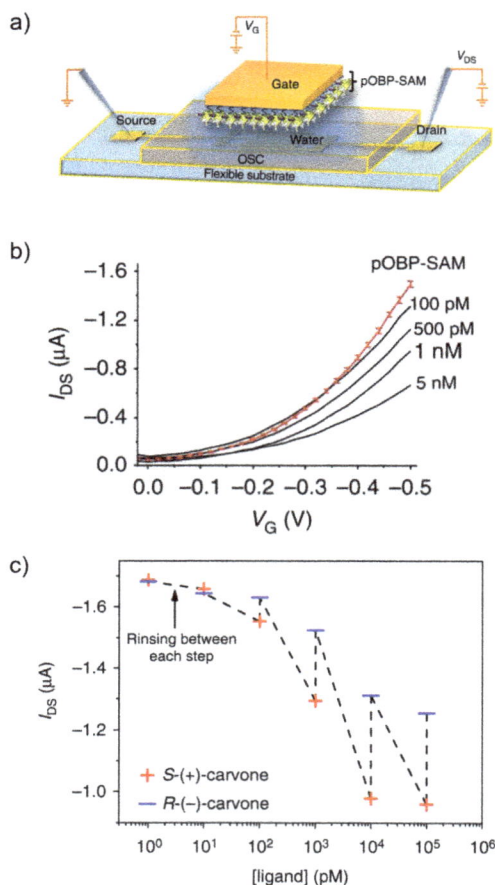

Figure 2.   *a) Device schematics, showing the OBP covalently bound to the gate electrode through the presence of a SAM. b) Transfer characteristics of the pristine device and upon exposure to increasing concentrations of (S)-(+)-carvone. c) Chiral differential detection of carvone enantiomers. Reproduced from ref. [10] under Creative Commons Attributions license.*

Organic Bioelectronics for Life Science and Healthcare   Materials Research Forum LLC
Materials Research Foundations **56** (2019) 71-96   doi: https://doi.org/10.21741/9781644900376-2

SAMs can also be formed on gold gate electrodes to functionalize them with –COOH or –NH$_2$ functionalities to be further exploited to bind covalently to amine- or carboxylic group-exposing aminoacids on a protein, respectively, with the well-established protocols employing a carbodiimide coupling agent and an auxiliary nucleophile. Mulla et al. [10] used this strategy to covalently bind a recombinant porcine odorant binding protein (OBP) to a gold gate functionalized with a 3-mercaptopropionic acid SAM (Fig. 2). The modified electrode surface was the gate of an EGOFET device, with p-type PBTTT-C14 as OSC, that could sense the small volatile compounds S-carvone down to a LOD of 50 pM and, most notably, discriminate the odorant from its enantiomer R-carvone. The authors explain the remarkable chiral differential detection on the basis of subtle differences in the changes occurring in the OBP upon binding neutral odorants S- or R-carvone. The binding of the odorants to the gate immobilized OBP causes a decrease of I$_{DS}$, dominated by the changes in the capacitance occurring at the OBP/electrolyte interface, and the purely capacitive nature of the transduction due to the full passivation of the gate by the SAM.

Besides SAM-forming molecules, the S affinity for Au can also be exploited for direct covalent immobilization of biomolecules on gold gate electrodes. When it comes to proteins, either surface exposed cysteines or intramolecular disulfide bridges can serve as a binding site: very often, protein engineering leads to recombinant proteins featuring a single surface exposed Cys, resulting in (ideally) uniform orientation of the biomolecule on the surface, although often with sub-monolayer coverage. This strategy is not suitable to all proteins though, as some undergo denaturation upon direct immobilization on Au. On the contrary, -SH terminated single stranded DNA molecules are quite stable when chemisorbed on gold. White et al. [22] used this strategy in the design of a floating gate transistor (FGT) working as a sensor for food toxin Ricin B. They adopt a two-step functionalization procedure of the right arm of the floating gate: i) incubation with thiolated aptamers selective towards the toxin, followed by ii) incubation with Poly(ethylene glycol) methyl ethyl thiol to passivate the surface spots not covered by the aptamers, thus minimizing non-specific binding to the floating gate. The authors developed the FGT configuration (Fig. 3) to prevent undesired interaction with the OSC, a potential issue in EGOFET biosensors that becomes more crucial when real world "dirty" samples, featuring several potential interfering agents, are to be investigated. The result is a sensor able to detect Ricin in test solutions down to 30 pM (1 ng/mL) and operating in a range that matches the clinically relevant levels of this toxin. The sensor is also able to quantify Ricin even in potable liquids (namely orange juice and 2% milk), although the LOD is about one order of magnitude higher in complex media than in PBS, as might be expected since the presence of ions and small molecules in the sample might interfere with the aptamer-Ricin recognition.

Single stranded thiolated nucleic acid oligomers can also be surface-immobilized on metallic gates for sensing complementary DNA in solution, as in the FGT biosensor developed by Frisbie's group [29]. They functionalized the floating gate with ssDNA and a 6-mercaptohexanol blocking layer and used P3HT as active material: upon DNA hybridization, a voltage shift was observed; the ΔV magnitude as a function of the complementary DNA concentration followed a Langmuir model and could be used to quantify DNA in the 10 nM-1 μM range.

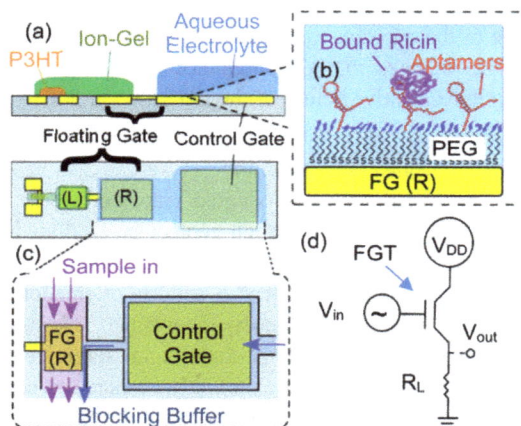

Figure 3.   Schematic of an FGT-based sensor for Ricin detection in potable liquids. Reprinted with permission from [22]. Further permissions related to the material excerpted should be directed to the ACS.

Gold electrodes are amenable to functionalization exploiting functional groups different from sulphur containing ones. One interesting case is represented by the polyhistidine tag, an amino acid motif characterized by the presence of 6 consecutive His residues, very common in recombinant proteins for purification in affinity chromatography, using polymeric resins functionalized with bivalent cations (typically Ni^{2+} or Co^{2+}). The polyhistidine tag can also be exploited to bind biomolecules to polycrystalline gold surfaces [38, 39]: as in the two-step strategy for functionalization of Au gate electrodes with antibodies with uniform orientation demonstrated by our group. The gold surface can be first functionalized with a His-tagged recombinant Protein G (PG) and then incubated with a solution containing (monoclonal) antibodies, which bind to the PG sub-monolayer thanks to the PG high affinity for the heavy chain of the Fc fragment of immunoglobulins. Through this gate derivatization scheme, Casalini et al. [16] demonstrated an EGOFET-based biosensor for detecting cytokine Interleukin 4 (IL4) at 5 nM concentration in aqueous test solutions. The uniform orientation imparted to the biorecognition unit through the PG layer was found to heavily impact on the performances of the EGOFET as a biosensor: when compared with gate electrodes functionalized with randomly oriented anti-IL4 antibodies, the His-tagged PG-antibody construct would lead to a higher surface density of antibodies active toward IL4 thus enhancing the probability of specific binding events (Fig. 4).

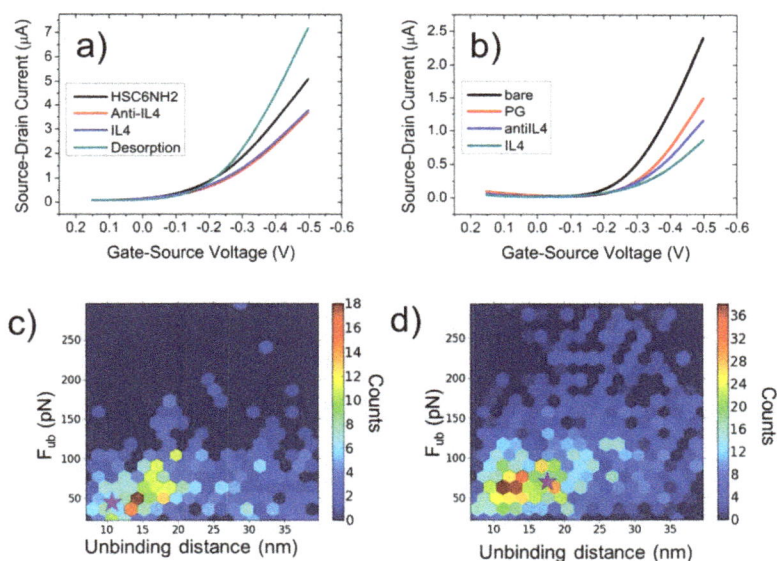

Figure 4. Transfer characteristics (a,b) and two-dimensional histograms of the unbinding distance and unbinding force obtained with single-force spectroscopy (c,d) for HSC6NH2- (a,c) and HisTagged Protein G- (b,d) based functionalization protocols, respectively. Adapted with permission from [16]. Copyright © 2015, American Chemical Society.

Such strategy was later employed by our group to develop an EGOFET pentacene biosensor for the detection of pro-inflammatory cytokine TNFα, by immobilizing His-tagged PG and anti-TNFα monoclonal antibodies on the gate surface in an *in situ* functionalization procedure within a PDMS microfluidics [28]. A dose curve was constructed by plotting the drain current changes ΔI normalized with respect to the current I_0 at [TNFα] = 0, (thus giving a signal S defined as S = -$\Delta I/I_0$) and a LOD of 100 pM was obtained (Fig. 5). Two response regimes could be identified, a superexponential rise in the sub-nM regime and a slowly-rising linear region at higher cytokine concentrations. The shape of the dose curve was attributed to the fact that cytokine binding changes the electrostatic potential of the gate electrode, causing a shift of the Fermi level in the tail of the pentacene density of states.

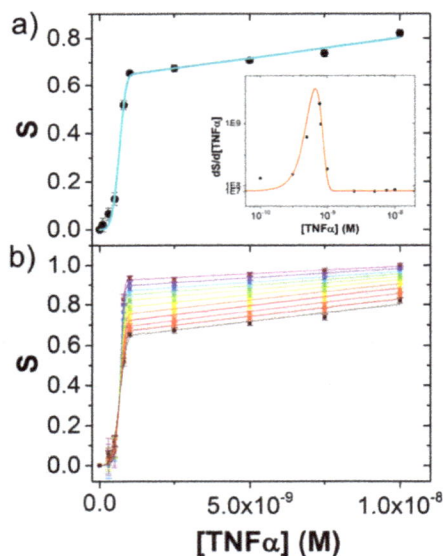

Figure 5. a) Dose curve for an EGOFET-based TNF α biosensor obtained at V_{GS} = -0.8 V, and b) Dose curves acquired at decreasing gate voltages from V_{GS} = −0.6 V (top curve, purple) to V_{GS} = −0.8 V (bottom curve, brown). Reprinted with permission from [28]. Copyright © 2016, American Chemical Society.

Our group further explored the sensing of inflammatory biomarkers by developing EGOFET peptide aptasensors, by functionalizing the gold gate electrode with Affimers specific to TNFα [14] or to Interleukin-6 (IL-6), a cytokine with context dependent pro- or anti-inflammatory action [13]. In both cases, Affimers were directly immobilized on the surface exploiting the presence of a polyhistidine tag at their C-terminus, in a simpler surface modification strategy if compared to the two-step functionalization involving PG and antibodies described above. Both the TNFα and IL-6 biosensors exhibited a LOD as low as 1 pM in PBS: this value is lower than the increased levels of both cytokines in the serum of patients suffering from inflammatory response to a range of pathologies. Therefore, the analytical performances of these devices make EGOFET peptide aptasensors very promising candidates for monitoring inflammatory processes. To further test the operability of the devices in real biological fluids, the TNFα biosensors were also operated using mammalian cell culture medium (enriched with 10% Fetal Bovine Serum) as electrolyte gating solution; the selectivity of the peptide aptasensor was checked by constructing a dose curve in this medium, spanning [TNFα] in the 1 pM – 10 nM range. From a technological point of view, it is more straightforward to develop a device that provides a current value as output. For this reason, most of the dose curves constructed to demonstrate the sensing capability of EGOFET biosensors rely on the normalized current change as a function of the analyte level. Nevertheless, the multi-parametric nature of the EGOFET response allows one to focus on concentration-dependent changes of other figures of merit, such as V_{th} or g_m. Their extraction requires numerical fitting from a full transfer curve, making them a less appealing output for device implementation, but they can be affected by much smaller device-to-device variability than drain current, although typically following a similar trend as a function of the analyte concentration. This is particularly true for g_m, whose value decreased monotonically, for both TNFα and IL-6 biosensors, following a trend that could be approximated to a simple Langmuir model and would enable for the construction of a calibration curve for both cytokines with extremely small standard deviation associated to each data point.

5.　EGOFET biosensors based on competitive binding

One further possibility to the construction of biointerfaces for sensing with EGOFET is the following: instead of observing binding events at the surface between a freely diffusing analyte and an immobilized biorecognition moiety, one might monitor the detachment of the weakly adsorbed biorecognition element from one of the sensing-relevant interfaces, when the bioanalyte is present in solution. This strategy was recently proposed by B. Piro and his co-workers.

Inspired by the work of Wijaya et al. [40] using SWCNT liquid-gated field effect transistor (LGFET), and by classical competitive binding experiments, the authors proposed a brilliant sensing strategy, which also overcomes the theoretical limitations imposed by the Debye screening length, which would in principle rule out the use of biomolecules as large as antibodies as recognition moieties in electronic immunoassays [12]. Their approach is based on the use of three ingredients: the target compound that needs to be quantified, an antibody molecule specific to the target, and a mime compound, also able to bind to the same Ab molecule, albeit with a lower affinity (basically, a poor inhibitor of the target molecule). The rationale behind their sensing strategy is the following: the mime compound must be adsorbed on one of the EGOFET interfaces relevant for sensing, then exposed to a solution containing the Abs, which will therefore bind to the surface. Once this biointerface is exposed to a solution containing the target compound, the latter will displace the equilibrium by binding the Ab, thus removing it from the surface. The dissociation of the Ab from the surface-bound target mime causes an increase in the capacitance at the OS/electrolyte interface, resulting in a current increase.

They first applied this innovative approach to sense Bisphenol A (BPA), a compound used to produce plasticware and in general polymer- and resin-based items. As a mime compound, they employed a BPA alkylated derivative alkBPA, non-covalently bound to the pBTTT semiconductor, but rather through the preparation of an alkBPA:pBTTT blend. LOD was estimated to be as low as ca. 2 pg mL^{-1} BPA, comparable to that exhibited by electrochemical approaches, and, most importantly, with the range of BPA concentrations that can be measured in tap or bottled mineral water.

The same principle was exploited one year later for the demonstration of an EGOFET biosensor for 2,4-dichlorophenoxyacetic acid (2,4-D), a potent herbicide widely used as weed killer in agriculture, albeit being banned as potentially carcinogenic [11]. At variance with the BPA sensor, this time the hapten compound was immobilized on the gate, covalently grafted via electroreduction of a diazonium salt. The choice of the OSC is also worth of mention: the authors used poly(N-alkyldiketopyrrolo-pyrrole dithienylthieno[3,2-b]thiophene) (DPP-DTT), rarely used in EGOFET, despite appealing structural (ordered compact lamellar structure) and functional features (hole mobility as high as about 1 cm^2 V^{-1} s^{-1}). Upon addition of 2,4-D in the electrolyte gating solution, the detachment of the antibody from the gate led to an increase in drain current; the calibration curve yielded a LOD as low as 2.5 fM, a value well below the accepted pollution level in drinking water. It is important to highlight that the performances that one can obtain with the competitive binding approach heavily depend on the relative affinity of the target and hapten compound for the antibody.

Materials Research Forum LLC
doi: https://doi.org/10.21741/9781644900376-2

Conclusions and future perspectives

The main goal of the overview of the most recent applications of EGOFET in biosensing, provided in the present chapter, was to make the reader aware of the high potential of this specific organic bioelectronics architecture for monitoring purposes in healthcare and life sciences. The tremendous sensitivity displayed by these devices, their versatility, the possibility of having two addressable interfaces (paving the way for diverse surface functionalization strategies) are among their most remarkable features. Nevertheless, there are still unsolved issues that are hindering the transition of these devices from laboratory prototypes to technological industrial products to be used routinely, replacing state-of-the-art commercial platforms. Some of these limiting factors are common to many other proposed biosensing schemes, especially to those that require surface confinement of a biomolecule. Some examples are the high risk of loss of biomolecular structure and/or functionality upon surface immobilization and the limited number of experimental techniques to monitor these processes, as well as the low reproducibility, hardly avoidable when dealing with complex, large molecules confined to environmental conditions rather different from physiological ones.

Other crucial points to be addressed are instead specific to EGOFETs. From a technological point of view, we see two major bottlenecks: i) avoiding unspecific response at the interface that does not constitute the core sensing unit and ii) developing more solution-stable, easily processable OSCs. The first point is particularly crucial for (bio)sensing, especially when measuring in complex matrices that contain a large portfolio of extremely diverse molecules and ions that can cause severe interference. Advances in microfluidic technologies surely represent an invaluable step forward to this end, ensuring selective exposure of the sensing interface to the sample. The most promising and effective solution probably is that of coupling *ad hoc* microfluidics with the implementation of a floating (or extended) gate [41–43], avoiding the contact between the OSC and the fluid sample, which could yield non-negligible unspecific binding and possibly perturb the OSC stability.

The final limitation concerns a still significant lack of understanding of the detailed working mechanism of EGOFET. The community is still far from reaching an unambiguous assignment of the molecular events that cause specific changes to the transistor figures of merit (threshold voltage, field effect mobility, capacitance), despite important contributions in the last years to this end [10, 35, 44, 45]. Moreover, evidences are being collected that the OSC in EGOFETs is not fully impermeable to ion penetration [3, 6] so that the working principle depicted in Fig. 1 might need to be somehow reconsidered.

Interestingly, applying the EGOFET architecture to the detection of a wide range of biologically-relevant analytes is also contributing to further investigating the fundamental mechanism underlying the operation of these devices, which possess all the mandatory features to become major player in biosensing in the upcoming future.

References

[1]   D. T. Simon, E. O. Gabrielsson, K. Tybrandt, and M. Berggren, "Organic Bioelectronics: Bridging the Signaling Gap between Biology and Technology," *Chem. Rev.*, vol. 116, no. 21, pp. 13009–13041, Nov. 2016. https://doi.org/10.1021/acs.chemrev.6b00146

[2]   A. M. Pappa, O. Parlak, G. Scheiblin, P. Mailley, A. Salleo, and R. M. Owens, "Organic Electronics for Point-of-Care Metabolite Monitoring," *Trends Biotechnol.*, vol. 36, no. 1, pp. 45–59, 2017. https://doi.org/10.1016/j.tibtech.2017.10.022

[3]   M. J. Panzer and C. D. Frisbie, "Exploiting ionic coupling in electronic devices: Electrolyte-gated organic field-effect transistors," *Adv. Mater.*, vol. 20, no. 16, pp. 3176–3180, 2008. https://doi.org/10.1002/adma.200800617

[4]   J. Rivnay, S. Inal, A. Salleo, R. M. Owens, M. Berggren, and G. G. Malliaras, "Organic electrochemical transistors," *Nat. Rev. Mater.*, vol. 3, no. 2, p. 17086, Jan. 2018.

[5]   S. H. Kim, K. Hong, W. Xie, K. H. Lee, S. Zhang, T. P. Lodge, and C. D. Frisbie, "Electrolyte-gated transistors for organic and printed electronics," *Advanced Materials*, vol. 25, no. 13. pp. 1822–1846, 04-Apr-2013. https://doi.org/10.1002/adma.201202790

[6]   M. Di Lauro, S. Casalini, M. Berto, A. Campana, T. Cramer, M. Murgia, M. Geoghegan, C. A. Bortolotti, and F. Biscarini, "The Substrate is a pH-Controlled Second Gate of Electrolyte-Gated Organic Field-Effect Transistor," *ACS Appl. Mater. Interfaces*, vol. 8, no. 46, pp. 31783–31790, Nov. 2016. https://doi.org/10.1021/acsami.6b06952

[7]   L. Torsi, M. Magliulo, K. Manoli, and G. Palazzo, "Organic field-effect transistor sensors: a tutorial review," *Chem. Soc. Rev.*, vol. 42, no. 22, p. 8612, 2013. https://doi.org/10.1039/c3cs60127g

[8]   F. Buth, D. Kumar, M. Stutzmann, and J. A. Garrido, "Electrolyte-gated organic field-effect transistors for sensing applications," *Appl. Phys. Lett.*, vol. 98, no. 15, pp. 2009–2012, 2011. https://doi.org/10.1063/1.3581882

[9]   S. Casalini, F. Leonardi, T. Cramer, and F. Biscarini, "Organic field-effect transistor for label-free dopamine sensing," *Org. Electron.*, vol. 14, no. 1, pp. 156–163, Jan. 2013. https://doi.org/10.1016/j.orgel.2012.10.027

[10] M. Y. Mulla, E. Tuccori, M. Magliulo, G. Lattanzi, G. Palazzo, K. Persaud, and L. Torsi, "Capacitance-modulated transistor detects odorant binding protein chiral interactions," *Nat. Commun.*, vol. 6, p. 6010, Jan. 2015. https://doi.org/10.1038/ncomms7010

[11] T. T. K. Nguyen, T. N. Nguyen, G. Anquetin, S. Reisberg, V. Noël, G. Mattana, J. Touzeau, F. Barbault, M. C. Pham, and B. Piro, "Triggering the Electrolyte-Gated Organic Field-Effect Transistor output characteristics through gate functionalization using diazonium chemistry: Application to biodetection of 2,4-dichlorophenoxyacetic acid," *Biosens. Bioelectron.*, vol. 113, no. April, pp. 32–38, 2018. https://doi.org/10.1016/j.bios.2018.04.051

[12] B. Piro, D. Wang, D. Benaoudia, A. Tibaldi, G. Anquetin, V. Noël, S. Reisberg, G. Mattana, and B. Jackson, "Versatile transduction scheme based on electrolyte-gated organic field-effect transistor used as immunoassay readout system," *Biosens. Bioelectron.*, vol. 92, no. November 2016, pp. 215–220, 2017. https://doi.org/10.1016/j.bios.2017.02.020

[13] C. Diacci, M. Berto, M. Di Lauro, E. Bianchini, M. Pinti, D. T. Simon, F. Biscarini, and C. A. Bortolotti, "Label-free detection of interleukin-6 using electrolyte gated organic field effect transistors," *Biointerphases*, vol. 12, no. 5, p. 05F401-6, 2017. https://doi.org/10.1116/1.4997760

[14] M. Berto, C. Diacci, R. D'Agata, M. Pinti, E. Bianchini, M. Di Lauro, S. Casalini, A. Cossarizza, M. Berggren, D. Simon, G. Spoto, F. Biscarini, and C. A. Bortolotti, "EGOFET Peptide Aptasensor for Label-Free Detection of Inflammatory Cytokines in Complex Fluids," *Adv. Biosyst.*, vol. 1700072, p. 1700072, 2017. https://doi.org/10.1002/adbi.201700072

[15] M. Magliulo, D. De Tullio, I. Vikholm-Lundin, W. M. Albers, T. Munter, K. Manoli, G. Palazzo, and L. Torsi, "Label-free C-reactive protein electronic detection with an electrolyte-gated organic field-effect transistor-based immunosensor," *Anal. Bioanal. Chem.*, vol. 408, no. 15, pp. 3943–3952, Jun. 2016. https://doi.org/10.1007/s00216-016-9502-3

[16] S. Casalini, A. C. Dumitru, F. Leonardi, C. A. Bortolotti, E. T. Herruzo, A. Campana, R. F. De Oliveira, T. Cramer, R. Garcia, and F. Biscarini, "Multiscale sensing of antibody-antigen interactions by organic transistors and single-molecule force spectroscopy," *ACS Nano*, vol. 9, no. 5, 2015. https://doi.org/10.1021/acsnano.5b00136

[17] P. Seshadri, K. Manoli, N. Schneiderhan-Marra, U. Anthes, P. Wierzchowiec, K. Bonrad, C. Di Franco, and L. Torsi, "Low-picomolar, label-free procalcitonin analytical detection with an electrolyte-gated organic field-effect transistor based

electronic immunosensor," *Biosens. Bioelectron.*, vol. 104, no. October 2017, pp. 113–119, Dec. 2017. https://doi.org/10.1016/j.bios.2017.12.041

[18]   P. Ko Ferrigno, "Non-antibody protein-based biosensors," *Essays Biochem.*, vol. 60, no. 1, pp. 19–25, 2016. https://doi.org/10.1042/EBC20150003

[19]   A. D. Keefe, S. Pai, and A. Ellington, "Aptamers as therapeutics.," *Nat. Rev. Drug Discov.*, vol. 9, no. 7, pp. 537–550, 2010. https://doi.org/10.1038/nrd3141

[20]   K. Škrlec, B. Štrukelj, and A. Berlec, "Non-immunoglobulin scaffolds: a focus on their targets," *Trends Biotechnol.*, vol. 33, no. 7, pp. 408–418, Jul. 2015. https://doi.org/10.1016/j.tibtech.2015.03.012

[21]   S. Reverdatto and D. S. B. and A. Shekhtman, "Peptide Aptamers: Development and Applications," *Current Topics in Medicinal Chemistry*, vol. 15, no. 12. pp. 1082–1101, 2015. https://doi.org/10.2174/1568026615666150413153143

[22]   S. P. White, S. Sreevatsan, C. D. Frisbie, and K. D. Dorfman, "Rapid, Selective, Label-Free Aptameric Capture and Detection of Ricin in Potable Liquids Using a Printed Floating Gate Transistor," *ACS Sensors*, vol. 1, no. 10, pp. 1213–1216, 2016. https://doi.org/10.1021/acssensors.6b00481

[23]   M. Magliulo, A. Mallardi, M. Y. Mulla, S. Cotrone, B. R. Pistillo, P. Favia, I. Vikholm-Lundin, G. Palazzo, and L. Torsi, "Electrolyte-Gated Organic Field-Effect Transistor Sensors Based on Supported Biotinylated Phospholipid Bilayer," *Adv. Mater.*, vol. 25, no. 14, pp. 2090–2094, Apr. 2013. https://doi.org/10.1002/adma.201203587

[24]   M. Magliulo, A. Mallardi, R. Gristina, F. Ridi, L. Sabbatini, N. Cioffi, G. Palazzo, and L. Torsi, "Part per Trillion Label-Free Electronic Bioanalytical Detection," *Anal. Chem.*, vol. 85, no. 8, pp. 3849–3857, Apr. 2013. https://doi.org/10.1021/ac302702n

[25]   S. Lai, F. A. Viola, P. Cosseddu, and A. Bonfiglio, "Floating gate, organic field-effect transistor-based sensors towards biomedical applications fabricated with large-area processes over flexible substrates," *Sensors (Switzerland)*, vol. 18, no. 3, pp. 1–12, 2018. https://doi.org/10.3390/s18030688

[26]   L. Kergoat, B. Piro, M. Berggren, M. C. Pham, A. Yassar, and G. Horowitz, "DNA detection with a water-gated organic field-effect transistor," *Org. Electron.*, vol. 13, no. 1, pp. 1–6, 2012.

[27]   C. Suspène, B. Piro, S. Reisberg, M.-C. Pham, H. Toss, M. Berggren, A. Yassar, and G. Horowitz, "Copolythiophene-based water-gated organic field-effect transistors for biosensing," *J. Mater. Chem. B*, vol. 1, no. 15, p. 2090, 2013. https://doi.org/10.1039/c3tb00525a

[28] M. Berto, S. Casalini, M. Di Lauro, S. L. Marasso, M. Cocuzza, D. Perrone, M. Pinti, A. Cossarizza, C. F. Pirri, D. T. Simon, M. Berggren, F. Zerbetto, C. A. Bortolotti, and F. Biscarini, "Biorecognition in Organic Field Effect Transistors Biosensors: The Role of the Density of States of the Organic Semiconductor," *Anal. Chem.*, vol. 88, no. 24, pp. 12330–12338, Dec. 2016. https://doi.org/10.1021/acs.analchem.6b03522

[29] S. P. White, K. D. Dorfman, and C. D. Frisbie, "Label-free DNA sensing platform with low-voltage electrolyte-gated transistors," *Anal. Chem.*, vol. 87, no. 3, pp. 1861–1866, 2015. https://doi.org/10.1021/ac503914x

[30] A. Operamolla and G. Farinola, "Molecular and Supramolecular Architectures of Organic Semiconductors for Field-Effect Transistor Devices and Sensors: A Synthetic Chemical Perspective," *European J. Org. Chem.*, vol. 2011, no. 3, pp. 423–450, Dec. 2011. https://doi.org/10.1002/ejoc.201001103

[31] L. Torsi, G. M. Farinola, F. Marinelli, M. C. Tanese, O. H. Omar, L. Valli, F. Babudri, F. Palmisano, P. G. Zambonin, and F. Naso, "A sensitivity-enhanced field-effect chiral sensor," *Nat. Mater.*, vol. 7, p. 412, Apr. 2008. https://doi.org/10.1038/nmat2167

[32] M. Magliulo, B. R. Pistillo, M. Y. Mulla, S. Cotrone, N. Ditaranto, N. Cioffi, P. Favia, and L. Torsi, "PE-CVD of hydrophilic-COOH functionalized coatings on electrolyte gated field-effect transistor electronic layers," *Plasma Process. Polym.*, vol. 10, no. 2, pp. 102–109, 2013. https://doi.org/10.1002/ppap.201200080

[33] P. Favia, E. Sardella, R. Gristina, and R. d'Agostino, "Novel plasma processes for biomaterials: Micro-scale patterning of biomedical polymers," *Surf. Coatings Technol.*, vol. 169–170, pp. 707–711, 2003. https://doi.org/10.1016/S0257-8972(03)00174-9

[34] M. Y. Mulla, P. Seshadri, L. Torsi, K. Manoli, A. Mallardi, N. Ditaranto, M. V. Santacroce, C. Di Franco, G. Scamarcio, and M. Magliulo, "UV crosslinked poly(acrylic acid): a simple method to bio-functionalize electrolyte-gated OFET biosensors," *J. Mater. Chem. B*, vol. 3, no. 25, pp. 5049–5057, 2015. https://doi.org/10.1039/C5TB00243E

[35] G. Palazzo, D. De Tullio, M. Magliulo, A. Mallardi, F. Intranuovo, M. Y. Mulla, P. Favia, I. Vikholm-Lundin, and L. Torsi, "Detection Beyond Debye's Length with an Electrolyte-Gated Organic Field-Effect Transistor," *Adv. Mater.*, vol. 27, no. 5, pp. 911–916, Feb. 2015. https://doi.org/10.1002/adma.201403541

[36] L. S. Wong, F. Khan, and J. Micklefield, "Selective covalent protein immobilization: Strategies and applications," *Chem. Rev.*, vol. 109, no. 9, pp. 4025–4053, 2009. https://doi.org/10.1021/cr8004668

[37]　S. Cotrone, M. Ambrico, H. Toss, M. D. Angione, M. Magliulo, A. Mallardi, M. Berggren, G. Palazzo, G. Horowitz, T. Ligonzo, and L. Torsi, "Phospholipid film in electrolyte-gated organic field-effect transistors," Org. Electron., vol. 13, no. 4, pp. 638–644, 2012. https://doi.org/10.1016/j.orgel.2012.01.002

[38]　G. D. Bachand and C. D. Montemagno, "Constructing Organic/Inorganic NEMS Devices Powered by Biomolecular Motors," *Biomed. Microdevices*, vol. 2, no. 3, pp. 179–184, Jun. 2000.

[39]　P. Ghisellini, M. Caiazzo, A. Alessandrini, R. Eggenhöffner, M. Vassalli, and P. Facci, "Direct electrical control of IgG conformation and functional activity at surfaces," *Sci. Rep.*, vol. 6, p. 37779, Nov. 2016. https://doi.org/10.1038/srep37779

[40]　I. P. Mahendra Wijaya, T. J. Nie, S. Gandhi, R. Boro, A. Palaniappan, G. W. Hau, I. Rodriguez, C. R. Suri, and S. G. Mhaisalkar, "Femtomolar detection of 2,4-dichlorophenoxyacetic acid herbicides via competitive immunoassays using microfluidic based carbon nanotube liquid gated transistor," *Lab Chip*, vol. 10, no. 5, pp. 634–638, 2010. https://doi.org/10.1039/B918566F

[41]　M. Demelas, S. Lai, A. Spanu, S. Martinoia, P. Cosseddu, M. Barbaro, and A. Bonfiglio, "Charge sensing by organic charge-modulated field effect transistors: application to the detection of bio-related effects," *J. Mater. Chem. B*, vol. 1, no. 31, pp. 3811–3819, 2013. https://doi.org/10.1039/c3tb20237b

[42]　A. Spanu, S. Lai, P. Cosseddu, M. Tedesco, S. Martinoia, and A. Bonfiglio, "An organic transistor-based system for reference-less electrophysiological monitoring of excitable cells," *Sci. Rep.*, vol. 5, p. 8807, Mar. 2015. https://doi.org/10.1038/srep08807

[43]　S. P. White, K. D. Dorfman, and C. D. Frisbie, "Operating and Sensing Mechanism of Electrolyte-Gated Transistors with Floating Gates: Building a Platform for Amplified Biodetection," *J. Phys. Chem. C*, vol. 120, no. 1, pp. 108–117, Jan. 2016. https://doi.org/10.1021/acs.jpcc.5b10694

[44]　M. S. Thomas, S. P. White, K. D. Dorfman, and C. D. Frisbie, "Interfacial Charge Contributions to Chemical Sensing by Electrolyte-Gated Transistors with Floating Gates," *J. Phys. Chem. Lett.*, vol. 9, no. 6, pp. 1335–1339, 2018. https://doi.org/10.1021/acs.jpclett.8b00285

[45]　T. Cramer, A. Campana, F. Leonardi, S. Casalini, A. Kyndiah, M. Murgia, and F. Biscarini, "Water-gated organic field effect transistors – opportunities for biochemical sensing and extracellular signal transduction," *J. Mater. Chem. B*, vol. 1, no. 31, p. 3728, 2013. https://doi.org/10.1039/c3tb20340a

Organic Bioelectronics for Life Science and Healthcare   Materials Research Forum LLC
Materials Research Foundations **56** (2019) 97-114   doi: https://doi.org/10.21741/9781644900376-3

Chapter 3

The Organic Charge-Modulated Field-Effect Transistor: a Flexible Platform for Application in Biomedical Analyses

S. Lai[1], A. Spanu[1,2], P. Cosseddu[1], A. Bonfiglio[1]*

[1]Dept. of Electrical and Electronic Engineering, University of Cagliari

[2]Center for Materials and Microsystems, FBK - "Bruno Kessler" Foundation

annalisa@diee.unica.it

Abstract

Organic device-based sensors are currently being extensively investigated as key elements in easy-to-use, portable platforms for life science and healthcare. Filling the gap between laboratory environment and real application scenarios poses several challenges that researchers must address in order to meet the requirements for the realization of low-cost and efficient devices for Point-of-Care applications. Here we report a specific device architecture, namely the Organic Charge-Modulated Field-Effect Transistor (OCMFET), that represents a convenient option for the development of several kinds of electronic biosensors and bio-interfaces. A complete description of the OCMFET working principle will be provided, as well as its peculiar properties, which make it a unique device in the (bio)sensing field. Application of OCMFET principle for biochemical and biophysical sensor will be also discussed.

Keywords

Organic Biosensors, Field-Effect-Based Biosensor, DNA Sensing, pH Sensing, Cell Interface, Pharmacology

Contents

1.   Introduction

Among the topics covered by bioengineering, bioelectronics is surely one with the highest growth potential in the next years. Taking advantage of the continuous improvements in electronic technologies, researchers are developing novel classes of devices meant to revolutionize the consolidated approaches in life science and healthcare. Chemical and physical methods currently represent the gold standard in these fields: however, these methods are complex, requiring sophisticated, costly and bulky instrumentation and the need of expert operators working in laboratory environment. Although the high accuracy and reliability ensured by these approaches, the current trend in biomedical sciences is to overcome their limitations, providing novel, user-friendly and low-cost instruments suited for Point-of-Care (PoC) applications. Silicon technology is one of the foundations of this new paradigm: the consolidated fabrication processes, the possibility of device miniaturization and the low costs for mass production are being widely exploited in a large number of applications, including portable instruments which are already on the market (e.g. glucose meters and portable Holter monitors), and more complex devices, such as Lab-on-Chips (LoCs). In particular, LoCs aim either at integrating standard methods into miniaturized, highly-parallel platforms [1, 2], or proposing novel working principles that can bring further advantages in terms of straightforwardness of analyses and integration with elaboration units [3-7].

Nonetheless, the exploration of novel technologies and materials is a fundamental element for a complete change of paradigm in life science and healthcare applications. Indeed, inorganic materials and their technologies have different drawbacks that could represent an insurmountable limitation to their actual employment in biomedical applications: biocompatibility is not always ensured, and reduced flexibility and softness, opacity and relevant cost for large area production must be taken into account in the development of novel devices. On these basis, it is possible to justify the increasing interest in the novel frontiers of electronics, where innovative materials and processes are taken into account to broaden the boundaries of current technologies.

In the field of bioelectronics, organic materials have been thoroughly considered in the latest decades. Organic electronic devices are conceived for being complementary to their inorganic counterparts on different aspects. Taking advantage of the intrinsic properties of organic materials, organic devices can be fabricated onto unconventional substrates, like plastic, paper and fabrics, with peculiar properties such as transparency, flexibility and light weight. Organic materials can be processed from liquid phase, thus enabling several innovative fabrication techniques suited for cost-effective fabrication over large areas. Moreover, organic materials can be tailored by chemical procedures for providing specific properties required by the investigated applications: therefore, more degrees of freedom both in fabrication and in applications are provided by organic devices if compared to the one given by inorganic materials.

So far, organic materials have actually revolutionized the field of optoelectronics, and can be found at a market size as a reliable technology. Besides photovoltaics, sensor applications in biomedical sciences is generally considered one of the future exploitation fields for organic electronics. As a matter of fact, in the last two decades researchers have reported several examples of organic electronic devices for biochemical and physical sensing applications. In particular, the employment of Organic Thin-Film Transistors (OTFTs) is being thoroughly explored. Transistors can be used for electronic transduction of the investigated phenomena, thus directly allowing the generation of electrical signals that can be available for the final elaboration unit. Transistors also have intrinsic amplification properties, and this further simplifies the signals readout. More than for their inorganic counterparts, the multi-parametric nature of organic transistors can be exploited in different transduction mechanism, making OTFTs suitable sensors for different kinds of application in biomedical sciences. In the last decades, OTFTs have been employed for the development of several kinds of sensors with potential application in healthcare and life science, including devices for biochemical sensing, such as ion-sensitive devices [8], genetic [9-11], enzymatic [12], protein [13-16] and immunological sensors [17], which can be employed into sensing platforms in different scenarios, including laboratory environment and field-measurement kits. Disposability enabled by low fabrication costs is a peculiar property of this class of devices if compared with inorganic transistor-based sensors.

Although organic devices have peculiar properties that can be effectively exploited for the development of different kind of sensors for biomedical applications, their diffusion in real application scenarios is still limited. As a matter of fact, bridging the gap between demonstration of device functionality in laboratory environment and diffusion of sensors based on organic devices requires facing several problems with no trivial resolution. Some of them are related to the intrinsic properties of organic materials in terms of degradation in ambient conditions, which significantly affects the life cycle of a possible product, moreover, such materials are generally characterized by a low electronic conductivity, resulting in high power-needs for actual devices operation that usually leads to reduced portability. Some other aspects are related to the reliability of the fabrication processes, which are surely lower than the ones used for standard electronics, further reducing the reproducibility of device performances. As in several cases device sensing performances are directly related to the electronic properties of the organic materials constituting the sensor, the sum of these criticalities would represent an insurmountable obstacle for the development of devices aiming at reaching industrial production.

In this sense, an increasing effort has been put by researchers in exploring possible solutions to the above-mentioned issues and limitations. Here, we report about a possible approach for the development of OTFT-based sensing devices where the peculiar properties of a specific device structure, namely Organic Charge-Modulated Field-Effect Transistor (OCMFET), are exploited for the development of a whole set of different sensors covering a wide range of applications in biomedical sciences. In the following, we will discuss about the working principle and main properties of the OCMFET platform, which has been conceived for the development of reliable biochemical and physical sensors. A demonstration of the functionality of OCMFET-based sensors in biomedical applications will be provided, thus allowing to foresee its actual exploitation in the development of innovative tools for life science and healthcare applications.

2. Organic Charge-Modulated Field-Effect Transistor: device structure and working principle

The basic structure of the OCMFET is reported in Figure 1. The core of the device is a standard OTFT structure, the Organic Field-Effect Transistor (OFET), developed in a bottom-gate/bottom-contact structure. In the OCMFET, the gate of the OFET structure is biased through a control capacitor (also referred as control gate): therefore, no bias is directly imposed (i.e. it is a floating gate) even if its voltage is actually fixed by several concurring effects. The floating gate is insulated from the environment by a dielectric layer, which both acts as the gate insulator and as a protecting layer impeding variations in the charge stored in it. If a part of the floating gate is left exposed, it is possible to influence the charge distribution inside it by an external stimulus, thus allowing the employment of the OCMFET as an extremely sensitive charge sensor.

Organic Bioelectronics for Life Science and Healthcare   Materials Research Forum LLC
Materials Research Foundations **56** (2019) 97-114   doi: https://doi.org/10.21741/9781644900376-3

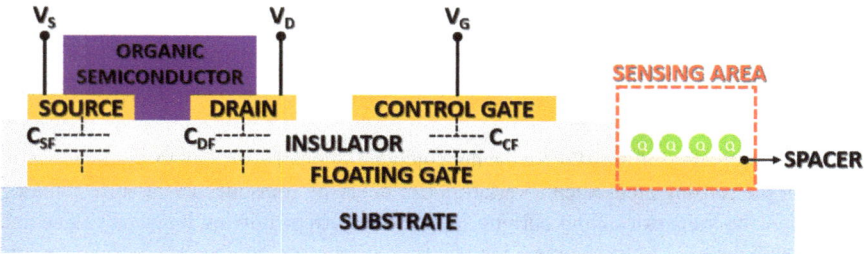

Figure 1: Cross section of a generic OCMFET structure. The OCMFET is a floating gate OTFT with a second gate called control gate that is needed to set the transistor working point. This device can be employed as a charge sensor by exposing the final part of the floating gate (called sensing area) to the measurement environment.

Its working principle has been firstly described by Barbaro et al. using a CMOS version of the device called Charge Modulated FET (CMFET) [18]. The charge inside the floating gate, Q_{TOT}, can be estimated taking into account the different voltage contributions in the device according to Gauss equation:

$$Q_{TOT} = C_{CF}\left(V_{FG} - V_G\right) + C_{DF}\left(V_{FG} - V_D\right) + C_{SF}\left(V_{FG} - V_S\right)$$

(1)

where C_{CF}, C_{DF} and C_{SF} are, respectively, the control capacitance and the parasitic capacitances related to the overlap between drain and source and the floating gate, V_G, V_D and V_S are the voltages applied to control capacitor, drain and source respectively and V_{FG} is the actual floating gate voltage. This last parameter can be thus written as

$$V_{FG} = \frac{C_{CF}}{C_{CF} + C_{DF} + C_{SF}}V_G + \frac{C_{DF}}{C_{CF} + C_{DF} + C_{SF}}V_D + \frac{C_{SF}}{C_{CF} + C_{DF} + C_{SF}}V_S + \frac{Q_{TOT}}{C_{CF} + C_{DF} + C_{SF}}.$$

(2)

If a charge Q_S is immobilized on top of the sensing area by interposing an insulating anchoring layer, Q_{TOT} can be written as $Q_0 + Q_i(Q_S)$, being Q_0 a constant amount of charge eventually incorporated in the floating gate during fabrication, and $Q_i(Q_S)$ the charge induced in the floating gate by Q_S. When the parasitic capacitances are negligible with respect to the control capacitance, the spacer is thinner than other insulating layer in the device (thus allowing perfect induction approximation, $Q_i(Q_S) \approx -Q_S$) and Q_0 is negligible, the last equation can be approximated as

Organic Bioelectronics for Life Science and Healthcare   Materials Research Forum LLC
Materials Research Foundations **56** (2019) 97-114   doi: https://doi.org/10.21741/9781644900376-3

$$V_{FG} \approx V_G - \frac{Q_S}{C_{CF}}.$$

(3)

Therefore, the value of the floating gate voltage is linearly related to the amount of charge on the sensing area. Such a variation can be easily transduced as a corresponding variation of the transistor output current. Indeed, the current flowing between source and drain is a function of the floating gate voltage according to the basic characteristic equations of OFETs [19]: for instance, if the device is maintained in the saturation regime ($|V_{DS}| \geq |V_{FG}-V_{TH}|$, being V_{TH} the transistor threshold voltage), the current can be expressed according to the relationship

$$I_{DS,SAT} = \frac{1}{2} \mu C_{INS} \frac{W}{L} \left(V_{FG} - V_{TH} \right)^2$$

(4)

where μ is the charge carrier mobility, C_{INS} is the capacitance per unit area of the floating gate insulator, W and L are the transistor's channel width and length respectively. If we substitute V_{FG} with the approximated relationship in (3), Equation 4 can be written as

$$I_{DS,SAT} \approx \frac{1}{2} \mu C_{INS} \frac{W}{L} \left[V_G - \left(\frac{Q_S}{C_{CF}} + V_{TH} \right) \right]^2.$$

(5)

A variation of the floating gate voltage is thus modeled as a shift in the transistor threshold voltage: if the charge Q_S varies of a quantity ΔQ_S, the threshold voltage shift is

$$\Delta V_{TH} = -\frac{\Delta Q_S}{C_{CF}}.$$

(6)

From the last equation, the sensitivity of the OCMFET can be derived as following:

$$\left| \frac{\Delta V_{TH}}{\Delta Q_S} \right| = \frac{1}{C_{CF}} = \frac{1}{C_{INS}} \frac{1}{A_{CC}} \approx \frac{1}{C_{INS}} \frac{1}{A_{TOT} - A_S - WL}$$

(7)

where the dependence of C_{CF} to the control capacitor area A_{CC} is made explicit, and this area is written as a function of the total floating gate area (A_{TOT}), which includes the control capacitor, the sensing area (A_S) and the transistor area (WxL).

A deeper insight into the device structure and the equation describing its working principle allows deriving a series of peculiar characteristics that make the OCMFET unique among other OTFT-based devices. Differently than other structures, like OFETs, Ion-Sensitive FET (ISFET)-like devices and Electrolyte-Gated OFETs (EGOFETs), there is no need of an external reference electrode set the working point in the measurement environment: this feature is particularly important for a proper integration of biochemical sensors into electronic platforms, since an external reference electrode is generally needed with the most employed architectures [8], [10-17]. Interestingly enough, the OCMFET sensing performance is substantially independent of the electrical properties of the transistor's active layer: indeed, the transistor is actually the transducing element for the actual sensing element in the OCMFET, i.e. the floating gate. This effective decoupling of transistor and sensing area brings several advantages in terms of device functionality: when the active layer is employed as sensing layer, as for instance in EGOFETs, the transistor is directly exposed to harsh environments (e.g., in the case of biochemical sensors [9], [11], [13-17]), thus leading to a rapid deterioration of the device performance or to serious damages of the structure. In the OCMFET structure, the organic semiconductor is physically separated from the sensing area, thus it can be encapsulated for ensuring a prolonged lifetime of the device. Moreover, when the active layer is employed as sensing element, the obtained sensing performances are reliant on the chosen material, and on the actual reliability of its electrical performance; this makes the derivation of proper design rules virtually impossible, thus seriously reducing the exploitation of the sensor at industrial level. On the contrary, the working principle of the OCMFET is virtually independent on the choice of the materials, allowing deriving specific equations which can be used to rule the sensor design. In this sense, it is possible to observe from Equation 7 that the sensitivity can be easily tuned by acting on the device geometry and that it barely depends on the design of the transistor. Indeed, transistor performances and sensing performances can be independently optimized: C_{INS}, W and L can be chosen to determine the basic transistor performances (e.g., a proper level for operating voltages, a specific output current in operating conditions), while the ratio between the overall dimensions of the floating gate and the sensing area can be changed to tune the sensitivity.

3.   OCMFET as sensor for biomedical applications

3.1   DNA sensing

DNA hybridization refers to the complex set of biochemical reactions that lead two complementary single-stranded oligonucleotides to interact forming the common double-stranded DNA structure; its (the DNA hybridization) detection is one of the most investigated biological reactions, employed in medical, pharmaceutical and forensic applications. The development of electronic sensors for label-free, direct DNA hybridization detection has been thoroughly explored over the years to overcome the main limitations of standard methods, which are mainly based on optical analyses, which requires bulky and costly instrumentation, highly-specialized operators as well as quite complex procedures for DNA labelling and amplification [20]. As a matter of fact, DNA hybridization detection through OTFT-based devices has been deeply investigated over the last decade [9-11]. Different OCMFET-based devices have been successfully tested as DNA hybridization sensors [21-23] and their working principle relies on the charge properties of DNA molecules. Indeed, DNA is a negatively-charged molecule, so if a single-stranded oligonucleotide is employed as a probe and anchored on the sensing area, hybridization with the complementary target sequence can be transduced as an increase of the negative charge immobilized on the sensing area, resulting in a floating gate voltage shift and thus in a variation of the output current of the transistor (as shown in Figure 2).

Starting from basic proof-of-concept demonstrated by Demelas et al. [21], a first complete characterization of the actual device functionality has been reported in Lai et al. [22]. In the latest version, low voltage operation was obtained by employing a nano-size, hybrid organic-inorganic dielectric [24], thus allowing the device to proper operate in aqueous environment. More recently, we reported an optimized DNA sensor based on a self-aligned OCMFET structure: in this particular layout, parasitic contributions coming from the transistor are minimized, thus allowing a considerable improvement of the device sensitivity [23]. By following the OCMFET design rules, both transistor and sensing performances have been further optimized: a detection limit in the range of 100 fM was obtained (see Figure 3), and the device was capable to correctly discriminate the single-nucleotide polymorphism (SNP). This performance represents a record result in terms of sensitivity and selectivity for DNA hybridization detection employing OTFT-based devices.

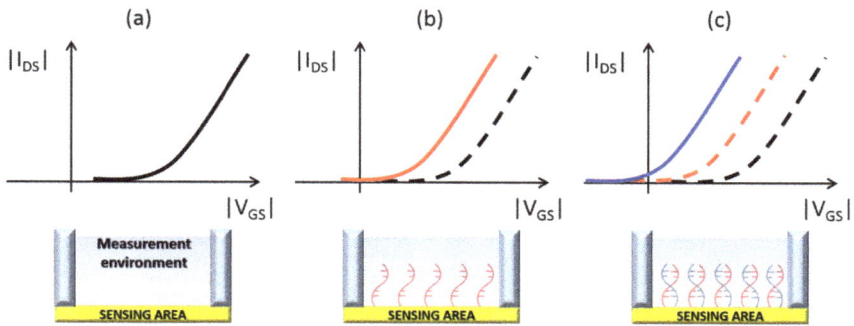

Figure 2: OCMFET as DNA sensor: (a) the transistor transfer characteristic curve before any interaction with oligonucleotides (measurement solution on the sensing area); (b) when single-stranded oligonucleotides are anchored on the sensing area (DNA probe), their negative charge determines a shift in the device threshold voltage; (c) a further shift of the threshold voltage is obtained when probes hybridize with complementary target oligonucleotides, thus determining an increase of the negative charge anchored on the sensing area; this last threshold voltage shift, and the related current variation, represent the detection mechanism for DNA hybridization.

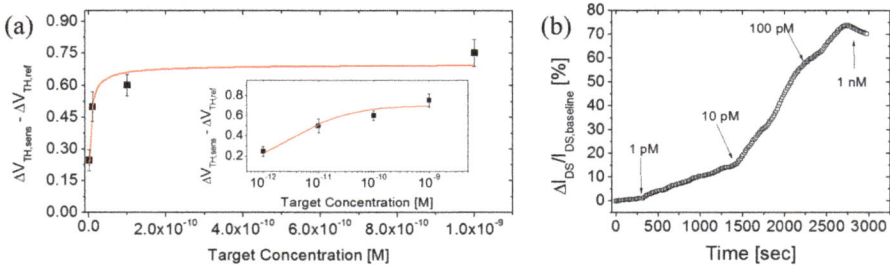

Figure 3: (a) Threshold voltage shift in linear and logarithmic scale (in the inset) as a function of target concentration; the shift is relative to threshold voltage shift evaluated in a reference device used for the compensation of unspecific voltage variations. Data points are fitted with a Sips isotherm, which represents absorption of molecules on surfaces; (b) percentage current variation as a function of time: arrows represent time of injection of target molecules on the sensing area at the indicated concentration. Reprinted from Sensors and Actuators B: Chemical, 223, S. Lai, M. Barbaro, A. Bonfiglio, Tailoring the sensing performances of an OFET-based biosensor, 314-319, Copyright (2016), with permission from Elsevier.

Interestingly enough, all these results have been obtained in measurement environments with relatively high ionic concentration ([NaCl] = 50 mM): this condition is considered prohibitive for proper device operation, as the intrinsic charge of the oligonucleotides is screened by counter ions in solution. In order to take into account this screening effect, the Debye length parameter is introduced: for monovalent ions, the Debye length can be written as

$$\lambda_D = \sqrt{\frac{\varepsilon_r \varepsilon_0 k_B T}{2 N_A e^2 I}}$$

(8)

where ε_r is the relative permittivity of the measurement environment, ε_0 is the free-space permittivity, k_B the Botzmann constant, T the absolute temperature, N_A the Avogadro number, e the electron charge and I the ionic strength of the media, the latter being directly proportional to the concentration of ions in solution. All charges beyond a distance λ_D from the sensing surface are screened by counter-ions, thus being impossible to detect. In the measurement conditions of [21-23], the Debye length is approximatively of 1.5 nm, while hybridization takes part from a distance of more than 4 nm, thus being in theory screened and thus not transduced by the OCMFET. Starting from the results of Rant and co-workers [25], field-effect induced tilting of DNA molecules was hypothesized to take place in OCMFET in normal operating conditions. Indeed, in order to reduce the bias stress, a pulse-shaped gate voltage is normally applied to the sensor: such a voltage determines a transient voltage drop between the floating gate and the solution that attracts the double-stranded DNA molecules towards the surface, repelling at the same time the counter ions and thus increasing the effective Debye length in the measurement environment. Such an effect was verified by means of fluorescence-quenching tests [26, 27]: Cyanine-3 labelled DNA strands were employed, and a significant and reversible reduction of emission in the bandwidth of the fluorochrome was observed during device operation (Figure 4). This effect is related to the tilting of the oligonucleotides, which brings the fluorochrome close to the metal sensing surface, where fluorescence quenching takes place.

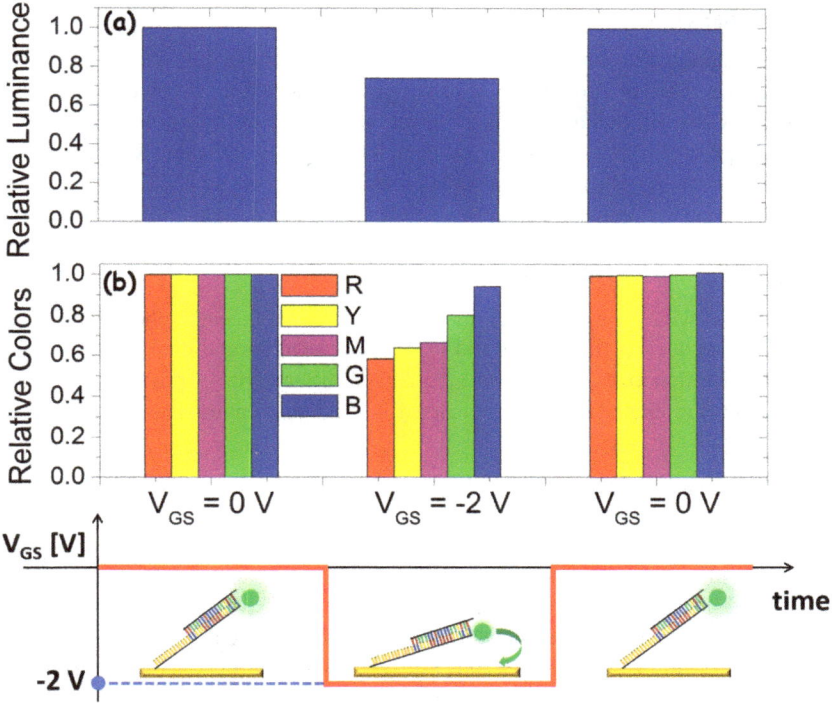

Figure 4: (a) Variation of relative luminance during fluorescence quenching tests: data are normalized to the value of luminance when the OCMFET is switched off (V_{GS} = 0 V); when the device is switched on, the electric field attracts oligonucleotides near to the sensing surface, thus determining a reduction of the luminance, which is then restored when the device is brought back to the off state. (b) Corresponding variation of the different wavelengths: the ones with the larger variation are red, yellow and magenta components, which are in the emission band of the Cyanine-3 fluorochrome; green (excitation) and blue components (out from emission band) are substantially constant. (c) Cartoon depicting the proposed mechanism that takes place on the sensing area during the experiment. Reprinted from S. Lai, M. Barbaro, A. Bonfiglio, Applied Physics Letters 107, 103301 (2015), with the permission of AIP Publishing.

Organic Bioelectronics for Life Science and Healthcare   Materials Research Forum LLC
Materials Research Foundations **56** (2019) 97-114   doi: https://doi.org/10.21741/9781644900376-3

3.2   Cellular electrical and pH sensing

A different application scenario that has gained considerable interest in the last two decades concerns the in vitro testing of electrogenic cells cultures. In fact, monitoring the electrical activity of living cells is of great importance in scientific fields such as pharmacology, (neuro)rehabilitation, and computational neuroscience. Recently, in vitro approaches involving the use of either microelectrode arrays or ISFET-like electronic devices have been employed for the investigation of fundamental cellular mechanisms [28] and in the disclosing of the insights of the way in which the action potential propagates in 2D and 3D neuronal networks [29-31].

Despite the wide employment in electrophysiology and pharmacology the aforementioned devices (the MEAs and ISFETs), present several disadvantages such as the high fabrication costs, the intrinsic rigidity of the materials, and the presence of a (usually bulky) reference electrode immersed in the liquid environment where the cells are cultured.

As previously highlighted, the OCMFET is a very convenient approach in all the applications where the detection of low charge variations in a liquid environment is involved. Particularly interesting features of this device for this specific application are the absence of a reference electrode during its operation (thanks to the presence of the control gate) and the elongated shape of the floating gate, which allows separating the active layer (i.e. the organic semiconductor) and the area where the actual sensing occurs (namely the sensing area). These features, together with its high charge sensitivity and relatively high cutoff frequency (up to 100 kHz), make the OCMFET a good candidate for the design of novel electrophysiological/pharmacological tools that can be both reference-less and low-cost, thus potentially having the capability to compete "market-wise" with the already existent and assessed MEA and ISFET technology. In order to meet the specific requirements of the electrophysiological application, a particular kind of OCMFET array (called Micro OCMFET Array - MOA) has been designed and fabricated, and its capability of transducing bio-electrical signals has been thoroughly investigated by using cardio-myocytes cultures. In fact, thanks to the optimal covering of the sensing area and their regular electrical activity in vitro, cardiomyocytes are a very good model to estimate the device sensitivity. As shown in figure 5, the MOA turned out to be capable of reliably monitoring the activity of rat cardiomyocytes in both basal conditions and upon thermal and chemical stimulation, thus confirming the suitability of the OCMFET approach for pharmacological testing [32].

Organic Bioelectronics for Life Science and Healthcare   Materials Research Forum LLC
Materials Research Foundations **56** (2019) 97-114   doi: https://doi.org/10.21741/9781644900376-3

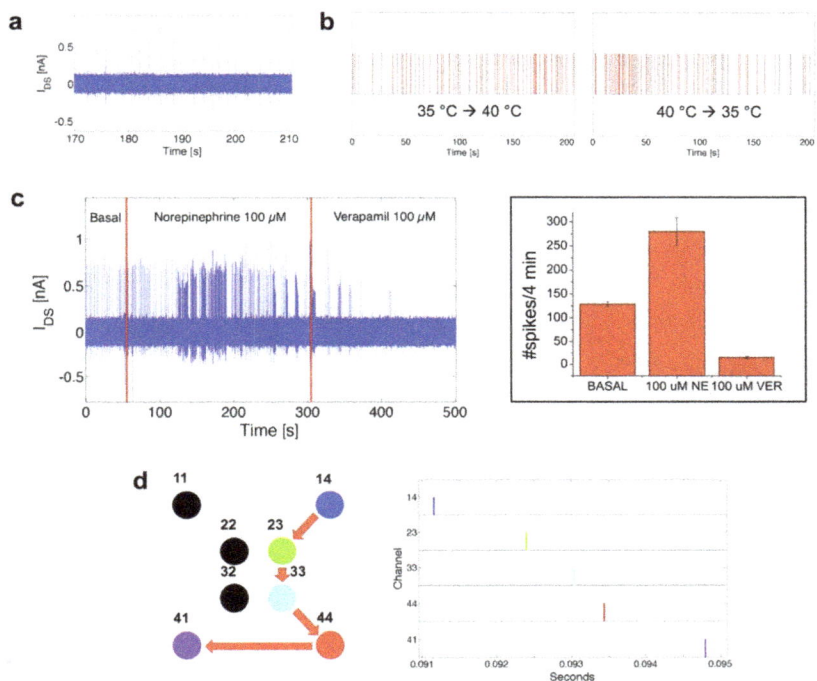

Figure 5: OCMFET validation with rat cardiac myocytes. (a) Basal activity measured with an OCMFET. (b) Raster plots of the thermal modulation of the culture activity. (c) Chemical tuning of the culture's activity. The spontaneous activity was accelerated by means of the addition of 100 mM of Norepinephrine and then suppressed with 100 mM of Verapamil. (c) (inset), Beating frequency modulation (statistics on 5 OCMFETs - average and standard deviation). (d) Multisite recording. The MOA allows to track electrical signals in the culture as shown in the raster plot of the spontaneous activity indicating a propagation of the signal from site 14 to site 41. The culture was maintained 8 days in vitro and measured at 37 °C . Adapted from Scientific Reports volume 5, Article number: 8807 (2015) [32].

Another important parameter when dealing with cell cultures is the cellular metabolic activity [33]. In fact, the pH of the medium where the cells are cultured influences the cells activity and, on the contrary, variations of cellular metabolic activity (either physiological or induced by means of chemical/electrical stimulation) causes relative pH variations of the medium. Thanks to its versatility, the OCMFET can be easily turned into a pH sensor by properly modify the sensing area. To this purpose, a simple pH-sensitive membrane, consisting in an oxygen plasma-activated Parylene C thin layer, has been employed. The transduction principle, as previously explained, is related to a variation of the transistor threshold voltage induced by the (pH-dependent) charge immobilized onto the sensing area. As shown in figure 6, such an OCMFET turned out to be a very sensitive pH sensor, thanks to the intrinsic charge amplification given by the peculiar structure of the device [34].

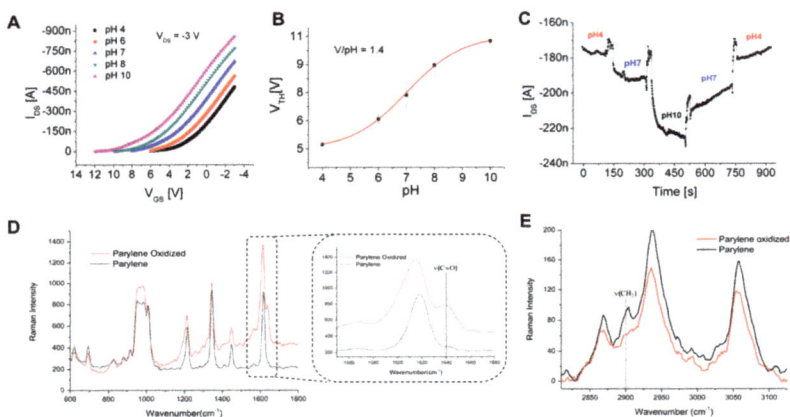

Figure 6: Characterization of an OCMFET for pH sensing. a) Transfer characteristics performed while the sensing area is exposed to buffer solutions at different pH. It is noticeable the gradual switching on of the transistor going from pH 4 to pH 10 due to the shift of the threshold voltage to a more positive value. b) Device calibration. The V_{TH} has been extrapolated from the trans-chars (three chars for each pH point) and plotted against the pH value as mean and standard deviation. The device shows a typical sigmoidal behavior and a relatively wide linear region (from pH 6 to pH 8). c) I_{DS} VS Time of another OCMFET device. As it can be noticed, the device (despite the current drift, which is mainly due to the bias stress) showed a fast response and a reproducible behavior. d) Raman spectra of pristine Parylene C (black) and plasma activated Parylene C (red). A silicon wafer has been employed as the carrier substrate. The additional band at 1640 cm^{-1} has been assigned to the vibration of the C=O stretching of the carboxyl group. (B) The vibration at 2900 cm^{-1}, related to the CH_2 stretching, decreased after the oxidation process with the related increase of the band at 1640 cm^{-1} [34]. Reprinted from Organic Electronics, 48, A. Spanu, F. Viola, S. Lai, P. Cosseddu, P. C. Ricci, A. Bonfiglio, A reference-less pH sensor based on an organic field effect transistor with tunable sensitivity, 188-193, Copyright (2017), with permission from Elsevier.

Such a super-nernstian and referenceless pH sensor can be employed in combination with the OCMFET for cell electrical activity monitoring for the realization of a multisensing platform, which can be able to simultaneously monitor different aspects of the cellular activity, thus opening up interesting perspectives in pharmacology and neural interface.

Conclusions

The technological approach that has been carried along by the introduction of the OCMFET represents a great leap forward in the biosensing field in that it offers the possibility to obtain a wide range of low-cost, referenceless and ultra-sensitive sensors and biosensors. The potentials in healthcare (and life science in general) of such a versatile device have been demonstrated for DNA hybridization sensing, for the detection of the electrical activity of living cells and for pH sensing beyond the Nerst limit, thus opening up new interesting perspectives in these fast-growing scientific fields. Further exciting developments are foreseen by the combination of different sensing capability into the same platform, thus paving the way to the development of low-cost and easy-to-fabricate multisensing devices to be applied in a variety of fields.

References

[1]   P.M. Levine, P. Gong, R. Levicky, K. L. Shepard, Active CMOS Sensor Array for Electrochemical Biomolecular Detection, IEEE J. Solid-State Circuits 43 (2008),1859–1871. https://doi.org/10.1109/JSSC.2008.925407

[2]   A. Manickam, A. Chevalier, M. McDermott, A. D. Ellington, A. Hassibi, A CMOS Electrochemical Impedance Spectroscopy (EIS) Biosensor Array. IEEE Trans. Biomed. Circuits Syst. 4 (2010), 379–390. https://doi.org/10.1109/TBCAS.2010.2081669

[3]   M. Schienle, C. Paulus, A. Frey, F. Hofmann, B. Holzapfl, P. Schindler-bauer, R. Thewes, A Fully Electronic DNA Sensor With 128 Positions and In-Pixel A/D Conversion, IEEE J. Solid-State Circuits, 39 (2004), 2438–2445. https://doi.org/10.1109/JSSC.2004.837084

[4]   K.-H. Lee, J. O. Lee, S. Choi, J.-B Yoon, G.-H. Cho, A CMOS label-free DNA sensor using electrostatic induction of molecular charges, Biosens. Bioelectron. 31(2011), 343–348. https://doi.org/10.1016/j.bios.2011.10.042

[5]   C. Stagni, C. Guiducci, L. Benini, S. Carrara, C. Paulus, M. Schienle, M. Augustyniak, R. Thewes, CMOS DNA Sensor Array With Integrated A/D Conversion

Based on Label-Free Capacitance Measurement, IEEE J. Solid-State Circuits 41 (2006), 2956–2964. https://doi.org/10.1109/JSSC.2006.884867

[6]   E. Anderson, J. Daniels, H. Yu, T. Lee, N. Pourmand, A Label-free CMOS DNA Microarray based on Charge Sensing. In I2MTC 2008 – IEEE Instrumentation and Measurement Technology Conference, 2008. https://doi.org/10.1109/IMTC.2008.4547305

[7]   J. M. Rothberg, et al., An integrated semiconductor device enabling non-optical genome sequencing, Nature 475 (2011), 348–352. https://doi.org/10.1038/nature10242

[8]   C. Bartic, B. Palan, A. Campitelli, G. Borghs, Monitoring pH with organic-based field-effect transistors, Sens. Actuator B-Chem 83 (2002), 115-122. https://doi.org/10.1016/S0925-4005(01)01053-X

[9]   Q. Zhang, V. Subramanian, DNA hybridization detection with organic thin film transistors: Toward fast and disposable DNA microarray chips, Biosens. Bioelectron. 22 (2007), 3182–3187. https://doi.org/10.1016/j.bios.2007.02.015

[10]   H. U. Khan, M. E. Roberts, O. Johnson, R. Förch, W. Knoll, Z. Bao, In Situ, Label-Free DNA Detection Using Organic Transistor Sensors, Adv. Mat. 22 (2010), 4452–4456. https://doi.org/10.1002/adma.201000790

[11]   L. Kergoat, B. Piro, M. Berggren, M.-C. Pham, A. Yassar, G. Horowitz, DNA detection with a water-gated organic field-effect transistor, Org. Electron. 13 (2012), 1-6. https://doi.org/10.1016/j.orgel.2011.09.025

[12]   C. Bartic, A. Campitelli, and S. Borghs, Field-effect detection of chemical species with hybrid organic/inorganic transistors, Appl. Phys. Lett. 82 (2003), 475-477. https://doi.org/10.1063/1.1527698

[13]   M. Y. Mulla, E. Tuccori, M. Magliulo, G. Lattanzi, G. Palazzo, K. Persaud, L. Torsi, Capacitance-modulated transistor detects odorant binding protein chiral interactions. Nat. Commun 6 (2015), 6010. https://doi.org/10.1038/ncomms7010

[14]   M. Y. Mulla, P. Seshadri, L. Torsi, K. Manoli, A. Mallardi, N. Ditaranto, M. V. Santacroce, C. Di Franco, G. Scamarcio M. Magliulo, UV crosslinked poly(acrylic acid): a simple method to bio-functionalize electrolyte-gated OFET biosensors, J. Mater. Chem. B 3 (2015), 5049–5057. https://doi.org/10.1039/C5TB00243E

[15]   M. D. Angione, S. Cotrone, M. Magliulo, A. Mallardi, D. Altamura, C. Giannini, N. Cioffi, L. Sabbatini, E. Fratini, P. Baglioni, G. Scamarcio, G. Palazzo, L. Torsi,

Proc. Natl. Acad. Sci. U.S.A. 109 (2012), 6429-6434.
https://doi.org/10.1073/pnas.1200549109

[16]   M. D. Angione, M. Magliulo, S. Cotrone, A. Mallardi, D. Altamura, C. Giannini, N. Cioffi, L. Sabbatini, D. Gobeljic, G. Scamarcio, G. Palazzo, L. Torsi, Volatile general anesthetic sensing with organic field-effect transistors integrating phospholipid membranes, Biosens. Bioelectron. 40 (2013) 303-307. https://doi.org/10.1016/j.bios.2012.07.068

[17]   M. Medina-Sánchez, C. Martínez-Domingo, E. Ramon, A. Merkoçi, An Inkjet-Printed Field-Effect Transistor for Label-Free Biosensing, Adv. Funct. Mater. 24 (2014), 6291–6302. https://doi.org/10.1002/adfm.201401180

[18]   M. Barbaro, A. Bonfiglio, L. Raffo, A Charge-Modulated FET for Detection of Biomolecular Processes: Conception, Modeling, and Simulation, IEEE Trans. Electron Devices 53 (2006), 158-166. https://doi.org/10.1109/TED.2005.860659

[19]   G. Horowitz, P. Lang, M. Mottaghi, H. Aubin, Extracting Parameters from the Current–Voltage Characteristics of Organic Field-Effect Transistors, Adv. Funct. Mater. 14 (2004), 1069-1074. https://doi.org/10.1002/adfm.200305122

[20]   W. J. Wall, "Techniques for DNA Analysis", in Ullmann's Encyclopedia of Industrial Chemistry, Wiley-VCH Verlag GmbH & Co. KGaA, 2000.

[21]   M. Demelas, S. Lai, G. Casula, E. Scavetta, M. Barbaro, A. Bonfiglio, "An organic, charge-modulated field effect transistor for DNA detection", Sens. Actuator B-Chem 171 (2012), 198-203. https://doi.org/10.1016/j.snb.2012.03.007

[22]   S. Lai, M. Demelas, G. Casula, P. Cosseddu, M. Barbaro, A. Bonfiglio, "Ultralow Voltage, OTFT-Based Sensor for Label-Free DNA Detection", Adv. Mater 25 (2013), 103–107. https://doi.org/10.1002/adma.201202996

[23]   S. Lai, M. Barbaro, A. Bonfiglio, "Tailoring the sensing performances of an OFET-based biosensor", Sens. Actuator B-Chem 233 (2016), 314–319. https://doi.org/10.1016/j.snb.2016.04.095

[24]   P. Cosseddu, S. Lai, M. Barbaro, A. Bonfiglio, Ultra-low voltage, organic thin film transistors fabricated on plastic substrates by a highly reproducible process, Appl. Phys. Lett. 100 (2012), 093305. https://doi.org/10.1063/1.3691181

[25]   S. Lai, M. Barbaro, A. Bonfiglio, The role of polarization-induced reorientation of DNA strands on organic field-effect transistor-based biosensors sensitivity at high ionic strength, Appl. Phys. Lett. 107 (2015), 103301. https://doi.org/10.1063/1.4930303

[26]   U. Rant, K. Arinaga, S. Fujita, N. Yokoyama, G. Abstreiter, M. Tornow, Dynamic electrical switching of DNA layers on a metal surface, Nano Lett. 4 (2004), 2441-2445. https://doi.org/10.1021/nl0484494

[27]   U. Rant, K. Arinaga, S. Fujita, N. Yokoyama, G. Abstreiter, M. Tornow, Electrical manipulation of oligonucleotides grafted to charged surfaces, Org. Biomol. Chem. 4 (2006), 3448-3455. https://doi.org/10.1039/b605712h

[28]   Cogollo, J. F. S., Tedesco, M., Martinoia, S., Raiteri, R. A new integrated system combining atomic force microscopy and micro-electrode array for measuring the mechanical properties of living cardiac myocytes. Biomed. Microdevices 13, 613–621 (2011). https://doi.org/10.1007/s10544-011-9531-9

[29]   Chiappalone, M., Bove, M., Vato, A., Tedesco, M., Martinoia, S. Dissociated cortical networks show spontaneously correlated activity patterns during in vitro development. Brain Res. 1093, 41–53 (2006). https://doi.org/10.1016/j.brainres.2006.03.049

[30]   Berdondini, L. *et al.* Active pixel sensor array for high spatio-temporal resolution electrophysiological recordings from single cell to large scale neuronal networks. Lab Chip 9, 2644–51 (2009). https://doi.org/10.1039/b907394a

[31]   M. Frega, M. Tedesco, P. Massobrio, M. Pesce, S. Martinoia, Network dynamics of 3D engineered neuronal cultures: a new experimental model for in-vitro electrophysiology. 1–14 (2014). https://doi.org/10.1038/srep05489

[32]   A. Spanu, S. Lai, P- Cosseddu, M. Tedesco, S. Martinoia, A. Bonfiglio, An organic transistor-based system for reference-less electrophysiological monitoring of excitable cells. Sci. Rep. 5, 8807 (2015). https://doi.org/10.1038/srep08807

[33]   S. Martinoia, N. Rosso, M. Grattarola, L. Lorenzelli, B. Margesin, M. Zen, Development of ISFET array-based microsystems for bioelectrochemical measurements of cell populations. Biosens. Bioelectron. 16, 1043–1050 (2001). https://doi.org/10.1016/S0956-5663(01)00202-0

[34]   A. Spanu, F. Viola, S. Lai, P. Cosseddu, P. C. Ricci, A. Bonfiglio, A reference-less pH sensor based on an organic field effect transistor with tunable sensitivity. Org. Electron. 48, 188–193 (2017). https://doi.org/10.1016/j.orgel.2017.06.010

Organic Bioelectronics for Life Science and Healthcare   Materials Research Forum LLC
Materials Research Foundations **56** (2019) 115-152   doi: https://doi.org/10.21741/9781644900376-4

Chapter 4

Graphene Based Field Effect Transistors for Biosensing: Importance of Surface Modification

Sabine Szunerits[1*], Rabah Boukherroub,[1] Alina Vasilescu,[2] Serban Peteu,[3]

[1] Uni. Lille, CNRS, Centrale Lille, ISEN, Uni. Valenciennes, UMR 8520-IEMN, F-59000 Lille, France

[2] International Center of Biodynamics, 1B Intrarea Portocalelor, Sector 6, Bucharest 060101, Romania

[3] Department of Chemistry, Michigan State University, 578 S Shaw Lane, East Lansing, MI 48824, United States

*sabine.szunerits@univ-lille.fr

Abstract

In the last decade the use of field-effect-based devices has become a basic structural element in a new generation of biosensors that allow label-free analysis. This field has been dominated for a long time by optically based readout techniques utilizing fluorescent markers or those requiring advanced spectroscopic equipment. While other fields have benefited from new technologies based upon advancements in semiconductor integrated circuit technology, chemical and biological sensors have remained dependent upon biochemical assays due to the challenges of achieving sensitivity and selectivity with semiconductor-based sensors. Silicon transistor-based readout sensors have been developed, but these devices suffered from poor sensitivity and selectivity due to fundamental shortcomings of the silicon structure. Recently, new electronic sensors have overcome the limitations of the current silicon sensors through the development of low dimensional materials, nanowires, nanotubes, and two-dimensional (2D) films. While sensors based upon one-dimensional (1D) structures, specifically carbon nanotubes (CNTs), have demonstrated excellent sensitivity and at least the promise of selectivity, the production of devices from 1D structures has proven difficult. Graphene offers the same performance opportunities as 1D structures along with the advantages of working with a planar film. The advancements in the production of graphene from the easier mechanical cleavage to more complex and higher quality constructions together with the synthesis of graphene on an industrial sale has been an important factor for considering graphene as element in biosensors. In this chapter, we show the advances made in the

Organic Bioelectronics for Life Science and Healthcare   Materials Research Forum LLC
Materials Research Foundations **56** (2019) 115-152   doi: https://doi.org/10.21741/9781644900376-4

integration of graphene and graphene composite materials into FET devices and present the surface chemistry strategies employed for GFET based biosensing. We conclude with some critical comments on the advantages and experimental challenges as well as some perspectives for the future research and development in this field.

Keywords

Graphene, Field Effect Transistor (FET), Biosensing, Surface Modification

Contents

Organic Bioelectronics for Life Science and Healthcare
Materials Research Forum LLC
Materials Research Foundations **56** (2019) 115-152
doi: https://doi.org/10.21741/9781644900376-4

1.   Introduction

1.1   From MOSFET to bioFET

Since the earliest works by Lilienfield in 1926 and Heil in 1935 on field effect transistors (FETs), these devices have become one of the key devices for the electronics industry. Their functioning is based on the concept that charge on a nearby object can attract charges within a semiconductor channel. The semiconductor determines the type of charge carriers that can accumulate or deplete in the channel, thus the current flow can either be the result of movement of holes ("p-type") or electrons ("n-type"). To measure this electrical field effect, a three-electrode system (source, drain, gate) with a semi-conductor channel in-between source and drain (Figure 1A) is used. A control electrode, called the gate, is capacitively coupled to the device through a thin dielectric layer, placed in very close proximity to the channel so that its electrical charge is able to affect the channel. In this way, the gate controls the flow of carriers, electrons or holes, from the source to the drain.   Throughout the last 50 years several FET concepts have been developed and implemented with the most widely employed being the metal-oxide-semiconductor field-effect transistor (MOSFET) structure using SiO_2 as channel material [1]. The classical silicon-based FET sensors suffer however from poor sensitivity and selectivity due to fundamental shortcomings of the silicon structure when I comes to sensing. To overcome the limitations of the current silicon-based field effect transistors, nanowires, nanotubes and two-dimensional films have been proposed as alternatives [2-3].

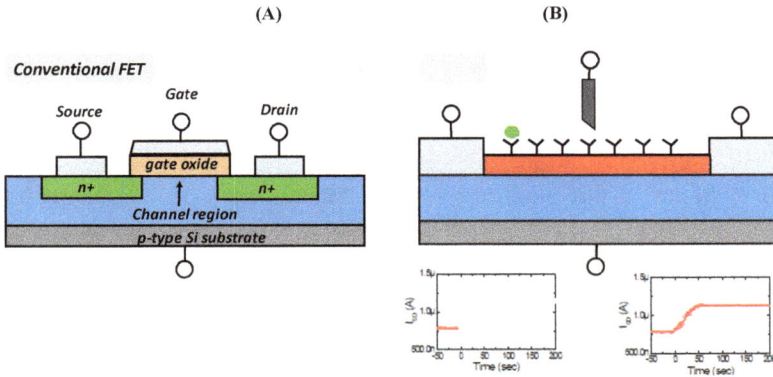

Materials Research Foundations **56** (2019) 115-152   doi: https://doi.org/10.21741/9781644900376-4

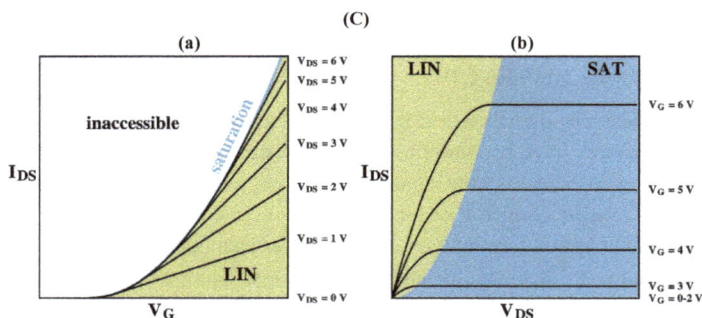

Figure 1: Different Field effect configurations: (A) Classical Si-based top-gated FET; (B) Schematic of a typical bioFET using a reference electrode as gate together with change in I_{DS} by the binding of analyte due to the additional electrical field induced buy the charged analyte; (C) (a) Example of a Transfer curve: Drain-source current (I_{DS}) as a function of gated voltage (V_G) for different drain-source voltages (V_{DS}) (b) I-V curves (I_{DS} vs. V_{DS}) for different gate voltages (downloaded on the 20th October 2017, http://www.stallinga.org/AcadActiv/Publications/theory/fet/index.html).

Field-effect transistors using silicon nanowires as channel materials have shown particular great promise as a potential platform for direct, label-fee detection of biomolecules with ultra-high sensitivity and scalability [4-5]. The structure of a biological FET (BioFET) is similar to that of a MOSFET, but instead of having a metal or polysilicon gate electrode, the gate (reference) electrode is a distance away from the dielectric, within an electrolyte solution (Figure 1B). A gate voltage is applied to the reference electrode to set the operating point of the device, and the conductance of the channel is measured by applying a drain to source voltage [6]. The gate material (silicon oxide, silicon nitride, aluminum oxide, etc) is in addition modified with a surface receptor specific to the analyte that one wishes to detect. Most biomolecules are charged when in solution, the charge being often dependent on the pH in addition. Therefore, when the analyte binds to the receptors, the field effect due to charges now bound close to the surface induces opposite charges within the semiconductor channel, which is then detected as a change in current or conductance (Figure 1B). The binding of positively charged target molecules on a p-type channel causes depletion of charge carriers (holes) and decrease in conductance, while in an n-type channel, positively charged molecules result in accumulation of charge carriers (electrons) and increase in conductance. The bioFET sensing characteristics can be also assessed by the transfer curve, which is a plot of the current across the drain-source electrodes (I_{DS}) as a function of gated voltage (V_G)

at a constant drain-source voltage (V_{DS}) (Figure 1C). A good bioFET based sensor results in measurable shifts in the I_{DS}/V_G characteristics.

The first realization of a bioFET date back to the 1970s with the concept of an ion sensitive field effect transistor (ISFET) [7]. However due to the large size of ISFTs, a lot of molecules were needed to achieve a good signal-to-noise ratio and most of the research focused on the detection of pH changes. Work performed on ISFETs eventually led to the pioneering work by Cui et al, who made use of CVD grown silicon nanowires as pH sensor as well as biosensor for detect streptavidin down to the picoMolar concentration range [8]. While this concept was further developed for DNA and even viruses [9-10], the bottom-up approach to creating such bioFETs suffers from two major drawbacks. Firstly, a large number of CNTs or Si NWs have to be integrated and aligned properly to decrease source/drain resistance, something far from being trivial with the need of unconventional and low yield methods. Also very little metal contact to the nanowires is achieved in general, since the nanotubes just lay on top of it and the contact area was therefore very small. Also, of in the case of CNT, due to their semiconducting nature, a Schottky barrier forms at the metal-semiconductor interface, increasing the contact resistance. A further challenge is the control of size, type, purity, chirality or the nanomaterial all of which uniquely determine its electrical properties.

1.2   From bioFET to GFET for sensing

In parallel to these different developments in the field of bioFETs, graphene, an atomically thin crystalline film formed by sp^2 hybridized carbon atoms in a hexagonal lattice (Figure 2A), attracts more recently a lot of attention as possible material for a future high-speed FET [11]. A typical SEM image of graphene sheets on a flat substrate is seen in Figure 2B. The color contrast in the SEM image is correlated to the number of graphene layers transferred onto the SPR chip [12]. The thicker the graphene, the lower is the number of secondary electrons, so that multilayer (or bilayer) graphene appears darker than monolayers. Raman analysis has in addition become one of the most important characterization tools for GFETs (Figure 2C) as it can provide useful information on the defects (D bands), in-plane vibration of sp^2 carbon atoms (G band) as well as the stacking order (2D bands) [13-15]. While single-layer graphene sheets present a single and sharp 2D band at 2679 cm^{-1}, for multi-layered graphene sheets the 2D band appears as a broadened peak with 19 cm^{-1} shift into higher wavenumbers.

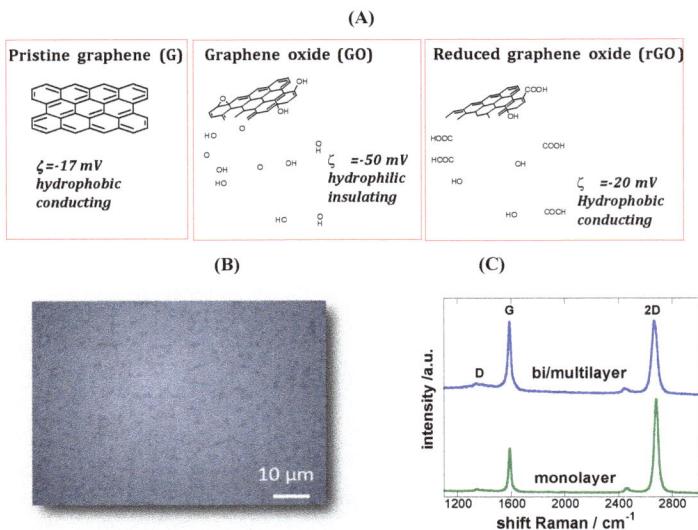

Figure 2. (A) Schematic image of graphene and its chemical derivatives often used in the construction of GFETS; (B) SEM image of pristine graphene on a flat surface; (C) Raman spectra of bright areas (monolayer) and darker areas (bi/multilayer) of the graphene-on-metal interface

Utilizing a 2D channel material has several advantages over classical Si-based MOSFET technology. For most semiconductor-based transistor sensors, local electric field changes at the channel surface have little effect deeper in the device channel, limiting the response sensitivity. With a GFET, the graphene channel is only one atom thick, meaning the entire channel is effectively on the surface and directly exposed to the environment. Any molecule attached to the surface of the channel impacts electronic transfer through the entire depth of the device. Near atomically thin silicon or other bulk semiconductors are not effective because at such a thickness, surface defects dominate the material characteristics. Next to the advantage of graphene having no dangling bonds to form defects, the most frequently stated advantage of graphene is its high carrier mobility at room temperature with motilities of 10.000-15.000 cm^2/V routinely measured for exfoliated graphene on SiO_2-covered silicon wafers at room temperature [16-17]. For large-area graphene grown on nickel and transferred to a substrate, motilities >3700 cm^2/V were measured [15]. Although the high motilities offered by graphene increases the speed of devices, it comes at the expense of making it difficult to switch devices off due to the zero-band gap of graphene. The other important property of graphene for its use as FET is a strong electric field effect which lead to an electrostatically tunable carrier density in the range of $n < 10^{14}$ cm^{-2}. The two-dimensional nature of graphene is another fact responsible for the rapid growth of its research, leaving behind other sp^2-carbon allotropes. There is still some disagreement in the literature regarding the point at which graphene becomes graphite ranging from structures composed of up to around ten graphitic sheets as proposed by Geim and Novoselov [18] to100 layers [19]. Most of the preparation methods of graphene actually result in few-layer graphene (3-10 layers). Next to pristine graphene (G) with almost no defects other graphene derivatives such as graphene oxide (GO) and notably reduced graphene oxide (rGO) have found interest for the construction of GFET (Figure 2A).

This review will discuss in details the milestones in the construction of GFETs and their use for the sensing of protein, DNA other biological relevant interactions. Before describing the different sensing concepts, some words about the synthesis of graphene and its derivatives as well as the different GFET device formations will be provided.

2.   Preparation of graphene, graphene oxide and reduced graphene and its transfer to surfaces for the formation of GFETs

The ability to make high-quality graphene is credited to Novoselov and Geim, who used Scotch tape to pull apart individual layers [17]. While mechanical exfoliation [17, 20] forms single, free standing G films, from which the real intrinsic properties of G could be derived, due to is low turnover this route is not well suited for the construction of sensors. Graphene synthesis *via* chemical vapor deposition (CVD) on nickel and copper films appears to be more ideally suited for applications with electrical sensors with regards to the prevalence of copious manufacturing volumes, large surface area, and uniform graphene sheets [21]. In this process, copper foils areas loaded into a furnace and heated to 1.000 °C in an argon/hydrogen reducing environment to remove any native oxide on the surface of the copper. A small flow of methane was added to the gas flow. Graphene formation begins with a few nucleation sites followed by lateral growth of the single atomic layer graphene crystals until the domains meet, completely coating the copper surface. The methane breaks down at the copper surface and the adsorbed carbon atoms travel on the surface until it meets up with it and adds to the graphene crystals. Continuous single atomic layer graphene (SLG) was formed after a short growth time of 5 to 30 minutes, depending primarily upon the gas flow ratio. Large-area graphene films, with about 95% single-layered graphene, in the order of centimeters can be nowadays grown on copper substrates by CVD using methane as carbon source [12, 22]. A further

Organic Bioelectronics for Life Science and Healthcare   Materials Research Forum LLC
Materials Research Foundations **56** (2019) 115-152   doi: https://doi.org/10.21741/9781644900376-4

appealing character of CVD grown graphene is that it can be transferred after its growth to any type of interface including silicon dioxide.

After the preparation of substrate gown graphene, the procedure for the fabrication of a GFET involves a cascade of steps including the fabrication of electrode holes though photolithography and lift-off process and deposition of source, drain and gate electrodes. The integration of graphene sheets onto silicon wafers with metal electrodes formed by thermal evaporation and patterned lithographically is *via* chemical transfer of CVD grown graphene on Cu or Ni sheets (Figure 3A). To perform the transfer, poly(methyl methacrylate) (PMMA) is spin-coated onto the graphene face of the copper substrate. The copper is separated from the PMMA/graphene by a mechanical separation and the graphene film placed on the surface, and baked to promote adhesion of the graphene to the interface. Then PMMA is removed by acetone washing or thermal annealing.

(A)

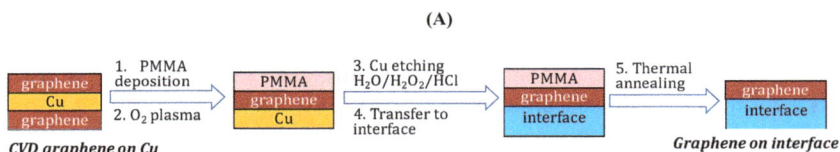

CVD graphene on Cu   *Graphene on interface*

(B)

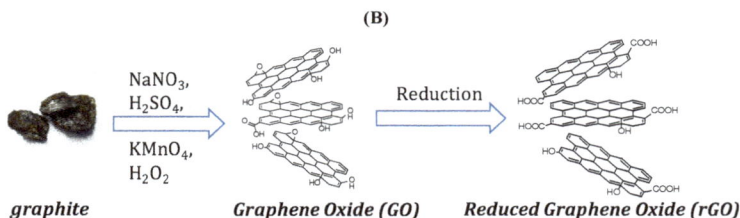

graphite   *Graphene Oxide (GO)*   *Reduced Graphene Oxide (rGO)*

(C)

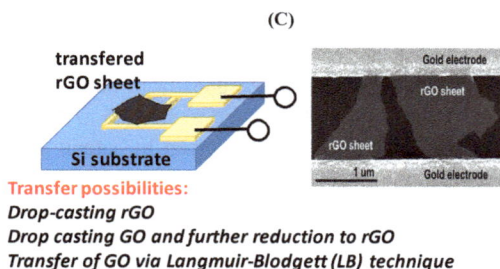

Transfer possibilities:
Drop-casting rGO
Drop casting GO and further reduction to rGO
Transfer of GO via Langmuir-Blodgett (LB) technique

Organic Bioelectronics for Life Science and Healthcare   Materials Research Forum LLC
Materials Research Foundations **56** (2019) 115-152   doi: https://doi.org/10.21741/9781644900376-4

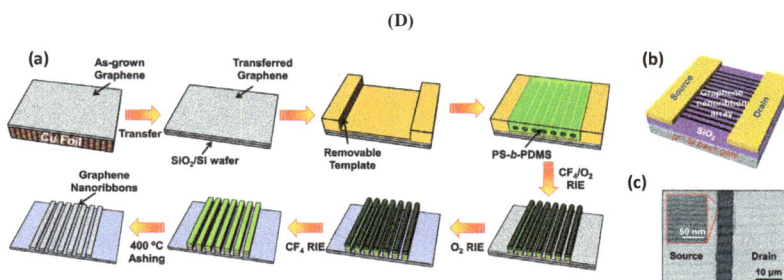

Figure 3: (A) A possible chemical transfer rout of CVD graphene grown on Cu sheets onto a flat interface in a GET device; (B) Chemically derived graphene nanostructures based on the Hummers method for the preparation of graphene oxide (GO) from graphite followed by reduction to rGO; (C) Possible transfer routs of rGO nanomaterials together with SEM image of rGO bridging a pair of gold electrode ; (D) Formation of graphene nanoribbons using lithographic patterning of self-assembled blockpolymers: (a) Schematics of fabrication process; (b) Schematics of multi-channel GFET, (c) SEM image of channel region of GFET (reprint with permission from Ref [23].)

While the transfer of CVD graphene has the advantage of using high quality graphene and offers a control over the number of graphene layers, the strategy implies that one has access to CVD grown graphene. Prior to the focus on graphene (G) there was extensive research on the preparation of graphite oxide and later graphene oxide (GO) [24-26]. The difference between G and GO is the presence of oxygen-containing functions bound to the carbon scaffold (Figure 3B). The Hummers method is probably the most widely used approach to GO and involves soaking of graphite in a solution of sulfuric acid and potassium permanganate [24]. Stirring or sonication of the formed GO is then performed to obtain single layers of GO. The formed GO consists of oxidized graphene sheets having the basal planes decorated mostly with epoxide and hydroxyl groups, in addition to carbonyl and carboxyl groups locates preferentially at the edge [27-28]. These oxygen functions render the GO layers hydrophilic and water molecules can readily intercalate into the interlayers. The negatively charged oxygen species help in addition to disperse GO as a single sheet by providing electrostatic repulsion and salvation. The drawback is that the presence of oxygen groups on the basal plane of graphene disrupts the π-conjugation, as a result, GO is insulating. The oxygen-containing functionalities on the GO surface can be removed through reducing procedures. This reduction of GO to reduced graphene oxide (rGO) (Figure 3B) can be achieved through chemically (e.g.

hydrazine monohydrate, sodium borohydride, hydroquinone), thermally, photochemically and other means [24, 29-46]. Hydrazine monohydrate is still most widely used, mainly due to its strong reduction activity to eliminate most oxygen-containing functional groups of GO and its ability to yield stable rGO aqueous dispersions [40]. However, with hydrazine as the reducing agent, its residual trace may strongly decrease the performance of rGO-based devices. In addition, hydrazine is a highly toxic and potentially explosive chemical. To avoid using hydrazine, many environmentally friendly and high-efficiency reductants have been developed and used for the reduction of GO, including vitamin C [47-48], amino acid [47], reducing sugar [46], alcohols [32], hydroiodic acid [49] reducing metal powder [35, 50], sodium citrate [51], tea [52], lysozyme [53], dopamine [37, 54], etc. However, independent of the reduction method used, the resulting graphene sheets differ from CVD graphene, since the structural integrity and hence the electronic properties are altered when compared to pristine graphene. The chemical derived rGO methods are however both versatile and scalable, offer the greatest ease for functionalization, and are adaptable to a wide variety of applications. Bottom-up organic synthesis method in solution have lately also been proposed to synthesize other morphologies next to sheets such as graphene nanoribbons (GNRs) [55-56]. While such bottom up approaches provide structurally defined graphene nanostructures, these nanomaterials have to be transferred to suitable substrates which is still a challenge. It is currently often simply based by drop-casting rGO or GO suspensions on the FET device (Figure 3C) [57-58]. In the case of GO further reduction to rGO is required which is achieved by exposure of the GO coated FET to hydrazine vapors followed by various durations of heat treatment at 400 °C in Ar atmosphere [58]. Another approach is the GO transfer onto SiO_2/Si substrates by Langmuir-Blodgett (LB) technique followed by reduction rGO by exposure to hydrazine vapors followed by various durations of heat treatment at 400 °C in Ar atmosphere [59].

Additional lithographic approaches can be and even allows the patterning of graphene. Self-assembled block copolymers have been used for example as etching masks to convert their pattern to graphene for the fabrication of GNR arrays. Spin coating poly(styrene-b-dimethoxysiloxane) (PS-b-PDMS) block polymers onto graphene coated Si wafers and annealing results in a phase separation into PS and PDMS domains (Figure 3D). The PS part can be removed using oxygen plasma treatment, leaving PDMS cylinders to serve as etching masks. CF_4 based RIE removes the unprotected polymer and underlying graphene resulting in GNRS of 10 nm in width [23].

Organic Bioelectronics for Life Science and Healthcare   Materials Research Forum LLC
Materials Research Foundations **56** (2019) 115-152   doi: https://doi.org/10.21741/9781644900376-4

3. Different graphene based transistor structures

Current designs of GFETs are shown in Figure 4A. The most often studied GFET structures are back-gated ones (Figure 4Aa), where the graphene flakes are connecting the source and drain electrodes, and the substrate, mostly doped Si, acts as a back gate [17]. Usually SiO_2 layers of about 300 nm in thickness are selected as substrate onto which graphene films are transferred, onto which electrode patterns are transferred by lithographic or electron beam etching using photoresist layers and metal electrodes are fabricated after electron beam evaporation and lift-off process. This configuration overcomes any doping effect in the graphene channel due to substrate, atmospheric or dielectric effects. Back-gated sensing GFETs has clear sensing advantages over top (liquid)-gated assemblies especially in conditions in which the analyte solution composition varies and in situation where the analyte is not in a solution matrix (e.g. vapor). Indeed, Kybert et al have recently reported the use of DNA modified back-gated GFETs for vapor sensing and demonstrated a significant positive shift in the $V_{G, min}$ after DNA deposition, explained as counteracting the negative field produced by the phosphate backbones of the adsorbed ssDNA [60]. Back-gated GFETs are less frequently employed however in liquid based bioFETs due to the difficulties associated with producing a device requiring a small gate voltage. By depositing a dielectric layer (SiO_2, Al_2O_3, etc) [61] on the top of graphene, and deposition of metal electrodes on the top of the gate dielectric, a top-gate configuration is achieved (Figure 4Ab) allowing both gate biases to control the charge concentration in the device channel [62]. These dual-gates GFETs have been widely use to study the i/V characteristics of the device.

Alternatively, by synthesizing graphene on silicon carbide (SiC) wafers, top-gated only devices can be fabricated (Figure 4Ac) [61, 63]. The carrier motilities in top-gated devices is much lower than in the back-gates structures as the top gate dielectric causes more electron scattering. Also graphene films might be more easily destroyed during the fabrication process.

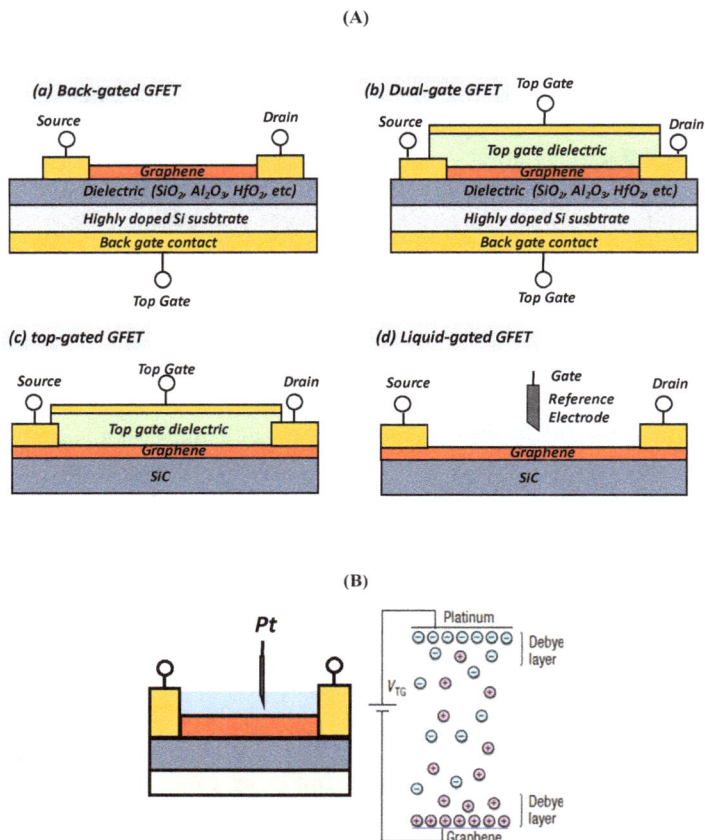

Figure 4: (A) Different GFET designs : (a) Back-gated GFET, (b) Dual-gated GFET; (c) Top-gated GFET, (d) liquid-gated GFET; (B) Formation of Debye ionic layer in liquid-gated GFET (reprint with permission from Ref. [64])

Liquid-gated GFETs (Figure 4 Ad), formed by placing a drop of buffer containing the analyte onto the surface of the graphene sheet and acting as the gate is one of the most widely employed configurations for biological GFET sensing. An important distinction between back and liquid gate GFETs is the potential for current to flow between the graphene channel and the gate electrode *via* the aqueous ionic solution separating the two conductive bodies. Gate current leakage is limited at low gate bias by the formation of an ionic double layer, Debye layer, formed along the surface of graphene and the gate electrode due to the voltage bias (Figure 4B) [64]. Higher gate bias potentials result however in the onset of electrolytic processes and high gate current.

4.   Integration of surface ligands onto GFET for selective sensing

Next to the construction of a GFET with low leakage current and optimized current/voltage characteristics, the integration of surface selective ligands is an ultimate prerequisite for the development of a sensitive, highly selective, direct, label-free biological GFET. When a receptor is immobilized onto the graphene channel and when target molecules bind to the receptor on the graphene surface, the redistribution of electronic charge will generate a change in the electric field across the GFET channel region, which changes the electronic conductivity in the channel and the overall device response as seen demonstrated in the transfer curve of a liquid-gated GFET for the analysis of DNA hybridization events (Figure 5A). Indeed, as a zero-bandgap semiconductor nanomaterial, the conductance of graphene is described by the transfer characteristic equation (equation 1)

$$I_{DS} = \mu \times (W/L) \times C \times V_{DS} \, (V_{GS} - V_{Dirac}) \tag{1}$$

where I_{DS} is the drain-source current, C is the top-gate capacitance per unit are, μ the carrier mobility, W and L the width and length of the graphene channel respectively, V_{GS} the applied gate-source voltage and V_{Dirac} the voltage indicating Dirac point, where I_{DS} achieves a minimum. Most importantly for sensing applications using GETs, the Dirac point is sensitive to the presence of immobilized DNA probes as well as to the hybridization of complementary target DNA. It shifts significantly to the left with the addition of DNA molecules, suggesting that DNA molecules n-dope the graphene film. Mismatched DNA had a non-significant impact on the Dirac point however. How can these receptors be incorporated onto the GET devise, thus?

Graphene generally does not react or bind with most materials; however, being composed of carbon, there are several chemistries that enable functionalization through formation of binding sites on the surface. Graphene shows generally increased adsorption of organic and biological molecules because their carbon-based ring structure enables π stacking interaction with the hexagonal cells of graphene. However, π-stacking interactions are not specific, and any organic molecule or biomolecule of appropriate aromatic structure will integrate onto the graphene matrix, a bit limitation for sensing applications. A number of procedures have been thus developed over time to allow addition of ligand into GFET devices (Figure 5B).

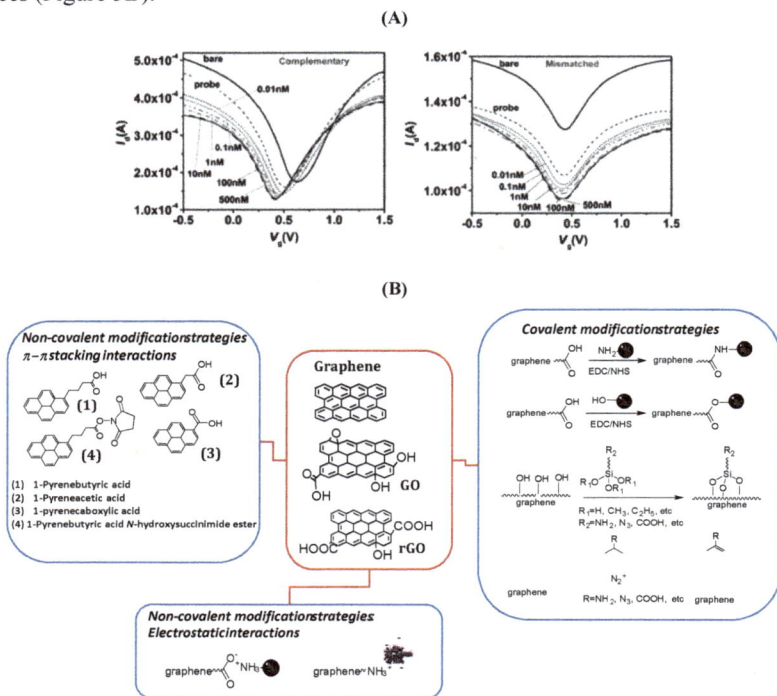

(A)

(B)

Figure 5: (A) Transfer characteristics for a GFET before adding DNA, after immobilization with probe DNA, and after reaction with complementary or one-base mismatched DNA molecules with the concentration ranging from 0.01 to 500 nM. (Reproduced with permission from Ref. [65] (B) Different surface strategies employed.

4.1   Non-covalent modification of graphene

The non-covalent modification of graphene has several advantages over covalent modification approaches [66]. The avoidance of damage to the graphene sp^2 network is essential to maintain many of its properties for FET based applications which is achieved using non-covalent functionalization. As these strategy is mainly based on hydrogen bonding and/or π–π stacking interaction no additional chemical need to be used for linking functional molecules to graphene. Indeed, the extended aromatic network of sp^2 carbon atoms in graphene makes this interaction important in graphene functionalization. An abundance of modifiers, mainly aromatic organic molecules (dopamine, 1,5-diaminonapthalene, tetrathiafulvene, pyrene derivatives) are available commercially and have found widespread use. By far the most common molecules used for π–π stacking are however pyrene derivatives such as 1-pyrenebutanoic acid, 1-pyreneacetic acid, 1-pyrenecarboxylic acid, 1-pyrenebutyric acid *N*-hydroxysuccinimide ester (Figure 5B) [67-69]. The presence of carboxylic acid and esters groups allows covalent integration of ligands carrying NH$_2$ functions *via* amide or ester bond formation. In order to functionalize the CCOH groups, in GO and rGO, they first have to be activated in order to increase their reactivity. The reagents used to do this include some commonly used for peptide synthesis such as 1-ethyl-3-[3-dimethylaminopropyl] carbodiimide/ N-hydroxysuccinimide (EDC/NHS) others more commonly found in synthetic organic chemistry such as thionyl chloride (SOCl$_2$). Once this activation has occurred the GO carboxylic acids readily react with molecules containing functionalities such amines to form amide bonds and hydroxyls to form esters.

Electrostatic interactions have to be added to the list of non-covalent modifications. As the carboxylic functions present on GO results in an overall negatively charged scaffold, physical and chemical strategies can be considered to allow the integration of negatively charged biomolecules such as bacteria. Treatment of GO and rGO with ammonia or nitrogen plasma or covalent linking of ethylenediamine (EDA) results furthermore in the amination of graphene, thus producing positively charged GO sheets.

4.2   Covalent modification of graphene

An alternative route to functionalized graphene is to use GO or rGO as a starting as the intrinsic presence of carboxylic functions allows for the direct coupling of amine-terminated ligands (Figure 5B). For example simple immersion of a amine-terminated DNA in the presence of an amine coupling reagent results in the covalent integration of DNA onto GO-based FETs [70]. While GO and rGO are ideal candidates to use for covalent functionalization due to its already high level of functionality, the exact identity and distribution of the oxygen containing groups on its surface is complex. It is widely believed that there are a range of epoxide, ether, aldehyde, ketone, alcohol, and carboxylic acid groups present on the sheets meaning that a variety of different reactions and chemistries can be used to covalently functionalize GO and rGO.

Carboxylic acid groups only make up part of the oxygen content of GO, and are generally believed to be located at the edges of sheets. Therefore, an alternative approach is to target the other functional groups present, most commonly the epoxide group.

Functionalization by the ring opening of epoxides can proceed *via* a number of methods including the use of sulphur nucleophiles, the conversion of epoxy groups to hydroxyl groups, and the use of malononitrile which adds to the epoxide group to leave pendent nitrile groups. Surprisingly none of these approaches have been so far employed in connection with GFETs and will this not be further elaborated here.

Another approach to the covalent functionalization of graphene oxide is the use of silanes as a modifier. Functionalization occurs by a reaction between hydroxyl groups on the GO/rGO surface and the trialkoxy groups present on the silanes.

Finally, diazonium chemistry based on the reduction of the 4-carboxybenzenediazonium cations to generate aryl radicals [71-73] can be used to link different functions. Such diazonium salts can be formed *in situ* upon addition of $NaNO_2/HCl$ to a solution of 4-aminobenzoic acid, where nitrogen is spontaneously eliminated, forming the 4-carboxyphenyl diazonium salts.

5.   Applications of GFETs for sensing of biomolecules

5.1   DNA sensors

GFETs DNA sensors are unique among other nucleic acid detection methods due to their label free character, allowing for the direct detection of oligonucleotides within a sample without any prior alteration of the sample. The first significant investigation between chemically-modified graphene and biomolecules was published by Mohanty and Berry in 2008 [70]. In this experiment, probe single-stranded (ss) DNA was anchored on the surface of graphene. When the DNA probe binds to the target strand, a double helix forms and the binding results in a measurable change in current and the electrical field of the GFET. Dong et al reported a liquid-gate GFET for DNA hybridization studies [65]. As shown before in Figure 5A, a negative shift in V_{Dirac} is observed upon the addition of DNA claimed to be due to n-doping of graphene by the π–π staking of electron-rich aromatic rings of the nucleic acids. Lin addressed this issue again and postulated that next to graphene doping, electrostatic gating and ionic impurity masking might take place [74]. As the phosphate groups of DNA are negatively charged, V_{Dirac} shots positively to overcome the opposing field while nucleotides n-dope in addition the graphene surface due to π–π taking interaction. The dsDNA structures thus control the level of doping by lifting the bases off the surface of graphene upon hybridization and following changes are expected: a right shift of V_{Dirac} with dsDNA and hybridization, but left shifts for sDNA. In a follow up work Dong et al used rGO and Au NPs and showed that the fabricated transistor exhibits higher hole mobility and conductance compared to that of bare graphene [75]. A decrease in drain current of about 13.5 % after hybridization with complementary was observed, while a single-mismatched DNA only made about 5.3 % current decrease.

An interesting DNA GFET was lately proposed by Zhu and co-workers though the linking of NH_2-modified DNA to thionine modified graphene using glutaraldehyde. Detection limit for complementary DNA of about 126 fM was shown [76]. More recently, a rGO based FET biosensor for the ultrasensitive detection (100 fM) of DNA *via* peptide nucleic acid (PNA)-DNA hybridization was shown.[77] Ping at all showed in addition that the sensitivity of DNA GFETs varies with the length of the oligonucleotides: a 60[mer] DNA show lower detection limits (1 fM) than a 40[mer] and 20[mer], where 100 fM and 100 pM detection limits were reported respectively [78].

Table 1: DNA sensing examples with GFET structures

Graphene	Modification	LOD	REF
rGO	-	10 nM	[79]
rGO	Use of amino-silane layer to stabilize rGO interface + EDA/glutaraldehye/NH_2-DNA	10 nM	[79]
G	ssDNA adsorption	100 pM	[65]
G	ssDNA adsorption	1 pM	[74]
G	-	1pM	[80]
G	DNA (20-40-60[mer])	100 pM-100 fM-1 fM	[78]
G	Thionine+NH_2-DNA	126 fM	[76]
G	-	100 fM	[81]
rGO	PNA	100 fM	[77]
G	Py-ester+PNA (array)	10 fM	[69]

EDA: ethylenediamine, py-ester: 1-pyrenebutaaoic acid succinimidyl ester

5.2   Protein sensors

The detection of protein biomarkers relevant for different medical conditions is one of the areas actively researched with GFETs due to their potential to achieve highly sensitive detection (Table 2). Protein sensing with GFETs is especially challenging given the constrains of complex, high ionic-strength buffered environments required. These ionic solutions form electronic double-layers that lower the apparent charge of proteins relative to a surface in a process called Debye screening [82]. The Debye screening length (λ_d), defined as the length scale between a surface and analyte in which charges on the analyte are screened by counter ions in solution, is highly dependent on solution composition. It has been experimentally validated that the performance of a GFET sensor where the analyte is close to the sensing substrate is improved as this close proximity does not allow ions to form a charge-neutralizing layer between the captured analyte and the sensing substrate [83]. Thus compared to large (10-15 nm) antibodies which are not well suited for field-effect applications, antibody Fab fragments of around 3 nm or DNA aptamers are advantageous as surface specific ligands as their smaller size allows sensing within the Debye layer (Figure 6A) [84].

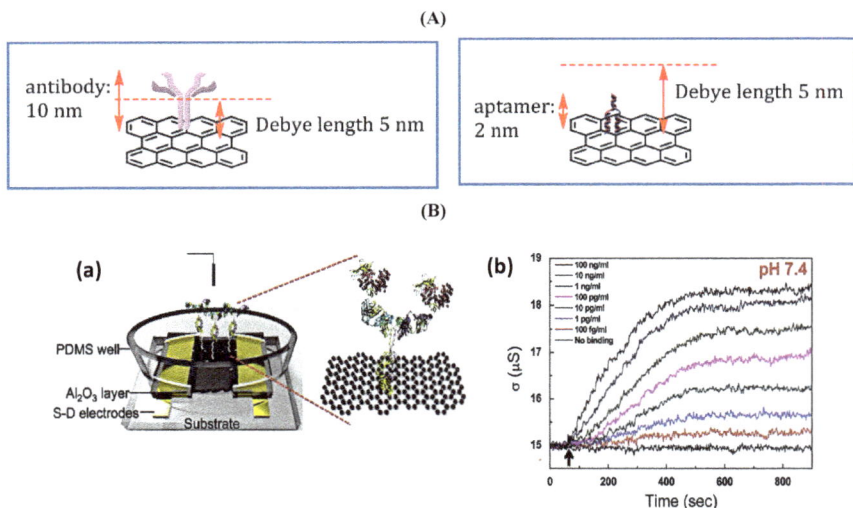

Figure 6: (A) Importance of use of surface specific ligands for sensitive protein sensing using GFETs to measure insure the about 5 nm thick Debye length [85], (B) PSA mAb antibody modified rGO FET immunosensor for the detection of PSA- ACT antigen complex, (a) Set up using a platinum reference electrode in the analyte solution and measuring of changes in the carrier density in the rGO channel due to the doping effects. (b) Change of electrical conductivity with upon addition of different concentrations of PSA-ACT (with kind permission from Ref. [86].

Table 2: Protein GFETs

Analyte	graphene	modification	Linear range	LOD	Ref.
Heat shock protein	G	Py-ester+ Fab	2.3 nM (K_D)	-	[84]
BNP	rGO/	Pt NPs modified with thioglycolic acid; BNP antibody attached by carbodiimide chemistry	100 fM-1 nM	100 fM	[87]
PSA-ACT	rGO	Py-ester+ PSA antibody	10 fg mL^{-1}-100 fg mL^{-1} (in PBS) 10 fg mL^{-1}-10 ng mL^{-1} (in serum)	10 fg mL^{-1}	[86]
VEGF	PPy-N-G	glutaraldehyde conjugated DAN + RNA aptamer	-	100 fM	[55]
IgE		IgE aptamer		47 nM (K_D)	[88]
PA	rGO	Py-ester + DNA-aptamer+ PA+ aptamer-AuNPs	12 aM - 120 fM	1.2 aM - 12 aM	[89]
thrombin	G	Py-ester+thrombin aptamer	-	170 nM (K_D)	[68]
carcinoembryonic antigen	G	Py-ester+antibody	-	100 pg mL^{-1} 6.35×10^{-11}M (K_D)	[90]
Norovirus protein (foodborne pathogen)	G	Inkjet printed G on a flexible Kapton substrate + adsorbed antibody	0.1 - 100 µg mL^{-1}	0.1 µg mL^{-1}	[91]
PSA	G	Py-acid/amine PEG+APT	1 - 1000 nM		[92]
BSA	rGO	Py-ester+antibody	-	66 ng mL^{-1}	[93]
Human IgG	thermally reduced GO	AuNPs-Antibody	-	0.2 ng mL^{-1}	[94]
IFN-γ	G	pyrene-modified IFN-γ aptamer	-	83 pM	[95]
cTnI	carboxylated graphene	anti-cTnI antibody bound by carbodiimide chemistry	-	1 pg mL^{-1}	[96]

BNP: brain natriuretic peptide; G: graphene: py-ester: 1-pyrenebutanoic acid succinimidyl ester; PSA-ACT: prostate specific antigen/α1-antichymotrypsin complex; VEGF: antivascular endothelial growth factor; PPy-N-G: polypyrrole-converted nitrogen-doped few-layer graphene; DAN: 1,5-diaminonaphthalene; PA: protective antigen ; py-acid: pyrene butyric acid; IFN-γ: Interferon-gamma; TnI: Troponin I

Applications in real samples are still rarely described and quite challenging considering the high ionic strength and complexity of real samples of blood, saliva etc. First steps to solve this are based on desalting the samples and getting rid of some potential interferences using filters, which resulted in the detection of brain natriuretic peptide, a biomarker in heart failure [87]. Kim et al. [86] reported in 2013 on a rGO bioFET for the ultrasensitive detection of a prostate cancer specific antigen/a1-antichymotrypsin(PSA-

ACT) complex (Figure 6B). The immunoreaction of PSA-ACT complexes with PSA monoclonal antibodies on the rGO channel surface caused a linear response in the shift of the gate voltage with a fM detection limit even when analyzing human serum samples.

5.3   Other molecules

Various small analytes ranging from odorant molecules to metal ions, glucose and mycotoxins (Table 3) have been in addition targeted by GFET biosensors. Depending on the analyte investigated, enzymes, aptamers or specific binding proteins were used as surface bioreceptors. For example, the μ-opioid receptor (MUR) is a G protein-coupled receptor (GPCR) with roles in pain and reward signaling pathways. This MUR protein with high binding affinity for opioids was immobilized on CVD graphene in a GFET and used to detect the opioid receptor antagonist naltrexone [97]. With regards to practical applications, the detection of Hg^{2+} ions in spiked lake water [98] and of aflatoxin B1 in spiked corn [99] were reported. Although not using a biorecognition element but an iron porphyrin to achieve selectivity, the GFET detecting nitric oxide in human umbilical vein endothelial cells (HUVECs) at single cell level [100] is also notable.

Table 3: Comparison of the analysis of small sized analytes by GFETs

Analyte	Graphene	Modification	Linear range	LOD	Ref.
Naltrexone	G	diazonium salt+ covalently attached MUR protein	$Ka = 7.8 \pm 1.6$ ng mL^{-1}	10 pg mL^{-1}	[97]
ATP	G	adsorbed APT	0-50 μM	10 pM	[101]
Atrazine	G	Urease (inhibition of urease activity by atrazine)	2×10^{-4} - 20 ppb	0.05 ppt	[102]
AFB1	rGO	polylysine/Ab	10^{-4} ppt - 1 ppt	0.1 fg ml^{-1}	[99]
Glucose	G	Py-ester+ glucose oxidase	0.1 - 1.0 mM	-	[103]
Glucose	G	Py-ester + glucose oxidase		0.1 mM	[104]
Glutamate	G	Py-ester +glutamic dehydrogenase	5 μM - 50 μM and 50 μM- 1.2 mM,	5 μM	[104]
Glucose	G	Pt NPs/Nafion-GOx-chitosan	0.5 μM - 1 mM	0.5 μM	[105]

Odorant molecules	rGO	Py-ester+ OBP14	4×10^{-6} M (K_D with homovanillic acid) 20×10^{-6} M (K_D with methyl vanilinate) 40×10^{-6} M (K_D with eugenol) 800×10^{-6} M (K_D with citral) 1400×10^{-6} M (K_D with Methyl eugenol); 3300×10^{-6} M (K_D with geraniol)	-	[6]
Nitric oxide	rGO	iron−porphyrin	1 pM - 100 nM	1 pM (in PBS) 10 pM (in the cell medium)	[100]
Ca^{2+} Mg^{2+} Hg^{2+} Cd^{2+}	rGO	Calmodulin (CaM) for Ca^{2+} and Mg^{2+}; Metallothionein type II protein for Hg^{2+} and Cd^{2+}	-	1 μM (Ca^{2+}) 1 nM (Hg^{2+} and Cd^{2+}	[98]
Glucose	rGO	carboxylated polypyrrole -nanotube (NT) + Glucose oxidase	1 nM -100 mM	1 nM	[106]
K^+ Pb^{2+}	G	TBA terminated by methylene blue molecule	100 μM- 50mM (K^+) 10 μM - 10mM (Pb^{2+})		[95]

ATP: (adenosine triphosphate), odorant-binding; protein 14 (OBP14), thrombin binding aptamer (TBA)

Among different analytes, a particular interest is in glucose sensing with GFETs, not surprisingly, considering the importance of glucose in diabetes but also as a model for biosensors. Glucose oxidase is employed as specific bioreceptor in most of the sensors as seen in Figure 7. The sensitivity of the sensor can be improved by co-modifying the graphene gates with Pt nanoparticles due to the enhanced electrocatalytic activity of the electrodes, as illustrated in Figure 7Ac. The sensing mechanism is attributed to the reaction of H_2O_2 generated by the oxidation of glucose near the gate. The glucose sensors showed detection limits down to 0.5 mM and good selectivity, that is sensitive enough for non-invasive glucose detections in body fluids. Also, the authors indicate that these devices can be conveniently fabricated with low cost, as a sensing platform for other types of biosensors with practical applications in the near future.

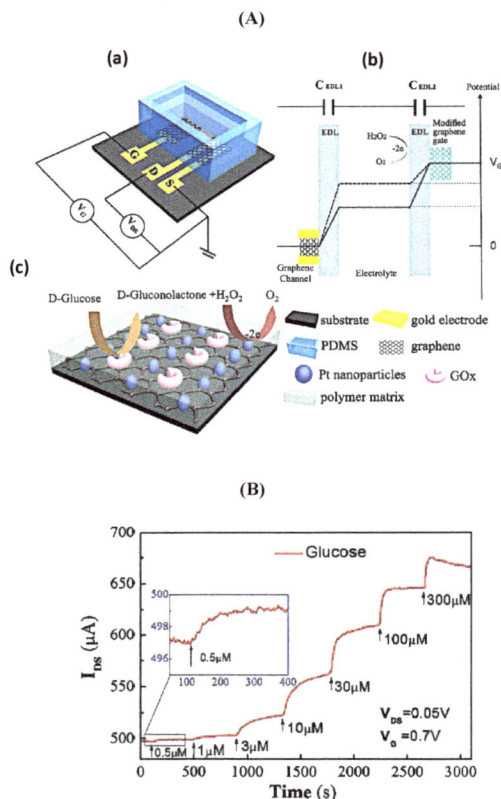

Figure 7. (A) (a) Schematic diagram of the glucose sensor; (b) Potential drops across the two electric double layers (EDLs) on the surfaces of graphene channel and gate. (c) The GOx catalyzed oxidation of glucose and the oxidation of H_2O_2 cycles on the GOx-CHIT/Nafion/PtNPs/graphene gate electrode; (B) The channel current responses to additions of glucose with different concentrations (with kind permission from Ref. [105]).

Another interesting application is the nitric oxide (NO) detection at cell level, reported by Xie et al. [100] and shown in Figure 8A. The NO sensing is based on iron porphyrin functionalized rGO FET, *via* the π−π stacking interaction. The highly catalytic specificity induced by metallo-porphyrin ensures an ultrasensitive and highly specific detection of NO in the range from 1 pM to 100 nM with the limit of detection as low as 1 pM in PBS and 10 pM in the cell medium, respectively. The real time monitoring of NO released from human umbilical vein endothelial cells (HUVECs) at single-cell level was also reported. This miniaturized, ultrasensitive, fast response FET biosensor is also described as a promising new platform for potential biological and diagnostic applications.

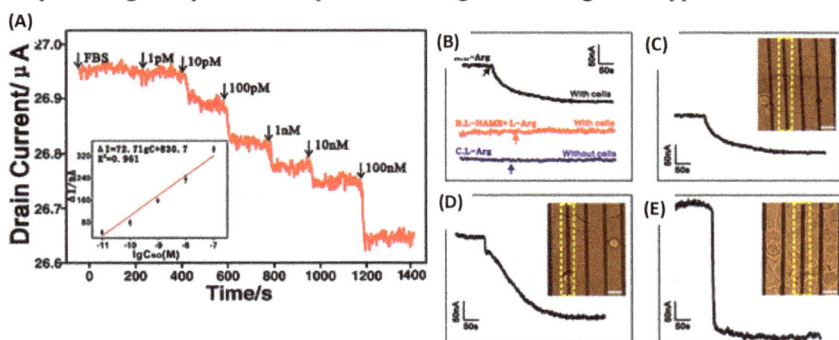

Figure 8. (A) Real-time electrical measurement at different concentrations of NO solution in the cell medium. Inset: the response of the RGO/FGPCs FET biosensors to NO at a series of concentrations (10, 100 pM and 1, 10, and 100 nM) in fetal bovine serum (FBS). The real-time monitoring of NO released from living cells. Typical response of the RGO/FGPCs FET sensor at different conditions: (B) stimulated by 3 mM L-Arg with cells; c, stimulated by 100 μM L-NAME + 3 mM L-Arg with cells; (D) stimulated by 3 mM L-Arg without cells. (C-E) Real-time responses of RGO/FGPCs FET sensor to NO sensor with different numbers of HUVECs cultured on the sensing surface (c, single cell; d, three cells; e, five cells). Scale bar: 20 μm (inset). The yellow dashed box shows the cells cultured inside the sensing channel. From reference (reprint permission from Ref. [100]).

Another distinctive example is the detection of odorant molecules reported by Knoll et al.[6] The olfactory biosensor was based on a rGO FET, functionalized by the odorant-binding protein 14 (OBP14) from the honey bee (*Apis mellifera*) and designed for the *in situ* real-time monitoring of a broad spectrum of odorants in aqueous solutions known to be attractants for bees. The results demonstrate that OBP14 is able to bind odorants even after immobilization on rGO and can discriminate between ligands binding within a range of dissociation constants. The strongest ligands, such as eugenol, and methyl vanillate, contain a hydroxy group seemingly important for the strong interaction with the protein.

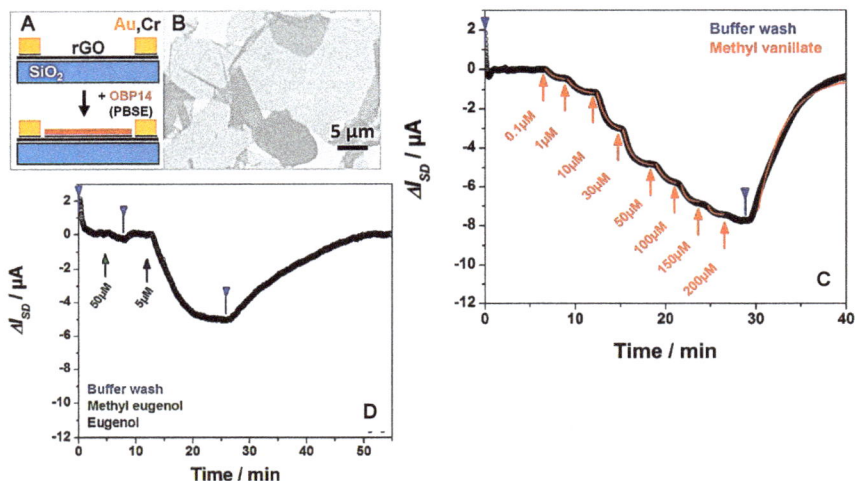

Figure 9. (A) Illustration of the OBP14 modified graphene biosensor device. (B) SEM image of the rGO-FET. (C) Real-time biosensor measurement of the binding of methyl vanillate to OBP14: Blue arrows indicate runs with pure buffer, red arrows indicate experiments with methyl vanillate solutions. Red curves are the fitting of the raw data by kinetic simulations of the association and dissociation processes based on the Langmuir model. (D) The real-time sensor response to the injection of a 50 uM solution of methyl eugenol, subsequent buffer wash, and the injection of 5 uM eugenol solution (with kind permission from Ref. [6]).

Another interesting examples are the detection of catecholamines epinephrine and dopamine. Mak et al. [107] described an epinephrine-sensitive biosensor prepared on a glass substrate using nafion and graphene or GO on the gate electrodes. The epinephrine is detected by oxidation at the gate electrode with a detection limit down to 0.1 nM, which makes it potentially useful for clinical applications.

Dopamine sensors were prepared by Zhang et al using solution-gated graphene transistors with graphene as both channel and gate electrodes were reported for the first time, with the detection limit down to 1 nM [108]. The sensing mechanism is attributed to the change of effective gate voltage induced by the electro-oxidation of dopamine at the graphene gate electrodes. The selectivity of the dopamine sensor was improved by modifying the gate electrode with a thin Nafion film by solution process.

5.4   Cells and bacteria

While still rare, large sized analytes such as cells and bacteria have been investigated with GFETs (Table 4). With cells, the interest is in flexible devices that can be interfaced with cultured cells in order to gather information about their action potential [109]. A study by Cohen-Karni et al. [110] on spontaneously beating embryonic chicken cardiomyocytes demonstrated that compared to classic SiNW FETs, graphene-based devices provide a lower spatial resolution, as the signal obtained is averaged from the several sources on the cell surface, according to the surface of the graphene flake in contact with the cells. The change in polarity across the Dirac point in GFETs offers them however an advantage over SiNW FET as the possibility for both n- and p-type recording enables to gather more information on the investigated cells.

With bacteria, the dream of detecting a single bacterium seems achievable considering the sensitivity of GFETs. So far, for *E. coli* the detection limits reached are 10 cfu mL^{-1} with antibodies as biorecognition elements. Remarkably, the G-FET were applied to detect the metabolic activities of bacteria in real time [111].

Table 4: Cell and bacteria based GFETs

Analyte	Graphene	Modification	LOD	References
E. coli	G	py-ester+ antibody	10 cfu mL^{-1}	[112]
E. coli	G	antibody	10 cfu mL^{-1}	[111]
ECC	G	ECC monolayer		[110]
cardiomyocyte-like cells	G	HL-1 cells		[109]

ECC: embryonic chicken cardiomyocytes

Conclusion and Perspectives

The advantages of large surface area and ease of specific functionalization of graphene and its derivatives together with the unique sensing mechanisms using GFETs, have resulted in an up-rise of these analytical devices in the field of biological sensing of DNA, proteins, biomarkers and small biological relevant molecules such as dopamine and glucose and others. While several steps are needed for the modification sensitive GFETs, these sensors have distinct fabrication advantages over devices fabricated with 1D materials, such as carbon nanotubes (CNTs) and nanowire. Graphene can be produced in uniform films, with uniform material characteristics. This uniformity results in limited variation of response characteristics and thus device-to-device consistency. With the advances of wafer-scale graphene film production by chemical vapor deposition and the amenability of these films to photolithographic techniques, GFETs are highly promising sensor platforms of commercial interest. Although some of the promises are clear, commercial GFETs are yet to come and there is still a need for breakthrough in sensor research for improving the sensing performance, lowering sensor costs and developing more adaptable and stable sensor fabrication methods. Among the challenges, improving the selectivity and miniaturization are both important for future applications. Moreover, attention has to be still given to long time stability of GFETs using aqueous solutions as well as biological fluids such as serum, saliva etc. Research is also needed to limit non-specific binding events in these more complex biological media. In fact, most tests were currently performed in non-biological fluidics, making the sensors of only limited use as point of care devices in hospital setting. The detection in physiological solutions using GFETs is thus still highly challenging due to the Debye screening limitation length. Some innovative work is still highly needed in this area to make GFETs future analytical devices.

Acknowledgements

Financial support from the Centre National de la Recherche Scientifique (CNRS), the University Lille 1, the Hauts-de-France region, the CPER "Photonics for Society", the Agence Nationale de la Recherche (ANR) through FLAG-ERA JTC 2015-Graphtivity are acknowledged.

References

[1]   Wong, H. S. P., Beyond the conventional transistor. *Solid-State Electronics* **2005,** *49* (5), 755-762. https://doi.org/10.1016/j.sse.2004.10.014

[2]   Elfström, N.; Karlström, A. E.; Linnros, J., Silicon Nanoribbons for Electrical Detection of Biomolecules. *Nano Letters* **2008,** *8* (3), 945-949. https://doi.org/10.1021/nl080094r

[3]   Vacic, A.; Criscione, J. M.; Stern, E.; Rajan, N. K.; Fahmy, T.; Reed, M. A., Multiplexed SOI BioFETs. *Biosensors and Bioelectronics* **2011,** *28* (1), 239-242. https://doi.org/10.1016/j.bios.2011.07.025

[4]   Li, Z.; Chen, Y.; Li, X.; Kamins, T. I.; Nauka, K.; Williams, R. S., Sequence-Specific Label-Free DNA Sensors Based on Silicon Nanowires. *Nano Letters* **2004,** *4* (2), 245-247. https://doi.org/10.1021/nl034958e

[5]   Stern, E.; Klemic, J. F.; Routenberg, D. A.; Wyrembak, P. N.; Turner-Evans, D. B.; Hamilton, A. D.; LaVan, D. A.; Fahmy, T. M.; Reed, M. A., Label-free immunodetection with CMOS-compatible semiconducting nanowires. *Nature* **2007**, *445* (7127), 519-522. https://doi.org/10.1038/nature05498

[6]   Larisika, M.; Kotlowski, C.; Steininger, C.; Mastrogiacomo, R.; Pelosi, P.; Schütz, S.; Peteu, S. F.; Kleber, C.; Reiner-Rozman, C.; Nowak, C.; Knoll, W., Electronic Olfactory Sensor Based on A. mellifera Odorant-Binding Protein 14 on a Reduced Graphene Oxide Field-Effect Transistor. *Angewandte Chemie International Edition* **2015**, *54* (45), 13245-13248. https://doi.org/10.1002/anie.201505712

[7]   Bergveld, P., Development of an ion-sensitive solid-state device for neurophysiological measurements. *IEEE Trans Biomed Eng* **1970**, *17* (1), 70-1. https://doi.org/10.1109/TBME.1970.4502688

[8]   Cui, Y.; Wei, Q.; Park, H.; Lieber, C. M., Nanowire nanosensors for highly sensitive and selective detection of biological and chemical species. *Science* **2001**, *293* (5533), 1289-92. https://doi.org/10.1126/science.1062711

[9]   Hahm, J.-i.; Lieber, C. M., Direct Ultrasensitive Electrical Detection of DNA and DNA Sequence Variations Using Nanowire Nanosensors. *Nano Letters* **2004**, *4* (1), 51-54. https://doi.org/10.1021/nl034853b

[10]   Zheng, G.; Patolsky, F.; Lieber, C. M., Multiplexed Electrical Detection of Single Viruses. *MRS Proceedings* **2011**, *828*, 79-84. https://doi.org/10.1557/PROC-828-A2.2

[11]   Schwierz, F., Graphene transistors. *Nature Nanotechnology* **2010**, *5*, 487–496. https://doi.org/10.1038/nnano.2010.89

[12]   Li, X.; Cai, W.; An, J.; Kim, S.; Nah, J.; Yang, D.; Piner, R.; Velamakanni, A.; Jung, I.; Tutuc, E.; Banerjee, S. K.; Colombo, L.; Ruoff, R. S., Large-area synthesis of high-quality and uniform graphene films on copper foils. *Science* **2009**, *324* (5932), 1312-4. https://doi.org/10.1126/science.1171245

[13]   Akhavan, O.; Ghaderi, E.; Esfandiar, A., Wrapping Bacteria by Graphene Nanosheets for Isolation from Environment, Reactivation by Sonication, and Inactivation by Near-Infrared Irradiation. *The Journal of Physical Chemistry B* **2011**, *115* (19), 6279-6288. https://doi.org/10.1021/jp200686k

[14]   Ferrari, A. C.; Meyer, J. C.; Scardaci, V.; Casiraghi, C.; Lazzeri, M.; Mauri, F.; Piscanec, S.; Jiang, D.; Novoselov, K. S.; Roth, S.; Geim, A. K., Raman Spectrum of Graphene and Graphene Layers. *Physical Review Letters* **2006**, *97* (18), 187401. https://doi.org/10.1103/PhysRevLett.97.187401

[15] Kim, K. S.; Zhao, Y.; Jang, H.; Lee, S. Y.; Kim, J. M.; Kim, K. S.; Ahn, J.-H.; Kim, P.; Choi, J.-Y.; Hong, B. H., Large-scale pattern growth of graphene films for stretchable transparent electrodes. *Nature* **2009,** *457* (7230), 706-710. https://doi.org/10.1038/nature07719

[16] Chen, J.-H.; Jang, C.; Xiao, S.; Ishigami, M.; Fuhrer, M. S., Intrinsic and extrinsic performance limits of graphene devices on SiO2. *Nature Nanotechnology* **2008,** *3* (4), 206-209. https://doi.org/10.1038/nnano.2008.58

[17] Novoselov, K. S.; Geim, A. K.; Morozov, S. V.; Jiang, D.; Zhang, Y.; Dubonos, S. V.; Grigorieva, I. V.; Firsov, A. A., Electric field in atomically thin carbon films. *Science* **2004,** *306* (5696), 666-669. https://doi.org/10.1126/science.1102896

[18] Geim, A. K.; Novoselov, K. S., The rise of graphene. *Nature Materials* **2007,** *6* (3), 183-191. https://doi.org/10.1038/nmat1849

[19] Pumera, M., Graphene-based nanomaterials and their electrochemistry. *Chemical Society reviews* **2010,** *39* (11), 4146-4157. https://doi.org/10.1039/c002690p

[20] Novoselov, K. S.; Jiang, D.; Schedin, F.; Booth, T. J.; Khotkevich, V. V.; Morozov, S. V.; Geim, A. K., Two-dimensional atomic crystals. *Proceedings of the National Academy of Sciences* **2005,** *102* (30), 10451-3. https://doi.org/10.1073/pnas.0502848102

[21] Shang, N. G.; Papakonstantinou, P.; McMullan, M.; Chu, M.; Stamboulis, A.; Potenza, A.; Dhesi, S. S.; Marchetto, H., Catalyst-Free Efficient Growth, Orientation and Biosensing Properties of Multilayer Graphene Nanoflake Films with Sharp Edge Planes. *Advanced Functional Materials* **2008,** *18* (21), 3506-3514. https://doi.org/10.1002/adfm.200800951

[22] Li, X.; Magnuson, C. W.; Venugopal, A.; Tromp, R. M.; Hannon, J. B.; Vogel, E. M.; Colombo, L.; Ruoff, R. S., Large-Area Graphene Single Crystals Grown by Low-Pressure Chemical Vapor Deposition of Methane on Copper. *Journal of the American Chemical Society* **2011,** *133* (9), 2816-2819. https://doi.org/10.1021/ja109793s

[23] Son, J. G.; Son, M.; Moon, K.-J.; Lee, B. H.; Myoung, J.-M.; Strano, M. S.; Ham, M.-H.; Ross, C. A., Sub-10 nm Graphene Nanoribbon Array Field-Effect Transistors Fabricated by Block Copolymer Lithography. *Advanced materials (Deerfield Beach, Fla.)* **2013,** *25* (34), 4723-4728. https://doi.org/10.1002/adma.201300813

[24] Hummers, W. S.; Offeman, R. E., Preparation of Graphitic Oxide. *Journal of the American Chemical Society* **1958,** *80* (6), 1339-1339. https://doi.org/10.1021/ja01539a017

[25]   Marcano, D. C.; Kosynkin, D. V.; Berlin, J. M.; Sinitskii, A.; Sun, Z.; Slesarev, A.; Alemany, L. B.; Lu, W.; Tour, J. M., Improved Synthesis of Graphene Oxide. *ACS Nano* **2010,** *4* (8), 4806-4814. https://doi.org/10.1021/nn1006368

[26]   Staudenmaier, L., Verfahren zur Darstellung der Graphitsäure. *Berichte der deutschen chemischen Gesellschaft* **1899,** *31* (2), 1481–1487. https://doi.org/10.1002/cber.18980310237

[27]   He, H.; Klinowski, J.; Forster, M.; Lerf, A., A new structural model for graphite oxide. *Chemical Physics Letters* **1998,** *287* (1), 53-56. https://doi.org/10.1016/S0009-2614(98)00144-4

[28]   Lerf, A.; He, H.; Forster, M.; Klinowski, J., Structure of Graphite Oxide Revisited. *The Journal of Physical Chemistry B* **1998,** *102* (23), 4477-4482. https://doi.org/10.1021/jp9731821

[29]   Allen, M. J.; Tung, V. C.; Kaner, R. B., Honeycomb Carbon: A Review of Graphene. *Chemical Reviews* **2010,** *110* (1), 132-145. https://doi.org/10.1021/cr900070d

[30]   Amarnath, C. A.; Hong, C. E.; Kim, N. H.; Ku, B.-C.; Kuila, T.; Lee, J. H., Efficient synthesis of graphene sheets using pyrrole as a reducing agent. *Carbon* **2011,** *49* (11), 3497-3502. https://doi.org/10.1016/j.carbon.2011.04.048

[31]   Compton, O. C.; Jain, B.; Dikin, D. A.; Abouimrane, A.; Amine, K.; Nguyen, S. T., Chemically Active Reduced Graphene Oxide with Tunable C/O Ratios. *ACS Nano* **2011,** *5* (6), 4380-4391. https://doi.org/10.1021/nn1030725

[32]   Dreyer, D. R.; Murali, S.; Zhu, Y.; Ruoff, R. S.; Bielawski, C. W., Reduction of graphite oxide using alcohols. *Journal of Materials Chemistry* **2011,** *21* (10), 3443-3447. https://doi.org/10.1039/C0JM02704A

[33]   Dreyer, D. R.; Park, S.; Bielawski, C. W.; Ruoff, R. S., The chemistry of graphene oxide. *Chemical Society reviews* **2010,** *39* (1), 228-40. https://doi.org/10.1039/B917103G

[34]   Fan, X.; Peng, W.; Li, Y.; Li, X.; Wang, S.; Zhang, G.; Zhang, F., Deoxygenation of Exfoliated Graphite Oxide under Alkaline Conditions: A Green Route to Graphene Preparation. *Advanced materials (Deerfield Beach, Fla.)* **2008,** *20* (23), 4490-4493. https://doi.org/10.1002/adma.200801306

[35]   Fan, Z.-J.; Kai, W.; Yan, J.; Wei, T.; Zhi, L.-J.; Feng, J.; Ren, Y.-m.; Song, L.-P.; Wei, F., Facile Synthesis of Graphene Nanosheets via Fe Reduction of Exfoliated Graphite Oxide. *ACS Nano* **2011,** *5* (1), 191-198. https://doi.org/10.1021/nn102339t

[36]   Fellahi, O.; Das, M. R.; Coffinier, Y.; Szunerits, S.; Hadjersi, T.; Maamache, M.; Boukherroub, R., Silicon nanowire arrays-induced graphene oxide reduction under UV irradiation. *Nanoscale* **2011,** *3* (11), 4662-4669. https://doi.org/10.1039/c1nr10970g

[37]   Kaminska, I.; Das, M. R.; Coffinier, Y.; Niedziolka-Jonsson, J.; Sobczak, J.; Woisel, P.; Lyskawa, J.; Opallo, M.; Boukherroub, R.; Szunerits, S., Reduction and Functionalization of Graphene Oxide Sheets Using Biomimetic Dopamine Derivatives in One Step. *ACS Applied Materials & Interfaces* **2012,** *4* (2), 1016-1020. https://doi.org/10.1021/am201664n

[38]   Kaminska, I.; Das, M. R.; Coffinier, Y.; Niedziolka-Jonsson, J.; Woisel, P.; Opallo, M.; Szunerits, S.; Boukherroub, R., Preparation of graphene/tetrathiafulvalene nanocomposite switchable surfaces. *Chemical Communications* **2012,** *48* (9), 1221-1223. https://doi.org/10.1039/C1CC15215G

[39]   Kaminska, I.; Barras, A.; Coffinier, Y.; Lisowski, W.; Roy, S.; Niedziolka-Jonsson, J.; Woisel, P.; Lyskawa, J.; Opallo, M.; Siriwardena, A.; Boukherroub, R.; Szunerits, S., Preparation of a Responsive Carbohydrate-Coated Biointerface Based on Graphene/Azido-Terminated Tetrathiafulvalene Nanohybrid Material. *ACS Applied Materials & Interfaces* **2012,** *4* (10), 5386-5393. https://doi.org/10.1021/am3013196

[40]   Li, D.; Muller, M. B.; Gilje, S.; Kaner, R. B.; Wallace, G. G., Processable aqueous dispersions of graphene nanosheets. *Nature Nanotechnology* **2008,** *3* (2), 101-105. https://doi.org/10.1038/nnano.2007.451

[41]   Moon, I. K.; Lee, J.; Ruoff, R. S.; Lee, H., Reduced graphene oxide by chemical graphitization. *Nature communications* **2010,** *1*, 73. https://doi.org/10.1038/ncomms1067

[42]   Park, S.; Ruoff, R. S., Chemical methods for the production of graphenes. *Nature Nanotechnology* **2009,** *4* (4), 217-224. https://doi.org/10.1038/nnano.2009.58

[43]   Shin, H.-J.; Kim, K. K.; Benayad, A.; Yoon, S.-M.; Park, H. K.; Jung, I.-S.; Jin, M. H.; Jeong, H.-K.; Kim, J. M.; Choi, J.-Y.; Lee, Y. H., Efficient Reduction of Graphite Oxide by Sodium Borohydride and Its Effect on Electrical Conductance. *Advanced Functional Materials* **2009,** *19* (12), 1987-1992. https://doi.org/10.1002/adfm.200900167

[44]   Stankovich, S.; Dikin, D. A.; Piner, R. D.; Kohlhaas, K. A.; Kleinhammes, A.; Jia, Y.; Wu, Y.; Nguyen, S. T.; Ruoff, R. S., Synthesis of graphene-based nanosheets via chemical reduction of exfoliated graphite oxide. *Carbon* **2007,** *45* (7), 1558-1565. https://doi.org/10.1016/j.carbon.2007.02.034

[45]   Tung, V. C.; Allen, M. J.; Yang, Y.; Kaner, R. B., High-throughput solution processing of large-scale graphene. *Nature Nanotechnology* **2009,** *4* (1), 25-9. https://doi.org/10.1038/nnano.2008.329

[46]   Zhu, C.; Guo, S.; Fang, Y.; Dong, S., Reducing Sugar: New Functional Molecules for the Green Synthesis of Graphene Nanosheets. *ACS Nano* **2010,** *4* (4), 2429-2437. https://doi.org/10.1021/nn1002387

[47]   Gao, J.; Liu, F.; Liu, Y.; Ma, N.; Wang, Z.; Zhang, X., Environment-Friendly Method To Produce Graphene That Employs Vitamin C and Amino Acid. *Chemistry of Materials* **2010,** *22* (7), 2213-2218. https://doi.org/10.1021/cm902635j

[48]   Fernández-Merino, M. J.; Guardia, L.; Paredes, J. I.; Villar-Rodil, S.; Solís-Fernández, P.; Martínez-Alonso, A.; Tascón, J. M. D., Vitamin C Is an Ideal Substitute for Hydrazine in the Reduction of Graphene Oxide Suspensions. *The Journal of Physical Chemistry C* **2010,** *114* (14), 6426-6432. https://doi.org/10.1021/jp100603h

[49]   Pei, S.; Zhao, J.; Du, J.; Ren, W.; Cheng, H.-M., Direct reduction of graphene oxide films into highly conductive and flexible graphene films by hydrohalic acids. *Carbon* **2010,** *48* (15), 4466-4474. https://doi.org/10.1016/j.carbon.2010.08.006

[50]   Fan, Z.; Wang, K.; Wei, T.; Yan, J.; Song, L.; Shao, B., An environmentally friendly and efficient route for the reduction of graphene oxide by aluminum powder. *Carbon* **2010,** *48* (5), 1686-1689. https://doi.org/10.1016/j.carbon.2009.12.063

[51]   Zhang, Z.; Chen, H.; Xing, C.; Guo, M.; Xu, F.; Wang, X.; Gruber, H. J.; Zhang, B.; Tang, J., Sodium citrate: A universal reducing agent for reduction / decoration of graphene oxide with au nanoparticles. *Nano Research* **2011,** *4* (6), 599-611. https://doi.org/10.1007/s12274-011-0116-y

[52]   Wang, Y.; Shi, Z.; Yin, J., Facile Synthesis of Soluble Graphene via a Green Reduction of Graphene Oxide in Tea Solution and Its Biocomposites. *ACS Applied Materials & Interfaces* **2011,** *3* (4), 1127-1133. https://doi.org/10.1021/am1012613

[53]   Yang, F.; Liu, Y.; Gao, L.; Sun, J., pH-Sensitive Highly Dispersed Reduced Graphene Oxide Solution Using Lysozyme via an in Situ Reduction Method. *The Journal of Physical Chemistry C* **2010,** *114* (50), 22085-22091. https://doi.org/10.1021/jp1079636

[54]   Xu, L. Q.; Yang, W. J.; Neoh, K.-G.; Kang, E.-T.; Fu, G. D., Dopamine-Induced Reduction and Functionalization of Graphene Oxide Nanosheets. *Macromolecules* **2010,** *43* (20), 8336-8339. https://doi.org/10.1021/ma101526k

[55]   Kwon, O. S.; Park, S. J.; Hong, J.-Y.; Han, A. R.; Lee, J. S.; Lee, J. S.; Oh, J. H.; Jang, J., Flexible FET-Type VEGF Aptasensor Based on Nitrogen-Doped Graphene Converted from Conducting Polymer. *ACS Nano* **2012,** *6* (2), 1486-1493. https://doi.org/10.1021/nn204395n

[56]   Xu, W.; Lee, T.-W., Recent progress in fabrication techniques of graphene nanoribbons. *Materials Horizons* **2016,** *3* (3), 186-207. https://doi.org/10.1039/C5MH00288E

[57]   Mao, S.; Cui, S.; Lu, G.; Yu, K.; Wen, Z.; Chen, J., Tuning gas-sensing properties of reduced graphene oxide using tin oxide nanocrystals. *Journal of Materials Chemistry* **2012,** *22* (22), 11009-11013. https://doi.org/10.1039/c2jm30378g

[58]   Wang, Z.; Eigler, S.; Halik, M., Scalable self-assembled reduced graphene oxide transistors on flexible substrate. *Applied Physics Letters* **2014,** *104* (24). https://doi.org/10.1063/1.4884064

[59]   Botcha, D. V.; Singh, G.; Narayanam, P. K.; Sutar, D. S.; Talwar, S. S.; Srinivasa, R. S.; Major, S. S., GO and RGO based FETs fabricated with Langmuir-Blodgett grown monolayers. *AIP Conference Proceedings* **2012,** *1447* (1), 327-328. https://doi.org/10.1007/s12274-013-0376-9

[60]   Kybert, N. J.; Han, G. H.; Lerner, M. B.; Dattoli, E. N.; Esfandiar, A.; Charlie Johnson, A. T., Scalable arrays of chemical vapor sensors based on DNA-decorated graphene. *Nano Research* **2014,** *7* (1), 95-103. https://doi.org/10.1073/pnas.1105113108

[61]   de Heer, W. A.; Berger, C.; Ruan, M.; Sprinkle, M.; Li, X.; Hu, Y.; Zhang, B.; Hankinson, J.; Conrad, E., Large area and structured epitaxial graphene produced by confinement controlled sublimation of silicon carbide. *Proceedings of the National Academy of Sciences* **2011,** *108* (41), 16900-16905. https://doi.org/10.1073/pnas.1105113108

[62]   Farmer, D. B.; Chiu, H.-Y.; Lin, Y.-M.; Jenkins, K. A.; Xia, F.; Avouris, P., Utilization of a Buffered Dielectric to Achieve High Field-Effect Carrier Mobility in Graphene Transistors. *Nano Letters* **2009,** *9* (12), 4474-4478. https://doi.org/10.1021/nl902788u

[63]   Faugeras, C.; Nerrière, A.; Potemski, M.; Mahmood, A.; Dujardin, E.; Berger, C.; de Heer, W. A., Few-layer graphene on SiC, pyrolitic graphite, and graphene: A Raman scattering study. *Applied Physics Letters* **2008,** *92* (1), 011914. https://doi.org/10.1063/1.2828975

[64]   Das, A.; Pisana, S.; Chakraborty, B.; Piscanec, S.; Saha, S. K.; Waghmare, U. V.; Novoselov, K. S.; Krishnamurthy, H. R.; Geim, A. K.; Ferrari, A. C.; Sood, A. K., Monitoring dopants by Raman scattering in an electrochemically top-gated graphene transistor. *Nature Nanotechnology* **2008**, *3* (4), 210-5. https://doi.org/10.1038/nnano.2008.67

[65]   Dong, X.; Shi, Y.; Huang, W.; Chen, P.; Li, L. J., Electrical detection of DNA hybridization with single-base specificity using transistors based on CVD-grown graphene sheets. *Advanced materials (Deerfield Beach, Fla.)* **2010**, *22* (14), 1649-53. https://doi.org/10.1002/adma.200903645

[66]   Johnson, D. W.; Dobson, B. P.; Coleman, K. S., A manufacturing perspective on graphene dispersions. *Current Opinion in Colloid & Interface Science* **2015**, *20* (5), 367-382. https://doi.org/10.1016/j.cocis.2015.11.004

[67]   Hwang, M. T.; Landon, P. B.; Lee, J.; Choi, D.; Mo, A. H.; Glinsky, G.; Lal, R., Highly specific SNP detection using 2D graphene electronics and DNA strand displacement. *Proceedings of the National Academy of Sciences* **2016**, *113* (26), 7088-7093. https://doi.org/10.1073/pnas.1603753113

[68]   Saltzgaber, G.; Wojcik, P., M. ; Sharf, T.; Leyden, M., R.; Wardini, J., L.; Heist, C., A. ; Adenuga, A., A. ; Remcho, V., T.; Minot, E., D. , Scalable graphene field-effect sensors for specific protein detection. *Nanotechnology* **2013**, *24* (35), 355502. https://doi.org/10.1088/0957-4484/24/35/355502

[69]   Zheng, C.; Huang, L.; Zhang, H.; Sun, Z.; Zhang, Z.; Zhang, G.-J., Fabrication of Ultrasensitive Field-Effect Transistor DNA Biosensors by a Directional Transfer Technique Based on CVD-Grown Graphene. *ACS Applied Materials & Interfaces* **2015**, *7* (31), 16953-16959. https://doi.org/10.1021/acsami.5b03941

[70]   Mohanty, N.; Berry, V., Graphene-Based Single-Bacterium Resolution Biodevice and DNA Transistor: Interfacing Graphene Derivatives with Nanoscale and Microscale Biocomponents. *Nano Letters* **2008**, *8* (12), 4469-4476. https://doi.org/10.1021/nl802412n

[71]   Balakrishnan, S.; Downard, A. J.; Telfer, S. G., HKUST-1 growth on glassy carbon. *Journal of Materials Chemistry* **2011**, *21* (48), 19207-19209. https://doi.org/10.1039/c1jm13912f

[72]   Eissa, S.; Zourob, M., A graphene-based electrochemical competitive immunosensor for the sensitive detection of okadaic acid in shellfish. *Nanoscale* **2012**, *4* (23), 7593-9. https://doi.org/10.1039/c2nr32146g

[73]   Pinson, J.; Podvorica, F., Attachment of organic layers to conductive or semiconductive surfaces by reduction of diazonium salts. *Chemical Society reviews* **2005,** *34* (5), 429-439. https://doi.org/10.1039/b406228k

[74]   Lin, C.-T.; Loan, P. T. K.; Chen, T.-Y.; Liu, K.-K.; Chen, C.-H.; Wei, K.-H.; Li, L.-J., Label-Free Electrical Detection of DNA Hybridization on Graphene using Hall Effect Measurements: Revisiting the Sensing Mechanism. *Advanced Functional Materials* **2013,** *23* (18), 2301-2307. https://doi.org/10.1002/adfm.201202672

[75]   Dong, X.; Huang, W.; Chen, P., In Situ Synthesis of Reduced Graphene Oxide and Gold Nanocomposites for Nanoelectronics and Biosensing. *Nanoscale Res Lett* **2010,** *6* (1), 60. https://doi.org/10.1007/s11671-010-9806-8

[76]   Zhu, L.; Luo, L.; Wang, Z., DNA electrochemical biosensor based on thionine-graphene nanocomposite. *Biosensors & bioelectronics* **2012,** *35* (1), 507-11. https://doi.org/10.1016/j.bios.2012.03.026

[77]   Cai, B.; Wang, S.; Huang, L.; Ning, Y.; Zhang, Z.; Zhang, G.-J., Ultrasensitive Label-Free Detection of PNA–DNA Hybridization by Reduced Graphene Oxide Field-Effect Transistor Biosensor. *ACS Nano* **2014,** *8* (3), 2632-2638. https://doi.org/10.1021/nn4063424

[78]   Ping, J.; Vishnubhotla, R.; Vrudhula, A.; Johnson, A. T. C., Scalable Production of High-Sensitivity, Label-Free DNA Biosensors Based on Back-Gated Graphene Field Effect Transistors. *ACS Nano* **2016,** *10* (9), 8700-8704. https://doi.org/10.1021/acsnano.6b04110

[79]   Stine, R.; Robinson, J. T.; Sheehan, P. E.; Tamanaha, C. R., Real-Time DNA Detection Using Reduced Graphene Oxide Field Effect Transistors. *Advanced materials (Deerfield Beach, Fla.)* **2010,** *22* (46), 5297-5300. https://doi.org/10.1002/adma.201002121

[80]   Chen, T.-Y.; Loan, P. T. K.; Hsu, C.-L.; Lee, Y.-H.; Tse-Wei Wang, J.; Wei, K.-H.; Lin, C.-T.; Li, L.-J., Label-free detection of DNA hybridization using transistors based on CVD grown graphene. *Biosensors and Bioelectronics* **2013,** *41* (Supplement C), 103-109. https://doi.org/10.1016/j.bios.2012.07.059

[81]   Xu, G.; Abbott, J.; Qin, L.; Yeung, K. Y. M.; Song, Y.; Yoon, H.; Kong, J.; Ham, D., Electrophoretic and field-effect graphene for all-electrical DNA array technology. *Nature communications* **2014,** *5*, 4866. https://doi.org/10.1038/ncomms5866

[82]   Stern, E.; Wagner, R.; Sigworth, F. J.; Breaker, R.; Fahmy, T. M.; Reed, M. A., Importance of the Debye Screening Length on Nanowire Field Effect Transistor Sensors. *Nano Letters* **2007,** *7* (11), 3405-3409. https://doi.org/10.1021/nl071792z

[83]   Vacic, A.; Criscione, J. M.; Rajan, N. K.; Stern, E.; Fahmy, T. M.; Reed, M. A., Determination of Molecular Configuration by Debye Length Modulation. *Journal of the American Chemical Society* **2011,** *133* (35), 13886-13889. https://doi.org/10.1021/ja205684a

[84]   Shogo, O.; Yasuhide, O.; Kenzo, M.; Koichi, I.; Kazuhiko, M., Immunosensors Based on Graphene Field-Effect Transistors Fabricated Using Antigen-Binding Fragment. *Japanese Journal of Applied Physics* **2012,** *51* (6S), 06FD08. https://doi.org/10.7567/JJAP.51.06FD08

[85]   Matsumoto, K.; Maehashi, K.; Ohno, Y.; Inoue, K., Recent advances in functional graphene biosensors. *Journal of Physics D: Applied Physics* **2014,** *47* (9), 094005. https://doi.org/10.1088/0022-3727/47/9/094005

[86]   Kim, D.-J.; Sohn, I. Y.; Jung, J.-H.; Yoon, O. J.; Lee, N. E.; Park, J.-S., Reduced graphene oxide field-effect transistor for label-free femtomolar protein detection. *Biosensors and Bioelectronics* **2013,** *41* (Supplement C), 621-626. https://doi.org/10.1016/j.bios.2012.09.040

[87]   Lei, Y.-M.; Xiao, M.-M.; Li, Y.-T.; Xu, L.; Zhang, H.; Zhang, Z.-Y.; Zhang, G.-J., Detection of heart failure-related biomarker in whole blood with graphene field effect transistor biosensor. *Biosensors and Bioelectronics* **2017,** *91* (Supplement C), 1-7. https://doi.org/10.1016/j.bios.2016.12.018

[88]   Ohno, Y.; Maehashi, K.; Matsumoto, K., Label-Free Biosensors Based on Aptamer-Modified Graphene Field-Effect Transistors. *Journal of the American Chemical Society* **2010,** *132* (51), 18012-18013. https://doi.org/10.1021/ja108127r

[89]   Kim, D.-J.; Park, H.-C.; Sohn, I. Y.; Jung, J.-H.; Yoon, O. J.; Park, J.-S.; Yoon, M.-Y.; Lee, N.-E., Electrical Graphene Aptasensor for Ultra-Sensitive Detection of Anthrax Toxin with Amplified Signal Transduction. *Small* **2013,** *9* (19), 3352-3360. https://doi.org/10.1002/smll.201203245

[90]   Zhou, L.; Mao, H.; Wu, C.; Tang, L.; Wu, Z.; Sun, H.; Zhang, H.; Zhou, H.; Jia, C.; Jin, Q.; Chen, X.; Zhao, J., Label-free graphene biosensor targeting cancer molecules based on non-covalent modification. *Biosensors and Bioelectronics* **2017,** *87* (Supplement C), 701-707. https://doi.org/10.1016/j.bios.2016.09.025

[91]　Xiang, L.; Wang, Z.; Liu, Z.; Weigum, S. E.; Yu, Q.; Chen, M. Y., Inkjet-Printed Flexible Biosensor Based on Graphene Field Effect Transistor. *IEEE Sensors Journal* **2016,** *16* (23), 8359 - 8364. https://doi.org/10.1109/JSEN.2016.2608719

[92]　Gao, N.; Gao, T.; Yang, X.; Dai, X.; Zhou, W.; Zhang, A.; Lieber, C. M., Specific detection of biomolecules in physiological solutions using graphene transistor biosensors. *Proceedings of the National Academy of Sciences* **2016,** *113* (51), 14633-14638. https://doi.org/10.1073/pnas.1625010114

[93]　Reiner-Rozman, C.; Larisika, M.; Nowak, C.; Knoll, W., Graphene-Based Liquid-Gated Field Effect Transistor for Biosensing: Theory and Experiments. *Biosensors & bioelectronics* **2015,** *70,* 21-27. https://doi.org/10.1016/j.bios.2015.03.013

[94]　Mao, S.; Yu, K.; Lu, G.; Chen, J., Highly sensitive protein sensor based on thermally-reduced graphene oxide field-effect transistor. *Nano Research* **2011,** *4* (10), 921. https://doi.org/10.1007/s12274-011-0148-3

[95]　Ke, X.; Xenia, M.; Barbara, M. N.; Eugene, Z.; Mitra, D.; Michael, A. S., Graphene- and aptamer-based electrochemical biosensor. *Nanotechnology* **2014,** *25* (20), 205501. https://doi.org/10.1088/0957-4484/25/20/205501

[96]　Tuteja, S. K.; Priyanka; Bhalla, V.; Deep, A.; Paul, A. K.; Suri, C. R., Graphene-gated biochip for the detection of cardiac marker Troponin I. *Analytica Chimica Acta* **2014,** *809* (Supplement C), 148-154. https://doi.org/10.1016/j.aca.2013.11.047

[97]　Lerner, M. B.; Matsunaga, F.; Han, G. H.; Hong, S. J.; Xi, J.; Crook, A.; Perez-Aguilar, J. M.; Park, Y. W.; Saven, J. G.; Liu, R.; Johnson, A. T. C., Scalable Production of Highly Sensitive Nanosensors Based on Graphene Functionalized with a Designed G Protein-Coupled Receptor. *Nano Letters* **2014,** *14* (5), 2709-2714. https://doi.org/10.1021/nl5006349

[98]　Sudibya, H. G.; He, Q.; Zhang, H.; Chen, P., Electrical Detection of Metal Ions Using Field-Effect Transistors Based on Micropatterned Reduced Graphene Oxide Films. *ACS Nano* **2011,** *5* (3), 1990-1994. https://doi.org/10.1021/nn103043v

[99]　Basu, J.; Datta, S.; RoyChaudhuri, C., A graphene field effect capacitive Immunosensor for sub-femtomolar food toxin detection. *Biosensors and Bioelectronics* **2015,** *68* (Supplement C), 544-549. https://doi.org/10.1016/j.bios.2015.01.046

[100]　Xie, H.; Li, Y.-T.; Lei, Y.-M.; Liu, Y.-L.; Xiao, M.-M.; Gao, C.; Pang, D.-W.; Huang, W.-H.; Zhang, Z.-Y.; Zhang, G.-J., Real-Time Monitoring of Nitric Oxide at Single-Cell Level with Porphyrin-Functionalized Graphene Field-Effect Transistor

Biosensor. *Analytical Chemistry* **2016,** *88* (22), 11115-11122.
https://doi.org/10.1021/acs.analchem.6b03208

[101] Mukherjee, S.; Meshik, X.; Choi, M.; Farid, S.; Datta, D.; Lan, Y.; Poduri, S.;
Sarkar, K.; Baterdene, U.; Huang, C. E.; Wang, Y. Y.; Burke, P.; Dutta, M.; Stroscio,
M. A., A Graphene and Aptamer Based Liquid Gated FET-Like Electrochemical
Biosensor to Detect Adenosine Triphosphate. *IEEE Trans Nanobioscience* **2015,** *14*
(8), 967-72. https://doi.org/10.1109/TNB.2015.2501364

[102] Thi Thanh, C.; Van Chuc, N.; Hai Binh, N.; Hung Thang, B.; Thi Thu, V.; Ngoc
Hong, P.; Bach Thang, P.; Le, H.; Maxime, B.; Matthieu, P.; Jean Louis, S.; Ngoc
Minh, P.; Dai Lam, T., Fabrication of few-layer graphene film based field effect
transistor and its application for trace-detection of herbicide atrazine. *Advances in
Natural Sciences: Nanoscience and Nanotechnology* **2016,** *7* (3), 035007.

[103] Viswanathan, S.; Narayanan, T. N.; Aran, K.; Fink, K. D.; Paredes, J.; Ajayan, P.
M.; Filipek, S.; Miszta, P.; Tekin, H. C.; Inci, F.; Demirci, U.; Li, P.; Bolotin, K. I.;
Liepmann, D.; Renugopalakrishanan, V., Graphene–protein field effect biosensors:
glucose sensing. *Materials Today* **2015,** *18* (9), 513-522.
https://doi.org/10.1016/j.mattod.2015.04.003

[104] Huang, Y.; Dong, X.; Shi, Y.; Li, C. M.; Li, L.-J.; Chen, P., Nanoelectronic
biosensors based on CVD grown graphene. *Nanoscale* **2010,** *2* (8), 1485-1488.
https://doi.org/10.1039/c0nr00142b

[105] Zhang, M.; Liao, C.; Mak, C. H.; You, P.; Mak, C. L.; Yan, F., Highly sensitive
glucose sensors based on enzyme-modified whole-graphene solution-gated transistors.
Scientific Reports **2015,** *5*, 8311.

[106] Park, J. W.; Lee, C.; Jang, J., High-performance field-effect transistor-type glucose
biosensor based on nanohybrids of carboxylated polypyrrole nanotube wrapped
graphene sheet transducer. *Sensors and Actuators B: Chemical* **2015,** *208* (Supplement
C), 532-537. https://doi.org/10.1016/j.snb.2014.11.085

[107] Mak, C. H.; Liao, C.; Fu, Y.; Zhang, M.; Tang, C. Y.; Tsang, Y. H.; Chan, H. L.
W.; Yan, F., Highly-sensitive epinephrine sensors based on organic electrochemical
transistors with carbon nanomaterial modified gate electrodes. *Journal of Materials
Chemistry C* **2015,** *3* (25), 6532-6538. https://doi.org/10.1039/C5TC01100K

[108] Zhang, M.; Liao, C.; Yao, Y.; Liu, Z.; Gong, F.; Yan, F., High-Performance
Dopamine Sensors Based on Whole-Graphene Solution-Gated Transistors. *Advanced
Functional Materials* **2014,** *24* (7), 978-985.

[109] Benno, M. B.; Martin, L.; Simon, D.; Andrea Bonaccini, C.; Karolina, S.; Lionel, R.; Gaëlle, L.; Jose, A. G., Flexible graphene transistors for recording cell action potentials. *2D Materials* **2016,** *3* (2), 025007. https://doi.org/10.1088/2053-1583/3/2/025007

[110] Cohen-Karni, T.; Qing, Q.; Li, Q.; Fang, Y.; Lieber, C. M., Graphene and Nanowire Transistors for Cellular Interfaces and Electrical Recording. *Nano Letters* **2010,** *10* (3), 1098-1102. https://doi.org/10.1021/nl1002608

[111] Huang, Y.; Dong, X.; Liu, Y.; Li, L.-J.; Chen, P., Graphene-based biosensors for detection of bacteria and their metabolic activities. *Journal of Materials Chemistry* **2011,** *21* (33), 12358-12362. https://doi.org/10.1039/c1jm11436k

[112] Akbari, E.; Buntat, Z.; Kiani, M. J.; Enzevaee, A.; Khaledian, M., Analytical model of graphene-based biosensors for bacteria detection. *International Journal of Environmental Analytical Chemistry* **2015,** *95* (9), 847-854. https://doi.org/10.1080/03067319.2015.1058930

Organic Bioelectronics for Life Science and Healthcare   Materials Research Forum LLC
Materials Research Foundations **56** (2019) 153-184   doi: https://doi.org/10.21741/9781644900376-5

Chapter 5

Graphene as an Organic and Bioelectronic Material

D. Kireev, A.Offenhäusser*

Institute of Complex Systems-Bioelectronics, Forschungszentrum Jülich, Germany

a.offenhaeusser@fz-juelich.de

Abstract

Graphene and graphene-based devices have recently shown a great potential in the field of bioelectronics and healthcare. Graphene-based electrodes and more complex transistors that comprise a new type of bioelectronic devices will be introduced in this chapter. Biocompatibility, stability, excellent and unique electronic properties, scalability, and pure two-dimensional structure make graphene the perfect material for bioelectronic applications. Usage of these devices was shown successful for *in vitro* studies of cardiac-like cell lines and cortical neuronal networks with signal to noise up to 100. Furthermore, *in vivo* applications of graphene based devices can lead to higher level of understanding of the brain.

Keywords

Graphene, GFETs, GMEAs, Extracellular Electrophysiology, Neuroprosthetics

Contents

1. Introduction

Undeniably, nowadays graphene and graphene-based devices form an exclusive field in science. After gaining the attraction to itself starting with discovery of its field effect and other unique physical effects in 2004 [1], it raised up into the adjacent subjects of biology and bioelectronics [2–5]. It was shown that graphene and graphene-like materials are inherently biocompatible and can be even more favorable for cellular outgrowth [6–8]. The unique transistor behavior, biocompatibility and possibility to implement liquid-gating of these transistors has pushed the devices to be used for interfacing with cells and electrophysiology.

In the field of electrophysiology, graphene has just recently gained influence in cellular interfacing since. Compared to silicon-based transistors, they are more sensitive, hence more favorable [9–11]. Moreover, intrinsic flexibility and biocompatibility of graphene has a promising future for this research, pointing out in the direction of fully flexible, soft and implantable neuroprosthetic applications. There, graphene can be used either actively as a transistor's active area [11,12] or passively as an electrode [13–16]. In the former case, the device requires at least two electrodes (source and drain) and complicated electrochemical gating and read-out schematics. In the latter case, however, there is just one feedline required to feed and record from a graphene electrode. Moreover, such graphene-based electrodes are comparably easy to fabricate, characterize and use.

As an allotrope of carbon, graphene is a single-layer, two-dimensional arrangement of carbon atoms. Each carbon molecule has six electrons that form one 2s and three 2p orbitals. These orbitals, when hybridized, form three sigma-bonds (strong in-plane bond) and one π-bond (weak out-of-plane bond). Graphite, that is a multi-layer arrangement of these single graphene layers is a typical example of sp^2 hybridization, is well known for its softness exactly due to these weak out-of-plane forces. However, the band structure of graphite shows that there is a direct intersection between its valence and conduction bands, and is therefore called a semi-metal and in general is a good conductor [17].

When graphene (a single layer of graphite) is considered, the sp^2 bonding provides remarkably new properties, such as absence of a bandgap at the K-points (also known as Dirac points) and linear energy dispersion behavior near these points. In some considerations (close to the Fermi level), it is possible to describe graphene as a "*two-*

Organic Bioelectronics for Life Science and Healthcare
Materials Research Foundations **56** (2019) 153-184

Materials Research Forum LLC
doi: https://doi.org/10.21741/9781644900376-5

dimensional gas of massless Dirac fermions". For comparison, typical for semiconductors energy-momentum relation is quadratic, and can be solved by the Schrödinger equation:

$$E = \frac{p^2}{2m_*}$$

where p is the momentum and m_* is mass of charge carriers.

As for graphene, the energy dispersion is linear near Fermi level, therefore charge carriers can be considered as massless and relativistic particles, and therefore must be solved by the Dirac equation, resulting in the following equation for energy dispersion:

$$E = v|p|,$$

where $c/v \cong 300$, and v is the Fermi velocity and c is the speed of light. This is a significantly different and unique dispersion as it does not depend on the mass of the charge carriers and linearly depends on momentum in the first order approximation [17].

In the recent years there has established a rather standard common strategy for improvement cell electrophysiology: going three-dimensional, increasing electrode's roughness, etc., to simply increase impedance and decrease cell-electrode cleft. However, graphene devices result in a slightly different situation. Since it is a purely two-dimensional material, with its charge carriers freely open to the surrounding environment, it is ultrasensitive to the surrounding environment regardless the low roughness and small impedance of the surface. Surprisingly due to this combination of graphene's excellent in-plane charge transfer, with its pure flatness and hydrophobicity, it can even outperform the 3D-structured electrodes. Considering the recent research, graphene might be even a superior material, especially for interfacing with neuronal cells [6–8], which have much smaller dimensions. With axonal sizes in the sub-micron range, it is easier to form a good coupling with (slightly hydrophobic) graphene surface rather than engulf a 3D electrode (see Figure 1). Moreover, graphene's transparency is already an advancement compared to the conventional gold or platinum MEAs, and allows to observe cellular outgrowth directly at-the-electrode through the substrate.

Organic Bioelectronics for Life Science and Healthcare

Materials Research Forum LLC

Materials Research Foundations **56** (2019) 153-184

doi: https://doi.org/10.21741/9781644900376-5

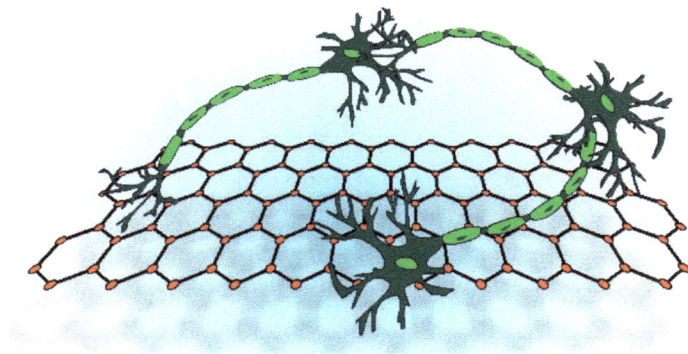

Figure 1. Schematics of a neuronal network grown on top of graphene surface. Not up to scale.

In order to fabricate the graphene-based transistor (GFETs) and graphene-based multielectrode arrays (GMEAs) it is important to use a single-layer graphene structure. There is nowadays a large variety of graphene production methods, starting from micromechanical exfoliation [1]. However, many methods provide either insufficiently small graphene flakes (micromechanical exfoliation), or non-uniform and uncontrollable layer distribution of graphene (liquid-phase exfoliation)[18], or difficulty to further transfer graphene (graphitization of SiC surface) [19]. Chemical vapor deposition is therefore selected and now commonly used for such purposes as it provides sufficiently large flake (grain) sizes up to cm-large, and it can be transferred to a rather arbitrary surface for further processing [20]. Quality-wise the CVD-grown graphene is nowadays almost on the same level with micromechanically exfoliated ones [21].

In this chapter, we will overview the organic properties of graphene and bioelectronics applications of the GFETs and GMEAs, review on their properties, advantages, drawbacks, and applications relevant for healthcare.

2. Bioelectronic devices based on graphene

2.1 Graphene transistors

Transistor is a semiconducting electrical component that is used to amplify or switch electrical signals via application of an electrical current or voltage [22]. Use of the two-dimensional carbon, graphene, as a transistor was first proposed and implemented by Geim and Novoselov in their Nobelprize winning work in 2004, showing that graphene is unique responsive to an applied external gate potential [1]. Further development and research had shown that indeed, theoretical predictions are not far from reality, and in some conditions the graphene-based transistors can exhibit extremely large mobilities (and consequently sensitivities), up to $1 \times 10^5 - 1 \times 10^6$ cm^{2}V^{-1}s^{-1}. However, these values are obtained for exfoliated graphene, suspended and measured in a close-to-zero temperature [23–25]. In the more ambient conditions, the mobility values are typically the range of 1000-50000 cm^{2}V^{-1}s^{-1}, depending on the quality of graphene [26]. Nevertheless, the values are rather large, especially if compared to that of silicon nanowires (~1000 cm^{2}V^{-1}s^{-1}) [27].

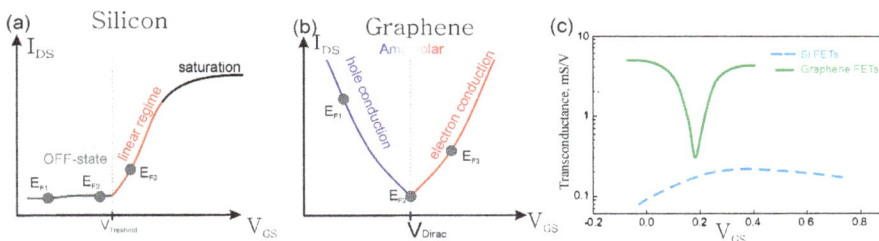

Figure 2. In (a) and (b) are schematically shown the transfer curves of Si-based unipolar FET (a) and ambipolar graphene-based FET (b). In the Si-devices, there is an energy gap therefore resulting I-V curve has an OFF state, and is unipolar, with a linear transfer regime and a saturation (due to a limited number of charge carriers and their mobility). Since graphene is a zero-gap semiconductor, a small gate potential shifts the charge carriers from hole (left, blue) into electron (right, red) conductivity, and therefore is an ambipolar transistor. (c) Shown comparison in transconductance values of typical GFETs (green) and Si-FETs (dashed blue).

As discussed above, graphene does not have a bandgap, therefore there is a finite OFF-state conductivity, that limits GFETs performance in logic gate operations. Nevertheless, for the application such as sensing, biosensing, high-frequency application, etc, graphene is probably one of the best candidates due to the steepness of its transfer curve [29]. For the biosensing applications, such as electrophysiology, the important parameter of the I-V sweep curve is the transconductance. Transconductance, g_m is the factor of change in the drain-source current (that is finitely recorded), upon changes in the gate potential ΔV_{GS} (that are caused by cells):

$$g_m = \frac{\Delta I_{DS}}{\Delta V_{GS}}$$

Interestingly, it was found that graphene transistors, biased through a liquid gate, exhibit a very impressive gating behavior [10,28,30,31]. See Figure 2 for comparison of the I-V curves of a typical Si-FET and a solution gated GFET. The change of reference electrode potential from zero to 500 mV is usually enough to move from p-type to n-type regime through the Dirac point (See Figure 3 for details). At the same time, the conductivity curve is very steep and the resulting transconductance, in some cases, can be as high as 11 mSV^{-1}, which is outperforming the Si-FETs by a factor of two [32].

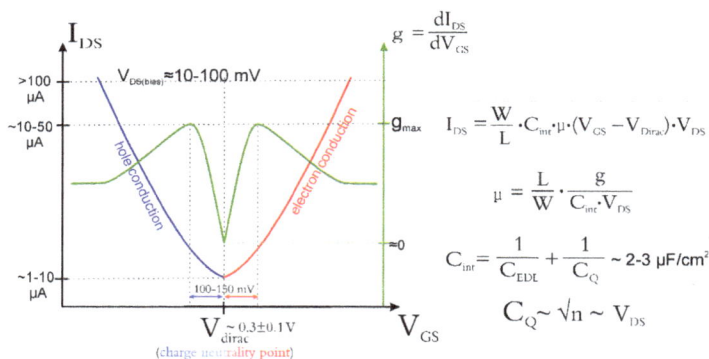

Figure 3. A representative transfer curve of a liquid-gated GFET. An I-V curve is shown in blue (hole conduction) and red (electron conduction) parabolic curve, while the first derivative of the plot, representing transconductance is shown in green. The values of applied potentials, flowing currents, and position of Dirac point and maximum transconductance points are given and are experimentally used/recorded. On the right are given the main formulas important for understanding of the transport properties of a liquid gated GFET.

A representative transfer curve of a liquid-gated GFET is shown in Figure 3, featuring a cone-like behavior of the current flowing through graphene channel upon applied electrical field to the gate dielectric or electrolyte. The current is typically at first dominated by holes, then it reaches minimum at the Dirac point, where the number of charge carriers is minimal, and then increases again, however the current is now dominated by electrons. The overall relation in the linear regimes of the plots can be expressed by the following well-known formula [26]:

$$I_{DS} = \frac{W}{L} \cdot C_{ox} \cdot \mu \cdot (V_{GS} - V_{Dirac}) \cdot V_{DS}$$

where I_{DS}, V_{DS}, and V_{GS} are the drain-source current, drain-source voltage and gate source voltage consecutively; V_{Dirac} is the Dirac potential, or charge neutrality point; μ is mobility, C_{ox} is the gate oxide's capacitance, W and L and the width and length of graphene's channel.

If all geometrical and measurement parameters are known, the two parameters, transconductance (g) and mobility (μ), can be computed:

$$g_m = \frac{\Delta I_{DS}}{\Delta V_{GS}},$$

$$\mu = \frac{L}{W} \cdot \frac{g}{C_{ox} \cdot V_{DS}}.$$

Assuming contact resistance negligible, and the V_{DS} potential to be dropping exactly on graphene channel, the transconductance of the devices, g, is simply calculated as first derivative of drain-source current over gate-source potential. Transconductance is the actual amplification factor, and, consequently the sensitivity factor and the figure of merit when the GFET is used for sensor applications. Mobility, on the other hand, is somewhat more complex property of a material, specially of such two-dimensional material. In a simple case, it is correlated to the interface oxide capacitance, or electrical double layer (EDL) capacitance in the case of liquid gating.

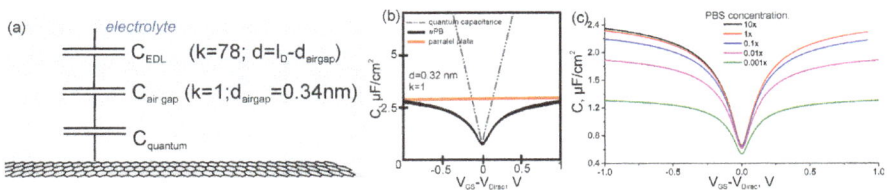

Figure 4. (a) The full capacitance models for a liquid gated GFET. (b) shows a model for the interfacial capacitance that considers these terms and extended Poisson-Boltzmann statistics is used. In (c) are given these capacitance dependencies on concentration of PBS electrolytes. (a) and (c) are reproduced in form with permission from [11].

Organic Bioelectronics for Life Science and Healthcare   Materials Research Forum LLC
Materials Research Foundations **56** (2019) 153-184   doi: https://doi.org/10.21741/9781644900376-5

Use of liquid-gated graphene field effect transistors for bioelectronics applications were first reported in 2010 by two groups at almost the same time [9,30]. As it turned out, the graphene is fully stable in an electrolyte environment, even upon application of sufficiently large electric fields (e.g. more than 1.4V in 1x PBS). When a gate potential is applied to the electrolyte, an electrical double layer (EDL) is then formed at the interface with graphene and plays a role of a dielectric/oxide (See Figure 4a). The EDL's structure, composition and dielectric properties are still under a careful examination, but as clear to the date, there are three main capacitances which are arranged in series (see Figure 4a) [11,28,30]:

1. Parallel plate double layer with a thickness equal to the Debye length, which is dependent on the electrolyte's molarity, M. Dielectric permittivity is considered to be equal $k = 78$, ε_0 is vacuum permittivity, and λ_D is the Debye length.

$$C_{EDL} = \frac{k_{EDL}\varepsilon_0}{\lambda_D}$$

$$\lambda_D = \frac{0.304}{\sqrt{M}} \ [nm]$$

2. Air gap capacitance that appears due to hydrophobicity of the graphene [30]. Dielectric permittivity is assumed to be that of air ($k_{air} = 1$), and of thickness, d, is estimated as graphene's interlayer distance, 0.3 nm.:

$$C_{airgap} = \frac{k_{air}\varepsilon_0}{d}$$

3. Quantum capacitance, which is an inherent property of graphene:

$$C_Q = \frac{2e^2}{\hbar v_F \sqrt{\pi}} \cdot \sqrt{|n_G| + |n^*|}$$

$$n_G = \left(\frac{eV_{GS}}{\hbar v_F \sqrt{\pi}}\right)^2$$

where e is the elementary charge, $\hbar$ is reduced Planck constant, $v_F = {}^c/_{300}$ is the Fermi velocity (c is the speed of light). n_G and n^* are the charge carrier concentrations, induced by gate potential and charged impurities, correspondingly. Since the charge carrier concentration, n_G, is quadratic proportional to the applied gate potential, V_{GS}, the quantum capacitance is linearly dependent on the applied potential ($C_Q \propto V_{GS}$), and the dependency is plotted in Figure 4b for one molarity value. Moreover, the number of charge carriers induced by charge impurities, n*, that is somewhat a factor of surface cleanness, and is known to varies from 1×10^{11} to 1×10^{12} cm^{-2}. When it is calculated numerically, it is clear that the n^* value gives difference only in the region in the close vicinity to the Dirac point (±50mV), while the influence can be neglected further from the point ($|V_{GS} - V_{Dirac}| > 100mV|$).

When all three terms are connected in series, the total interface capacitance can be calculated by the following equation:

$$\frac{1}{C_{total}} = \frac{1}{C_Q} + \frac{1}{C_{airgap}} + \frac{1}{C_{EDL}}$$

Knowing the actual interface capacitance is important to correct estimation of GFET's parameters, such as mobility and transconductance. The overall plots of such capacitances, which are dependent on electrolyte's ionic strength, are shown in Figure 4c. Since the interface capacitance of the EDL on top of graphene gated via electrolyte is proportional to V_{GS}, the mobility itself is then a value of the gate bias. Nevertheless, in order to simply estimate the mobility values, it is commonly arranged to use value of approximately 2 μF/cm^2 as the value of merit.

2.2   Graphene microelectrodes

Graphene microelectrode arrays (GMEA) can be produced with a significantly less complex fabrication process as compared to GFETs, with three to four fabrication steps allowing a wafer-scale process when using specific graphene transfer [34]. Typically, rigid SiO2/Si, quartz, or borofloat substrates are employed for the device fabrication. Borofloat wafers are used because of their transparency, which aids the long-term monitoring of the cell cultures, particularly in combination with the inherent transparency of graphene. For the fabrication of GMEAs, CVD grown graphene is used due to the need for large crystal sizes. Similarly to GFETs, polyimide or SU-8 photoresists are commonly utilized as passivation polymers. In agreement to previously developed MEAs and in order to utilize the same multichannel measurement schematics, the GMEA chips are typically 24 × 24 mm^2 in size and exhibit an array of 64 electrodes. They follow the same layout as typical metal MEAs with the active electrode area made up of graphene rather than a metal layer as shown in Figure 5a-b.

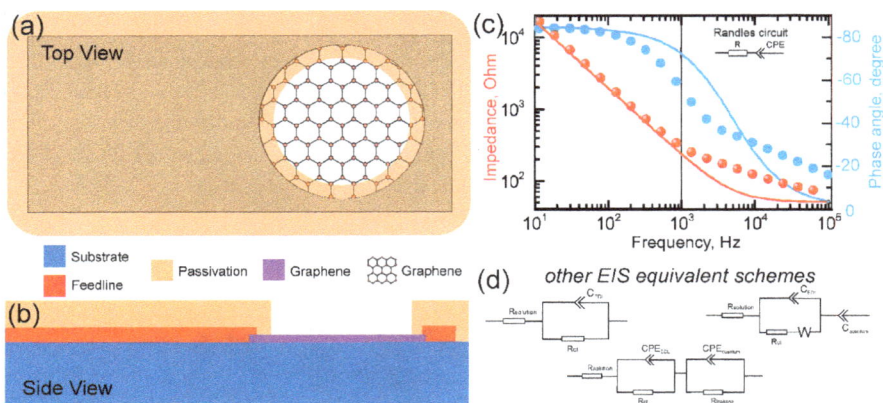

Figure 5. (a)-(b) Shows schematic layout of the fabricated GMEA chips in a top view (a) and side view (b). The geometry is rather classical, with metal feedline contacting graphene and passivation opening. In (c) are shown representative data for the Impedance and phase in response to different applied frequency. Reproduced with permissions from [72]. (d) features the three other common equivalent circuit that have been used to model the GMEA's behavior, showing complexity as well as diversity of the recorded signals.

The GMEAs are typically analyzed using electrical impedance spectroscopy (see Figure 5c-d). In comparison to gold electrodes (where 40 μF/cm^2 is expected [35]) of the same dimensions, the graphene-based electrodes exhibit slightly larger impedance, that is in the range of several MΩ at 1 kHz for 10-20 μm electrode openings. There are, however, some ways to still improve impedance response of graphene electrodes. One way is to induce doping into the graphene layer [13]. Another way is to increase the number of graphene monolayer [14,36] or even create macro porous graphene [37].

3.   Graphene devices for healthcare

3.1   GFET biosensing

Graphene transistors, as already introduced above are highly sensitive to the changes of the environment, specially to the change in the gate-source potential, and surface charge. The main drawback of graphene towards use as a biosensor can be mentioned its inertness of the surface. As reported previously, perfect graphene layer is highly insensitive to potential changes [38]. Therefore, one must engineer some kind of receptors on top of the graphene surface. The most common way of doing so is use of π-π interaction and use of molecules with pyrene-groups, which can be attached to graphene via this π-π interaction. Such interaction would typically introduce a shift in the surface potential (see Figure 6a), and therefore a shift in position of Dirac point, as well as any further molecular interaction would induce such shifts in the V_{Dirac} position (see Figure 6b), which in their turn can be simply detected via monitoring of the I_{DS} shift over time (see Figure 6c). The schematics of GFET biosensor is depicted in Figure 6, with the typical V-shaped transfer curve. For a more detailed review on graphene transistors for biosensing please refer to the following reviews [39,40].

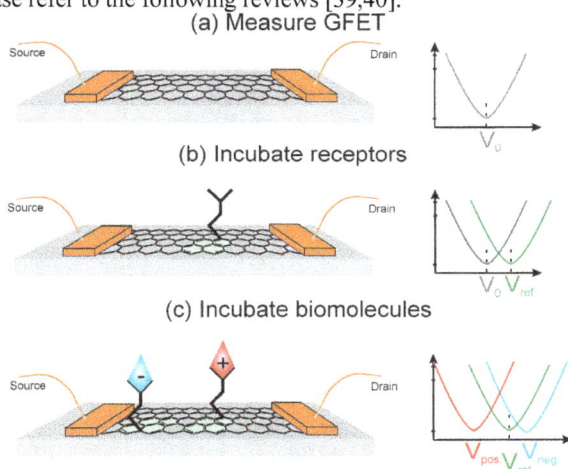

Figure 6. Schematics of a typical liquid-gated GFET biosensor. (a) Schematics and transfer curve of an untreated graphene transistor. (b) Schematics and the shift in transfer curve of a GFET modified with molecular receptors via π-π interaction. (c) Schematics and shift of GFET's performance under following exposure to biomolecules. Typically, positively charged molecules (red) would result in shifting the charge neutrality point to the left, and vice versa for the negatively charged biomolecules.

3.2   GFET electrophysiology

Sensing the cellular activity is similar to the described above, in some way even simpler, in some way more complex. If a complete capacitive coupling between the cell and transistor is assumed, the changes in the graphene's drain-source current (of potential drop over graphene electrode) can be easily transformed into changes of the cellular membrane potential and consequently into the intracellular potential.

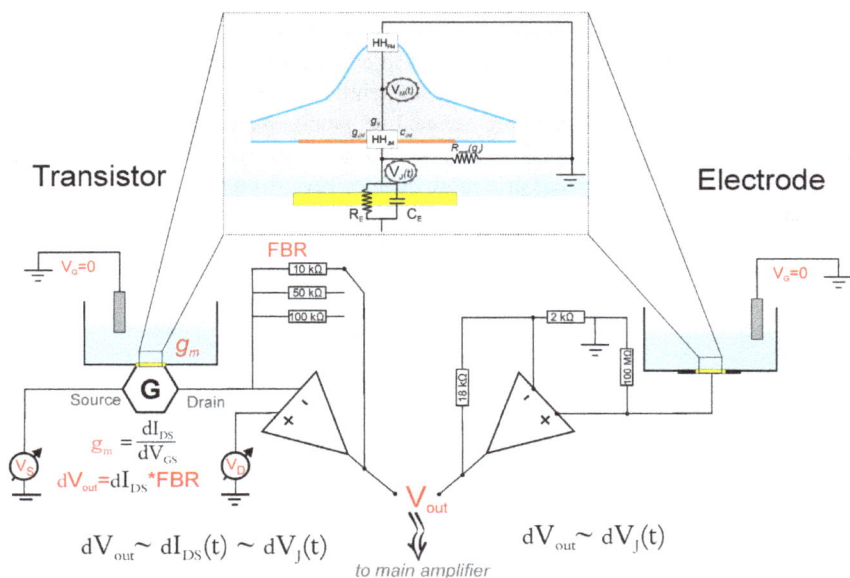

Figure 7. Read-out schematics of a GFET array (left) and GMEA array (right). Graphene is depicted in yellow. The pre-amplification schematics given are thus for BioMAS-based amplifier system [41,42]. Inset on top gives a simplified circuit of an electrogenic cell on top of graphene surface, Vj and Vm are encircled since the former is the one we are detecting and the latter is the one that happens inside the cell.

Typically, the extracellular measurement set-up consists of several layers of amplification and more importantly, a several constrictions, such as requirement of reference electrode to be grounded, $V_g=0V$, that somewhat complicates the schematics of applied potentials; This is also important for combination of extracellular recordings with patch-clamp experiment [43].

A typical write-read-out schematics of the GFETs and GMEAs is shown in Figure 7. An electrogenic cell (see Figure 7 inset), upon firing an action potential, would typically result in the change of the membrane potential over time, $V_M(t)$. That membrane potential will result in its turn in the change of junction potential, $V_J(t)$, that is further detected by a transistor or an electrode. R_{seal} is the junction resistance, or simply the factor of quality of the sealing between the cell and electrode. Typical R_J values are in the range of 10^5-10^6 Ω [44]. In the case of transistor, the finite change ΔV_J would transform into a change in drain-source current, ΔI_{DS}, that is recorded.

Based on this principle, different groups have used graphene transistors and their arrays to record cellular action potentials, their propagations and moreover, performed relative biochemical studies [9,11,12,31,32,45–47]. In particular, electrical activities of the cell lines as HEK-293 (kidney), HL-1 (cardiac) have been successfully measured with the GFETs.

While at first, the high quality exfoliated graphene was used to produce the devices, and the cell culture were not grown directly on top of the chips, it was difficult to study the graphene-cell interaction in full [9,47]. Therefore, it became important to fabricate transistor and electrode arrays, such as you can find in Figure 8 a-c [32,48,49]. Arraying the transistors helps in understanding propagation of the cellular activity and extracellular potentials through the culture.

When the cardiomyocyte cells (HL-1 cell line), are cultured on the chips' surface (see Figure 8d), they form a continuous and conformably beating layer. A typical timetrace recording of the HL-1 APs(Action Potentials) from one of the GFETs is shown in Figure 8d. The cells are "beating", producing repetitive APs that propagate through the membranes of the cellular layer, with rate of 23 ± 3 beats per minute (bpm) and amplitude of 1.2 ± 0.2 mV in this case. Typically, the beating rate of HL-1 cell culture may vary from 15 to 150 bpm and depends on many environmental and chemical conditions. It is also possible and important to study shape of the APs in order to understand what mechanisms are lying behind each of the spikes. An example if shown in Figure 8d, where more than a hundred consecutive spikes are averaged. The shape of the action potential, as in agreement with previous works represent a very good sealing between a cell and transistor [50,51].

Neuronal recordings with GFETs have also been reported to be possible: *in vitro* [11,49] as well as *in vivo* [52,53]. When cultured *in vitro*, cortical neurons must be cultures for at last DIV14 before they start producing spontaneous and large amplitude action potentials that can propagate through the network. Compared to HL-1, the neuronal extracellular action potentials are one to two orders of magnitude smaller. One representative neuronal action potential recording is presented in Figure 8e, with a well-defined bursting pattern. Averaged and reconstructed, the APs have comparably high amplitude of 600 µV, but a rather unusual and low SNR of the recordings due to large effective gate noise of up to 100-200 µV. Such noise level consequently impedes with possibility of recording ultra low amplitude neuronal signals (typically below 500 µV).

Furthermore, to show applicability of the devices as well as to prove the healthcare applications of the devices, HL-1 cell culture was chosen to perform the following tests. At first, when HL-1 cell mature enough to form a confluent cell layer, it makes possible to track down the dynamics of AP generation. These APs are repetitive and propagate through the whole cellular layer, and typically have a starting point, an origin – the pacemaker [54,55]. Since measurements of the whole chip array is done simultaneously via multichannel measurement system, it is possible to post-process the recorded signals in order to locate the pacemaker cell position as well as to calculate the signal propagation velocity. You can see the result of such data processing at the Figure 8f, where the time delay between the APs is used to evaluate the approximate location of the pacemaker cell. Further, in order to prove the biological origin of the signals, biochemical treatments could be performed on the cultures. An example of such test is shown in Figure 8g, when during the timetrace recording, norepinephrine (NorA), a well-known drug for heart rate stimulation, is added to the medium. Addition of the NorA to the medium, almost doubles the average concentration of the NorA in a "normal" culture, therefore stimulates the cells to produce AP more often, and almost doubles the beating frequency (Figure 8g insets). Furthermore, a concentrated sodium dodecyl sulfate (SDS, a surfactant) solution was later added to the same culture in order to perforate and dissolve the cellular layer, remove the cells from the chip's surface. This results in frequent but evanescent spikes that disappear completely in the span of minutes. Third test, as shown in Figure 8h, providing data that is unique for graphene transistors. As you remember from Figure 3, the GFET's I-V curve is V-shaped and the corresponding transconductance (amplification factor) curve typically have two maximums (one in hole regime and one in electron regime) and goes to its minimum at the Dirac point. Typically, the hole and electron regimes have a reversed sign in transconductance. Therefore, when an AP is recorded on top of such transistor, one can simply sweep the operation point (changing the set V_{GS}), and the resulting shape and amplitude of the action recorded AP is shown in Figure 8h. In that case, you can see the AP has a minimal amplitude at Dirac point (+0.1V at this case) and the APs have a different polarity if measured in a hole-regime or electron-regime. Therefore, in order to precisely understand the cellular AP origin it is important to keep in mind the operation point of the transistor and any possible factors that can reverse the signal.

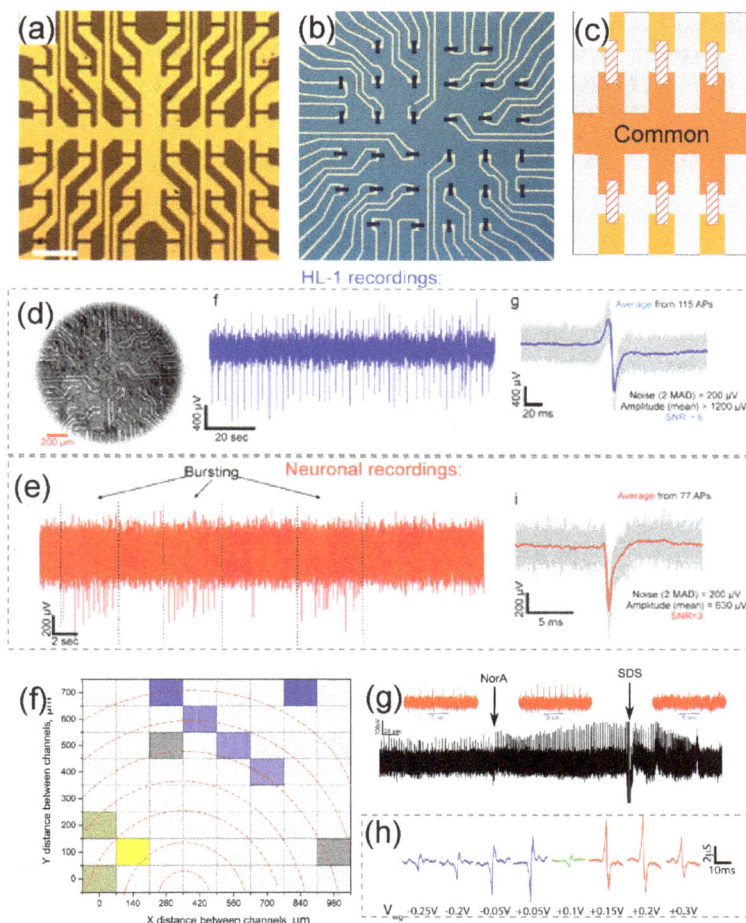

Figure 8. (a)-(c) Different geometrical arrangements of the graphene transistor arrays. In (a) and (c) are shown common-source layouts and 64 electrodes per chip, while (b) features somewhat more classical layout. (d) An optical micrograph of an HL-1 cell culture on top of a GFET chip; A typical time trace recording of HL-1 activity recorded with a graphene transistor; and the averaged HL-1 spikes (blue) from 115 individual consecutive spikes from the chip. (e) The neuronal recording time trace with the inherent neuronal feature of bursting, when the neurons exhibit alternating periods of high

Organic Bioelectronics for Life Science and Healthcare   Materials Research Forum LLC
Materials Research Foundations **56** (2019) 153-184   doi: https://doi.org/10.21741/9781644900376-5

frequency (bursts) and low frequency, intermittent, spiking; The averaged AP (red) from 77 individual APs (gray) from the neuronal time series. (f) Shows the propagation map of HL-1 signal with dotted circles featuring location of the pacemaker cell and radial propagation of the signal. (g) Recording, over 10 minutes long, from HL-1 cells where initially APs frequency has doubled up upon addition of NorA into the culture medium (from 0.3H to 0.5 Hz), further addition of SDS starts to perforate and dissolve the cellular layer, decreasing the amplitude and eventually ceasing the firing (third inset). (h) Unique for graphene dependency of recorded AP shape on the set Vgs potential. Blue show APs that were recorded at the hole conduction of the GFET, red – electron conduction, green - near the Dirac point of the GFET.

Additionally, based on many *in vitro* neuronal and HL-1 recordings and their analysis, there has been proposed a new way to passivate the GFETs. Up to date, the most common way of passivating the transistors for extracellular recordings consists of covering the whole chip's surface (Figure 9a) and having just tiny openings for graphene. Kireev et al. proposed the new kind of passivation, called "feedline follower" that covers only the area over the metallic feedlines [11]. Whilst the former type of the passivation has been in use and been proven to be important to create good and highly resistive sealing [56], it might be useful only to record from large cells, such as HL-1s [54,57]. Neuronal networks, on the other hand, do not form confluency, but try to grow out in a dense network of rather small and sparse neurites. The cell body sizes of neurons are much smaller than the openings themselves. The "feedline follower" passivation seems to be optimal for neuronal interfacing, (see Figure 9 for details), and there are several main points to that. First of all, graphene is a purely two-dimensional material with its charge carriers open to the surrounding environment, therefore it is important to bring the cell as close to the graphene's surface as possible. Second of all, as known and discussed, graphene comes a little bit hydrophobic [30]. As many works have already shown, this hydrophobicity of graphene can be used for supreme interaction with lipid bilayers (that are also hydrophobic) [58,59]. Cell membranes, in a simplified schematics is nothing else but lipid bilayer, and such graphene-lipid bilayer interaction might create a better change of sealing graphene and cell. Additionally, there is also a trade-off between low-noise recording interface and good cell-sensor coupling. Remembering that the GFET's multichannel measurement already feature large noise, there is no room for increasing the noise even more by creating highly resistive sealing.

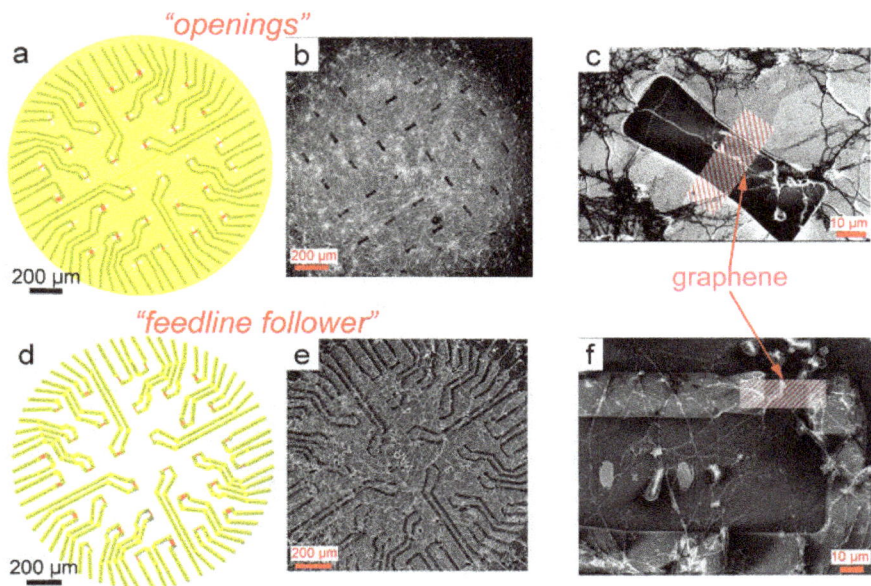

Figure 9. (a) and (d) represent the passivation openings and feedline follower technique respectively. The passivation is shown in yellow, openings/substrate in white and graphene in red. (b) and (e) represent SEM images of both kinds of chips with a neuronal network cultured on top, and zoom-ins into one GFET area are given in (c) and (f) for passivation openings and feedline follower technique respectively. Details on fixation and imaging are given in Methods. Due to the sputtered layer of Pt (required to visualize non-conducting neurons and dielectrics), graphene is not visible. In (b) and (c), due to a large layer of passivation, the openings look dark in the SEM image, while opposite situation in (e) and (f), where passivation covers the metal feedlines and are represented darker. The images are reproduced in parts with permission from [11].

Organic Bioelectronics for Life Science and Healthcare   Materials Research Forum LLC

Materials Research Foundations **56** (2019) 153-184   doi: https://doi.org/10.21741/9781644900376-5

3.3   GMEA electrophysiology

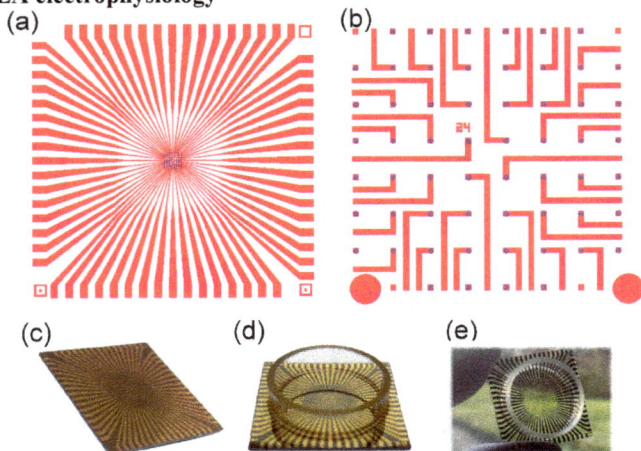

Figure 10. GMEA design overview. a) The design of a GMEA chip, 24 mm × 24 mm in size. b) Enlargement of the middle electrode array, of 1.4 mm × 1.4 mm in size. c) An optical image of the fabricated GMEA chip (on SiO₂/Si substrate). d) An optical image of the SiO₂/Si-based GMEA chip after encapsulation. e) An optical image of the borofloat-based GMEA chip after encapsulation. The images are reproduced in parts with permission from [16].

Similarly to the GFETs, the initial tests were performed with use of HL-1 cell culture. A live-dead image of a not-yet confluent layer of HL-1 cells on top of the GMEA chip can be seen in Figure 11a. Corresponding timetrace recording from some of the chips are given in Figure 11b, with a visible time-delay between the APs from different chips, that helps to estimate the propagation rate. The recorded beating frequency in this case is in the range of 1 ± 0.5 Hz. A more complex signal propagation map, recorded with one of the GMEAs is shown in Figure 11c, where the AP seem to propagate either in U-shape or there are two pacemakers cells that compete with each other. The recorded action potential amplitudes and their shapes vary from chip to chip (culture effect) and from electrode to electrode (sealing effect). A large number of HL-1 action potentials from various time series were analyzed in order to compare them with the experimental and simulated data [16]. The resulting pie-chart depicting the different AP shapes observed during this study is shown in Figure 11d.

As already discussed, many factors contribute to the large variety of shapes of the APs recorded with extracellular electrodes. Regardless of the impedance of the electrode itself, there are other physical and physiological parameters that affect the way the signal will be recorded and seen. First of all, the more mature the culture, the larger and more stable the APs [60]. The second important parameter is the sealing between the cellular layer and the electrode [51]. For the measurements conducted as part of this study [16], the action potential shapes A and B were observed with the highest occurrence (see Figure 11d). According to previously simulated data [51], spikes of type A result from a large sealing resistance, large sodium peak, and large amplitude. The difference in pre- and post-spike overshoot, their amplitude, and duration can be modeled by variations in the sealing resistance and current flows of Na+, Ca2+, and K+ ions [50,51,61]. Signal to noise ratio of such recordings, is typically reported to be above 30-50, often reaching values up to 100 and more [11,62].

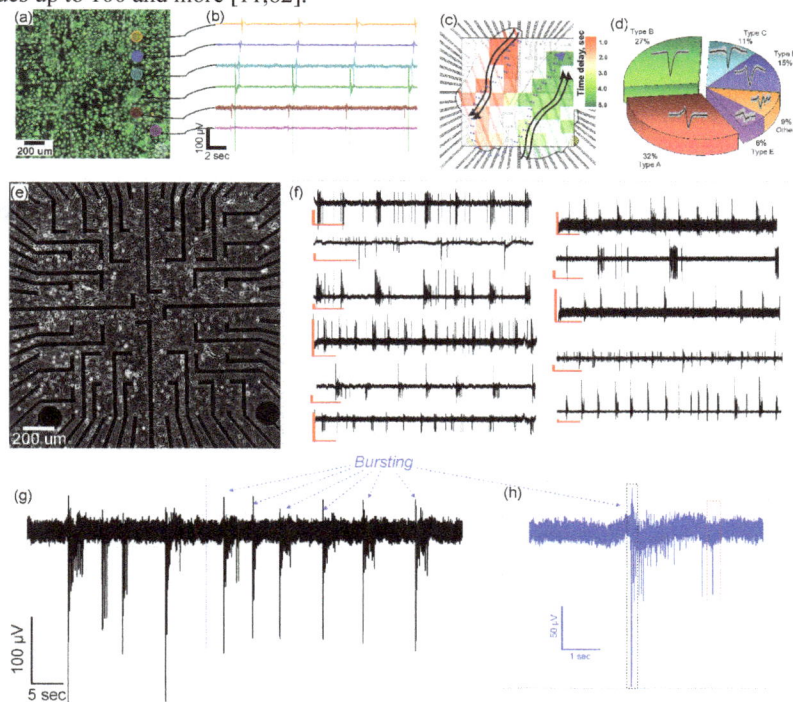

Figure 11. a) A picture of a Calcein/Ethidium Homodimer stained HL-1 culture on a GMEA chip. Live cells fluoresce green, dead cells fluoresce red. b) Timetraces recording from different channels from one GMEA chip, showing the repetitiveness of the spikes and clear propagation of the signal. The color code corresponds to the electrodes circled

in (a). (c) HL-1 signal propagation showing that either the signal is propagating in a U-shape (blue path) or there are two pacemakers, depicted as two separate black solid paths. (d) The result of analysis of over 13.000 recorded APs, that shows distribution of different signal shapes. The large occurrence of spikes of type A shows that good cell-electrode sealing can be achieved with the developed GMEA devices. (e) Microscope image of the neuronal culture on a GMEA chip. (f) A variety of neuronal activity patterns recorded from different GMEA channels. In (g) is given one single representative timetrace with zoom into one burst (h), showing that each time there is a small number of very large spikes, followed by a series of spikes with gradually decreased amplitude. The images are reproduced in parts with permission from [16].

Further, embryonic cortical neurons are cultured on top of the GMEAs until mature (see Figure 11e for an example). Typically, at DIV 14-21, such neuronal cultures are mature enough to produce spontaneous electrical activity that is propagating through the network. The variety of spontaneous spiking-bursting activities, recorded with GMEA chips are shown in Figure 11f. These patterns, resemble some kind of neuronal clustering, thinking and signal propagation mechanisms. These are very similar to the patters typically recorded with planar MEAs [63–65]. Figure 11g show one timetrace with distinguishable bursting and non-bursting spikes. As introduced by Droge et al. in 1986 [66], a criteria for defining bursting is following: the inter-burst intervals should exceed the inter-spike intervals. As it is clear from Figure 11g and zoom-in images of the bursting spikes (see Figure 11h), in average the bursts happen every 5 to 15 seconds (with dependency on a culture), each burst containing a series of rather high amplitude spikes (often as high as 800 µV) followed by a series of consequently evanescent spikes. The results are in accordance with the previously published data for standard planar inorganic MEAs [63–65]. A comprehensive signal propagation through neuronal cultures on top of GMEA chips has not been done yet due to limitations in signal processing and requirement of very high density electrodes in order to correctly evaluate this complex signal propagation mechanism.

Lastly, the proof-of-biological- origin of the signals, the neuronal cultures on top of GMEAs are treated with different chemicals, that can induce/eliminate the firing activity or even kill the whole culture. Therefore, sodium dodecyl sulfate (SDS), potassium chloride (KCl), and tetrodotoxin (TTX) were used in the study [16]. The effect of the above-mentioned chemicals on neuronal culture was observed in real-time recordings from the GMEA chips (see Figure 12). SDS, a surfactant, permeabilizes the cell membrane and dissolves cellular material from the surface. Less than 20 seconds is enough to produce an irreversible detachment of the cellular layer (see Figure 12a). Potassium chloride, a classical ion channel depolarizing agent, was added to the medium of the other chip to reach a final concentration of around 10-20 µM. This elevation of extracellular potassium concentration depolarizes the cell above threshold and prevents re-polarization (see Figure 12b). This results in the cells firing and then remaining in the depolarization block. TTX, is a neurotoxin which blocks voltage-gated sodium channels (Na_V). Addition of it to the cell culture slowly lowers APs amplitude, as well as their frequency as the toxin binds to more and more channels preventing their function, until no more signals are seen (Figure 12c). The effect is slightly different in time for different channels, which shows complexity of the whole neuronal networks [16].

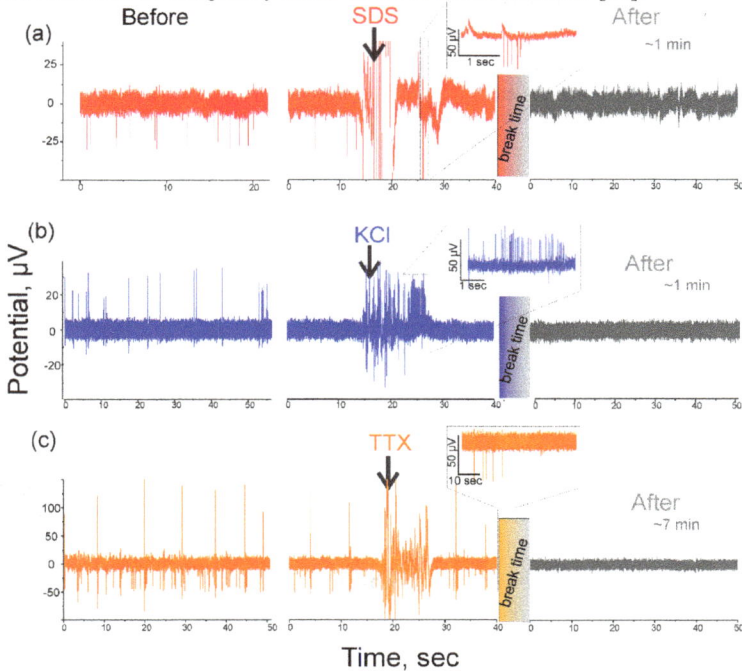

Figure 12. Chemical treatments of the neuronal networks with (a) SDS, (b) KCl, and (c) TTX. The left panels are the recordings before any treatment. The large signal fluctuations in the recordings in the middle panels are from addition of the chemicals and is due to mechanical perturbations and a slow mixing (pipetting in/out) of the liquid to distribute the substance evenly to the cells. The right panels are recordings 1 min (for SDS and KCl), and 7 min (for TTX) after the treatments. In the middle are the 40 s of the timetraces when the chemical was added into the medium. Small insets show the immediate effects when the chemicals alter the normal neuronal bursting–spiking activity. The images are reproduced in parts with permission from [16].

4.   Towards flexible bioelectronics

Combination of MEAs and flexible substrates has gained lots of attention recently, specifically when combined with carbon-based materials, like carbon nanotubes (CNT), which exhibit extreme performance and flexibility [67,68]. Graphene, regardless other promising features, such as biocompatibility and excellent electrical properties, possesses the feature of inherent flexibility and softness at the same time. In this section of the chapter, we would like to focus on fabrication and use of flexible graphene-based microelectrode arrays on biocompatible polymeric substrates. Surprisingly, such devices, even after a severe mechanical deformations, are stable, usable, and could record from *in vitro* and *ex vivo* extracellular cultures multiple times.

Fabrication of flexible graphene microelectrode arrays is typically standard photolithography based technology, with just one or two additional steps. The steps are: i) fabrication of a sacrificial layer; ii) fabrication of the flexible base; iii) release of the flexible device by means of releasing/etching the sacrificial layer. There are currently a variety of sacrificial layer structures as well as methodologies to release the wafers. Typically, polyimide and parylene-C are the most common flexible polymers that are used as substrates, both featuring extremely good film stability and biocompatibility.

Once the devices are fabricated, the sacrificial layer is etched, releasing the flexible chips (see Figure 13a), in order to perform *in vitro* test, connected to a carrier (see Figure 13c). This is an important step towards stable characterization and cell culture measurements. Due to the flexibility of the chips, a standard *in vitro* measurement process would be difficult or even impossible to perform. Therefore, soldering to a carrier (Figure 13c) helps to improve the in vitro compatibility of the devices and the long-term stability. Furthermore, Kireev et al. performed an experiment when one of the GMEA chips was tested for its mechanical stability: the flexible chip was crumpled severely (see Figures 13b) before soldering [62]. Nonetheless, the chip was still further used for recording from ex vivo heart tissue signals and *in vitro* HL-1 cells as visible in Figure 13d, eleven traces recorded from different channels (electrodes) on the same chip are shown. The AP's amplitudes and shapes are different from channel to channel, but stays persistent in one channel. Lastly, signal-to-noise ratios of these recordings were evaluated to be in the range of 20 ± 10.

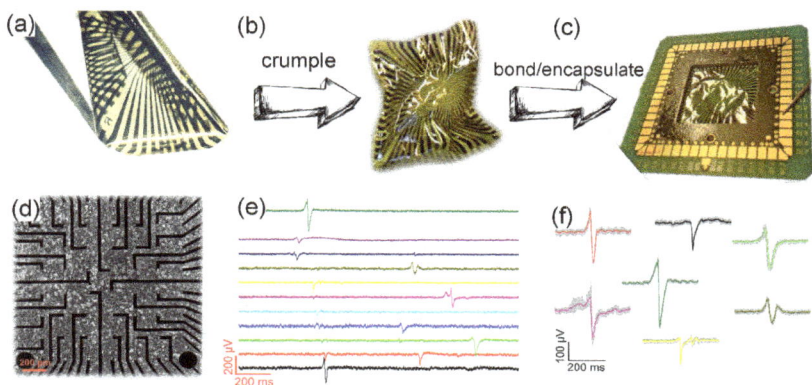

Figure 13. (a) one flexible chip, which was crumpled (b), then bonded and encapsulated (c). (d) a differential interference contrast (DIC) picture of HL-1 cells grown on top of a GMEA surface. (e) time trace recordings of HL-1 cells from eleven channels on one GMEA chip showing a time delay in recording of different electrodes that reflects spatial propagation. (f) the variety of different HL-1 action potential shapes recorded with the GMEA due to differences in cell–chip coupling. The images are reproduced in parts with permission from [62].

The first *in vivo* based GMEAs were built in 2011, with making use of just one rather large (0.31-0.59 mm^2) graphene piece on PDMS [69,70]. Further, such *in vivo* based devices were fabricated in a rather CMOS-compatible technology, with a simple use of polyimide (see Figure 14a) or parylene-C (see Figure 14b) as transparent and biocompatible substrate [13,14]. Shaped in a specific way, the devices can be easily fabricated for either cortical or on-dura µ-ECoG measurements. Ultrathin polymer base allows the probes to get conformal and tight contact with the tissue if placed on top, and to avoid fatal glia response in the case of through-tissue probes (see Figure 14b).

It is obvious that GMEA devices are extremely robust, low noise and functional devices. Additional advantages such devices possess are their flexibility and transparency [69,70]. As shortly introduced before, nowadays the direct combination of electrical and optical brain mapping is required. Graphene multielectrode arrays in this regard are the panacea. Simple flexible GMEAs, built for in vitro studies have shown robustness towards mechanical folding, while possessing ability to record cellular signals [62]. More advanced, fully flexible and implantable graphene-based *in vivo* sensors (see Figure 14) have been built and tested [13,14], as well as their implementation towards more complex neuroelecrical measurements have been recently shown [36,71].

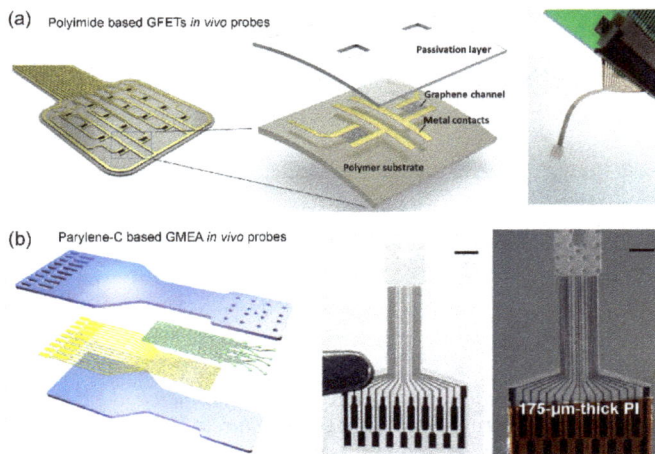

Figure 14. Two exemplary in vivo designs. (a) the in vivo GFET probe based on polyimide substrates. (b) the GMEAs in vivo probe based on Parylene-C. The images are reproduced in parts with permission from [14,53,71].

5.   Conclusions and outlook

Graphene bioelectronics is a rapidly growing field, offering unique opportunities and development of entirely new devices for interfacing with bio- and specially neuronal tissues. While *in vitro* studies gave us a general understanding of the ionic channels and cellular firing on a deep, even sub-cellular level, a further study of whole organs (e.g. brain or eye) activity is required in order to understand their functions and dysfunctions. The *in vivo* studies can lead to higher level of understanding of the brain's work, as well as to find treatments to such neurological diseases as Parkinson's.

Graphene-based electrode arrays (GMEAs) and more complicated graphene field effect transistors (GFETs) comprise a new type of bioelectronic device application. Biocompatibility, stability, excellent and unique electronic properties, scalability, and pure two-dimensional structure make graphene the perfect material for bioelectronic applications. The devices, used for extensive *in vitro* studies of a cardiac-like cell line and cortical neuronal networks, show excellent ability to extracellularly detect action potentials with signal to noise ratios in the range up to 100.

Research seem to be complete on the basic properties of graphene and a proof of concept devices have been realized. The current state of research at the moment is at the phase of optimization of the devices, in the search for a comprehensive design that shall use the advantages of graphene and neutralize its drawbacks. Noise in the graphene-based transistors has been extensively studied, yet solutions to decrease the noise in real application multichannel devices must be found. In terms of graphene microelectrode arrays, their applications to bioelectronics appear more and more often, yet the impedance of these devices vary greatly from a lab to a lab, possibly due to source of material and fabrication details. Therefore, further material standardization is required.

References

[1]   Novoselov K. S. S., Geim A. K. K., Morozov S. V. V, Jiang D., Zhang Y., Dubonos S. V. V, Grigorieva I. V. V and Firsov A. A. A. Electric field effect in atomically thin carbon films., Science 306 (2004) 666–9. https://doi.org/10.1126/science.1102896

[2]   Kostarelos K. and Novoselov K. S. Graphene devices for life, Nat. Nanotechnol. 9 (2014) 744–5. https://doi.org/10.1038/nnano.2014.224

[3]   Servant a, Bianco a, Prato M. and Kostarelos K. Graphene for multi-functional synthetic biology: the last "zeitgeist" in nanomedicine., Bioorg. Med. Chem. Lett. 24 (2014) 1638–49. https://doi.org/10.1016/j.bmcl.2014.01.051

[4]   Li N., Cheng Y., Song Q., Jiang Z., Tang M. and Cheng G. Graphene meets biology, Chinese Sci. Bull. 59 (2014) 1341–54. https://doi.org/10.1007/s11434-014-0158-0

[5]   Duan X., Fu T.-M., Liu J. and Lieber C. M. Nanoelectronics-biology frontier: From nanoscopic probes for action potential recording in live cells to three-dimensional cyborg tissues, Nano Today 8 (2013) 351–73. https://doi.org/10.1016/j.nantod.2013.05.001

[6]   Veliev F., Briançon-Marjollet A., Bouchiat V. and Delacour C. Impact of crystalline quality on neuronal affinity of pristine graphene, Biomaterials 86 (2016) 33–41. https://doi.org/10.1016/j.biomaterials.2016.01.042

[7]   Fabbro A., Scaini D., León V., Vázquez E., Cellot G., Privitera G., Lombardi L., Torrisi F., Tomarchio F., Bonaccorso F., Bosi S., Ferrari A. C., Ballerini L. and Prato M. Graphene-Based Interfaces Do Not Alter Target Nerve Cells, ACS Nano 10 (2016) 615–23. https://doi.org/10.1021/acsnano.5b05647

[8]   Bendali A., Hess L. H., Seifert M., Forster V., Stephan A., Garrido J. a and Picaud S. Purified Neurons can Survive on Peptide-Free Graphene Layers, Adv. Healthc. Mater. 2 (2013) 929–33. https://doi.org/10.1002/adhm.201200347

[9]   Cohen-Karni T., Qing Q., Li Q., Fang Y. and Lieber C. M. Graphene and Nanowire Transistors for Cellular Interfaces and Electrical Recording, Nano Lett. 10 (2010) 1098–102. https://doi.org/10.1021/nl1002608

[10]   Hess L. H., Hauf M. V., Seifert M., Speck F., Seyller T., Stutzmann M., Sharp I. D. and Garrido J. A. High-transconductance graphene solution-gated field effect transistors, Appl. Phys. Lett. 99 (2011) 33503. https://doi.org/10.1063/1.3614445

[11]   Kireev D., Brambach M., Seyock S., Maybeck V., Fu W., Wolfrum B. and Offenhäusser A. Graphene transistors for interfacing with cells: towards a deeper understanding of liquid gating and sensitivity, Sci. Rep. 7 (2017) 6658. https://doi.org/10.1038/s41598-017-06906-5

[12]   Hess L. H., Jansen M., Maybeck V., Hauf M. V, Seifert M., Stutzmann M., Sharp I. D., Offenhäusser A. and Garrido J. a Graphene transistor arrays for recording action potentials from electrogenic cells., Adv. Mater. 23 (2011) 5045–9, 4968. https://doi.org/10.1002/adma.201102990

[13]   Kuzum D., Takano H., Shim E., Reed J. C., Juul H., Richardson A. G., de Vries J., Bink H., Dichter M. A., Lucas T. H., Coulter D. A., Cubukcu E. and Litt B. Transparent and flexible low noise graphene electrodes for simultaneous electrophysiology and neuroimaging, Nat. Commun. 5 (2014) 5259. https://doi.org/10.1038/ncomms6259

[14]   Park D.-W., Schendel A. A., Mikael S., Brodnick S. K., Richner T. J., Ness J. P., Hayat M. R., Atry F., Frye S. T., Pashaie R., Thongpang S., Ma Z. and Williams J. C. Graphene-based carbon-layered electrode array technology for neural imaging and optogenetic applications, Nat. Commun. 5 (2014) 5258. https://doi.org/10.1038/ncomms6258

[15]   Du X., Wu L., Cheng J., Huang S., Cai Q., Jin Q. and Zhao J. Graphene microelectrode arrays for neural activity detection, J. Biol. Phys. 41 (2015) 339–47. https://doi.org/10.1007/s10867-015-9382-3

[16]   Kireev D., Seyock S., Lewen J., Maybeck V., Wolfrum B. and Offenhäusser A. Graphene Multielectrode Arrays as a Versatile Tool for Extracellular Measurements, Adv. Healthc. Mater. 6 (2017) 1601433. https://doi.org/10.1002/adhm.201601433

[17]   Castro Neto A. H., Guinea F., Peres N. M. R., Novoselov K. S. and Geim A. K. The electronic properties of graphene, Rev. Mod. Phys. 81 (2009) 109–62. https://doi.org/10.1103/RevModPhys.81.109

[18]   Sham A. Y. W. and Notley S. M. A review of fundamental properties and applications of polymer–graphene hybrid materials, Soft Matter 9 (2013) 6645. https://doi.org/10.1039/c3sm00092c

[19]   Norimatsu W. and Kusunoki M. Epitaxial graphene on SiC{0001}: advances and perspectives, Phys. Chem. Chem. Phys. 16 (2014) 3501. https://doi.org/10.1039/c3cp54523g

[20]   Mattevi C., Kim H. and Chhowalla M. A review of chemical vapour deposition of graphene on copper, J. Mater. Chem. 21 (2011) 3324–34. https://doi.org/10.1039/C0JM02126A

[21]   Wu T., Zhang X., Yuan Q., Xue J., Lu G., Liu Z., Wang H., Wang H., Ding F., Yu Q., Xie X. and Jiang M. Fast growth of inch-sized single-crystalline graphene from a controlled single nucleus on Cu–Ni alloys, Nat. Mater. 15 (2015) 43–7. https://doi.org/10.1038/nmat4477

[22]   Pierce J. R. The naming of the transistor, Proc. IEEE 86 (1998) 37–45. https://doi.org/10.1109/5.658756

[23]   Bolotin K. I., Sikes K. J., Jiang Z., Klima M., Fudenberg G., Hone J., Kim P. and Stormer H. L. Ultrahigh electron mobility in suspended graphene, Solid State Commun. 146 (2008) 351–5. https://doi.org/10.1016/j.ssc.2008.02.024

[24]   Banszerus L., Schmitz M., Engels S., Dauber J., Oellers M., Haupt F., Watanabe K., Taniguchi T., Beschoten B. and Stampfer C. Ultrahigh-mobility graphene devices from chemical vapor deposition on reusable copper, Sci. Adv. 1 (2015) e1500222. https://doi.org/10.1126/sciadv.1500222

[25]   Morozov S. V., Novoselov K. S., Katsnelson M. I., Schedin F., Elias D. C., Jaszczak J. A. and Geim A. K. Giant Intrinsic Carrier Mobilities in Graphene and Its Bilayer, Phys. Rev. Lett. 100 (2008) 16602. https://doi.org/10.1103/PhysRevLett.100.016602

[26]   Schwierz F. Graphene transistors., Nat. Nanotechnol. 5 (2010) 487–96. https://doi.org/10.1038/nnano.2010.89

[27]   Ramayya E. B., Vasileska D., Goodnick S. M. and Knezevic I. Electron Mobility in Silicon Nanowires, IEEE Trans. Nanotechnol. 6 (2007) 113–7. https://doi.org/10.1109/TNANO.2006.888521

[28]   Hess L. H., Seifert M. and Garrido J. a. Graphene Transistors for Bioelectronics, Proc. IEEE 101 (2013) 1780–92. https://doi.org/10.1109/JPROC.2013.2261031

[29]   Kim K., Choi J.-Y., Kim T., Cho S.-H. and Chung H.-J. A role for graphene in silicon-based semiconductor devices., Nature 479 (2011) 338–44. https://doi.org/10.1038/nature10680

[30]   Dankerl M., Hauf M. V., Lippert A., Hess L. H., Birner S., Sharp I. D., Mahmood A., Mallet P., Veuillen J.-Y. Y., Stutzmann M. and Garrido J. a. Graphene Solution-

Gated Field-Effect Transistor Array for Sensing Applications, Adv. Funct. Mater. 20 (2010) 3117–24. https://doi.org/10.1002/adfm.201000724

[31]   Cheng Z., Li Q., Li Z., Zhou Q. and Fang Y. Suspended graphene sensors with improved signal and reduced noise, Nano Lett. 10 (2010) 1864–8. https://doi.org/10.1021/nl100633g

[32]   Kireev D., Zadorozhnyi I., Qiu T., Sarik D., Brings F., Wu T., Seyock S., Maybeck V., Lottner M., Blaschke B., Garrido J., Xie X., Vitusevich S., Wolfrum B. and Offenhausser A. Graphene field effect transistors for in vitro and ex vivo recordings, IEEE Trans. Nanotechnol. 17 (2016) 1–1. https://doi.org/10.1109/TNANO.2016.2639028

[33]   Drieschner S., Guimerà A., Cortadella R. G., Viana D., Makrygiannis E., Blaschke B. M., Vieten J. and Garrido J. A. Frequency response of electrolyte-gated graphene electrodes and transistors, J. Phys. D. Appl. Phys. 50 (2017) 95304. https://doi.org/10.1088/1361-6463/aa5443

[34]   Kireev D., Sarik D., Wu T., Xie X., Wolfrum B. and Offenhäusser A. High throughput transfer technique: Save your graphene, Carbon N. Y. 107 (2016) 319–24. https://doi.org/10.1016/j.carbon.2016.05.058

[35]   Bard A. J. and Faulkner L. R. *Electrochemical Methods: Fundamentals and Applications* (Wiley).

[36]   Park D.-W., Ness J. P., Brodnick S. K., Esquibel C., Novello J., Atry F., Baek D.-H., Kim H., Bong J., Swanson K. I., Suminski A. J., Otto K. J., Pashaie R., Williams J. C. and Ma Z. Electrical Neural Stimulation and Simultaneous in Vivo Monitoring with Transparent Graphene Electrode Arrays Implanted in GCaMP6f Mice, ACS Nano 12 (2018) 148–57. https://doi.org/10.1021/acsnano.7b04321

[37]   Lu Y., Lyu H., Richardson A. G., Lucas T. H. and Kuzum D. Flexible Neural Electrode Array Based-on Porous Graphene for Cortical Microstimulation and Sensing, Sci. Rep. 6 (2016) 33526. https://doi.org/10.1038/srep33526

[38]   Fu W., Nef C., Knopfmacher O., Tarasov A., Weiss M., Calame M. and Schönenberger C. Graphene transistors are insensitive to pH changes in solution, Nano Lett. 11 (2011) 3597–600. https://doi.org/10.1021/nl201332c

[39]   Kuila T., Bose S., Khanra P., Mishra A. K., Kim N. H. and Lee J. H. Recent advances in graphene-based biosensors, Biosens. Bioelectron. 26 (2011) 4637–48. https://doi.org/10.1016/j.bios.2011.05.039

[40]   Fu W., Jiang L., van Geest E. P., Lima L. M. C. and Schneider G. F. Sensing at the Surface of Graphene Field-Effect Transistors, Adv. Mater. 29 (2017) 1603610. https://doi.org/10.1002/adma.201603610

[41]   Ecken H., Ingebrandt S., Krause M., Richter D., Hara M. and Offenhäusser A. 64-Channel extended gate electrode arrays for extracellular signal recording, Electrochim. Acta 48 (2003) 3355–62. https://doi.org/10.1016/S0013-4686(03)00405-5

[42]   Krause M., Ingebrandt S., Richter D., Denyer M., Scholl M., Sprössler C. and Offenhäusser A. Extended gate electrode arrays for extracellular signal recordings, Sensors Actuators, B Chem. 70 (2000) 101–7. https://doi.org/10.1016/S0925-4005(00)00568-2

[43]   Neher E. and Sakmann B. Single-channel currents recorded from membrane of denervated frog muscle fibres., Nature 260 (1976) 799–802. https://doi.org/10.1038/260799a0

[44]   Rutten W. L. C. Selective electrical interfaces with the nervous system., Annu. Rev. Biomed. Eng. 4 (2002) 407–52. https://doi.org/10.1146/annurev.bioeng.4.020702.153427

[45]   Hess L. H., Becker-Freyseng C., Wismer M. S., Blaschke B. M., Lottner M., Rolf F., Seifert M. and Garrido J. A. Electrical Coupling Between Cells and Graphene Transistors, Small 11 (2015) 1703–10. https://doi.org/10.1002/smll.201402225

[46]   Brown M. A., Crosser M. S., Leyden M. R., Qi Y. and Minot E. D. Measurement of high carrier mobility in graphene in an aqueous electrolyte environment, Appl. Phys. Lett. 109 (2016) 93104. https://doi.org/10.1063/1.4962141

[47]   Cheng Z., Hou J., Zhou Q., Li T., Li H., Yang L., Jiang K., Wang C., Li Y. and Fang Y. Sensitivity Limits and Scaling of Bioelectronic Graphene Transducers, Nano Lett. 13 (2013) 2902–7. https://doi.org/10.1021/nl401276n

[48]   Blaschke B. M., Lottner M., Drieschner S., Calia A. B., Stoiber K., Rousseau L., Lissourges G. and Garrido J. A. Flexible graphene transistors for recording cell action potentials, 2D Mater. 3 (2016) 25007. https://doi.org/10.1088/2053-1583/3/2/025007

[49]   Cheng J., Wu L., Du X.-W., Jin Q.-H., Zhao J.-L. and Xu Y.-S. Flexible Solution-Gated Graphene Field Effect Transistor for Electrophysiological Recording, J. Microelectromechanical Syst. 23 (2014) 1311–7. https://doi.org/10.1109/JMEMS.2014.2312714

[50]   Sprössler C., Denyer M., Britland S., Knoll W. and Offenhäusser a Electrical recordings from rat cardiac muscle cells using field-effect transistors., Phys. Rev. E. Stat. Phys. Plasmas. Fluids. Relat. Interdiscip. Topics 60 (1999) 2171–6. https://doi.org/10.1103/PhysRevE.60.2171

[51]   Schottdorf M., Hofmann B., Kätelhön E., Offenhäusser A. and Wolfrum B. Frequency-dependent signal transfer at the interface between electrogenic cells and

nanocavity electrodes, Phys. Rev. E - Stat. Nonlinear, Soft Matter Phys. 85 (2012) 31917. https://doi.org/10.1103/PhysRevE.85.031917

[52]   Blaschke B. M., Tort-Colet N., Guimerà-Brunet A., Weinert J., Rousseau L., Heimann A., Drieschner S., Kempski O., Villa R., Sanchez-Vives M. V. and Garrido J. A. Mapping brain activity with flexible graphene micro-transistors, 2D Mater. 4 (2017) 25040. https://doi.org/10.1088/2053-1583/aa5eff

[53]   Hébert C., Masvidal-Codina E., Suarez-Perez A., Calia A. B., Piret G., Garcia-Cortadella R., Illa X., Del Corro Garcia E., De la Cruz Sanchez J. M., Casals D. V., Prats-Alfonso E., Bousquet J., Godignon P., Yvert B., Villa R., Sanchez-Vives M. V., Guimerà-Brunet A. and Garrido J. A. Flexible Graphene Solution-Gated Field-Effect Transistors: Efficient Transducers for Micro-Electrocorticography, Adv. Funct. Mater. 1703976 (2017) 1703976. https://doi.org/10.1002/adfm.201703976

[54]   Claycomb W. C., Lanson N. A., Stallworth B. S., Egeland D. B., Delcarpio J. B., Bahinski A. and Izzo N. J. HL-1 cells: a cardiac muscle cell line that contracts and retains phenotypic characteristics of the adult cardiomyocyte, Proc Natl Acad Sci USA 95 (1998) 2979–84. https://doi.org/10.1073/pnas.95.6.2979

[55]   White S. M., Constantin P. E. and Claycomb W. C. Cardiac physiology at the cellular level: use of cultured HL-1 cardiomyocytes for studies of cardiac muscle cell structure and function., Am. J. Physiol. Heart Circ. Physiol. 286 (2004) H823–9. https://doi.org/10.1152/ajpheart.00986.2003

[56]   Spira M. E. and Hai A. Multi-electrode array technologies for neuroscience and cardiology, Nat. Nanotechnol. 8 (2013) 83–94. https://doi.org/10.1038/nnano.2012.265

[57]   Czeschik A., Rinklin P., Derra U., Ullmann S., Holik P., Steltenkamp S. S., Offenhäusser A. and Wolfrum B. Nanostructured cavity devices for extracellular stimulation of HL-1 cells, Nanoscale 7 (2015) 9275–81. https://doi.org/10.1039/C5NR01690H

[58]   Wang Y. Y., Pham T. D., Zand K., Li J., Burke P. J. and Al W. E. T. Charging the Quantum Capacitance of Graphene with a Single Biological Ion Channel., ACS Nano 8 (2014) 4228–38. https://doi.org/10.1021/nn501376z

[59]   Ang P. K., Jaiswal M., Lim C. H. Y. X., Wang Y., Sankaran J., Li A., Lim C. T., Wohland T., Barbaros O. and Loh K. P. A Bioelectronic Platform Using a Graphene−Lipid Bilayer Interface, ACS Nano 4 (2010) 7387–94. https://doi.org/10.1021/nn1022582

Organic Bioelectronics for Life Science and Healthcare
Materials Research Foundations **56** (2019) 153-184

Materials Research Forum LLC
doi: https://doi.org/10.21741/9781644900376-5

[60] Xie C., Lin Z., Hanson L., Cui Y. and Cui B. Intracellular recording of action potentials by nanopillar electroporation, Nat. Nanotechnol. 7 (2012) 185–90. https://doi.org/10.1038/nnano.2012.8

[61] Fromherz P., Offenhäusser A., Vetter T. and Weis J. A neuron-silicon junction: a Retzius cell of the leech on an insulated-gate field-effect transistor., Science 252 (1991) 1290–3. https://doi.org/10.1126/science.1925540

[62] Kireev D., Seyock S., Ernst M., Maybeck V., Wolfrum B. and Offenhäusser A. Versatile Flexible Graphene Multielectrode Arrays, Biosensors 7 (2016) 1. https://doi.org/10.3390/bios7010001

[63] Gross G. W. Simultaneous Single Unit Recording in vitro with a Photoetched Laser Deinsulated Gold Multimicroelectrode Surface, IEEE Trans. Biomed. Eng. BME-26 (1979) 273–9. https://doi.org/10.1109/TBME.1979.326402

[64] Gross G. W., Williams A. N. and Lucas J. H. Recording of spontaneous activity with photoetched microelectrode surfaces from mouse spinal neurons in culture, J. Neurosci. Methods 5 (1982) 13–22. https://doi.org/10.1016/0165-0270(82)90046-2

[65] Thomas C. A., Springer P. A., Loeb G. E., Berwald-Netter Y. and Okun L. M. A miniature microelectrode array to monitor the bioelectric activity of cultured cells, Exp. Cell Res. 74 (1972) 61–6. https://doi.org/10.1016/0014-4827(72)90481-8

[66] Droge M. H., Gross G. W., Hightower M. H. and Czisny L. E. Multielectrode analysis of coordinated, multisite, rhythmic bursting in cultured CNS monolayer networks., J. Neurosci. 6 (1986) 1583–92. https://doi.org/10.1523/JNEUROSCI.06-06-01583.1986

[67] Bareket-Keren L. and Hanein Y. Carbon nanotube-based multi electrode arrays for neuronal interfacing: progress and prospects, Front. Neural Circuits 6 (2013) 122. https://doi.org/10.3389/fncir.2012.00122

[68] Yi W., Chen C., Feng Z., Xu Y., Zhou C., Masurkar N., Cavanaugh J., Cheng M. M.-C. and Ming-Cheng Cheng M. A flexible and implantable microelectrode arrays using high-temperature grown vertical carbon nanotubes and a biocompatible polymer substrate, Nanotechnology 26 (2015) 125301. https://doi.org/10.1088/0957-4484/26/12/125301

[69] Chen C. H., Lin C. T., Chen J. J., Hsu W. L., Chang Y. C., Yeh S. R., Li L. J. and Yao D. J. A graphene-based microelectrode for recording neural signals *2011 16th International Solid-State Sensors, Actuators and Microsystems Conference* (IEEE) pp 1883–6. https://doi.org/10.1109/TRANSDUCERS.2011.5969794

[70] Chen C. H., Lin C. Te, Hsu W. L., Chang Y. C., Yeh S. R., Li L. J. and Yao D. J. A flexible hydrophilic-modified graphene microprobe for neural and cardiac

Organic Bioelectronics for Life Science and Healthcare   Materials Research Forum LLC
Materials Research Foundations **56** (2019) 153-184   doi: https://doi.org/10.21741/9781644900376-5

recording, Nanomedicine Nanotechnology, Biol. Med. 9 (2013) 600–4.
https://doi.org/10.1016/j.nano.2012.12.004

[71]   Park D., Brodnick S. K., Ness J. P., Atry F., Krugner-Higby L., Sandberg A.,
Mikael S., Richner T. J., Novello J., Kim H., Baek D., Bong J., Frye S. T., Thongpang
S., Swanson K. I., Lake W., Pashaie R., Williams J. C. and Ma Z. Fabrication and
utility of a transparent graphene neural electrode array for electrophysiology, in vivo
imaging, and optogenetics, Nat. Protoc. 11 (2016) 2201–22.
https://doi.org/10.1038/nprot.2016.127

[72]   D. Kireev and A. Offenhäusser, "Graphene & two-dimensional devices for
bioelectronics and neuroprosthetics," 2D Mater., vol. 5, no. 4, p. 042004, Sep. 2018.
https://doi.org/10.1088/2053-1583/aad988

Organic Bioelectronics for Life Science and Healthcare   Materials Research Forum LLC
Materials Research Foundations **56** (2019) 185-242   doi: https://doi.org/10.21741/9781644900376-6

Chapter 6

Graphene based Materials for Bioelectronics and Healthcare

Satish Kumar[1], Tetiana Kurkina[2†], Sven Ingebrandt[1], Vivek Pachauri[1]*

[1]Institute of Materials in Electrical Engineering 1, RWTH Aachen University, Sommerfeldstrasse 24, 52074 Aachen, Germany

[2]Department of Physics, Technical University Kaiserslautern, Erwin-Schroedinger-Strasse, Building 46, 67663 Kaiserslautern, Germany

*pachauri@iwe1.rwth-aachen.de

Abstract

In the last decade, graphene based materials (GBMs) have received special interest in physical sciences owing to their unique material properties at nanoscale. Extraction of two-dimensional lattice forms of carbon has allowed study of new physical phenomena at the molecular scales and allowing further miniaturization in electronics, which have tremendous implications for future technologies. Development of high-performance bioelectronics platforms is one such area where the use of GBMs is expected to yield cutting-edge sensor platforms with far reaching consequences for the advancement of life sciences and healthcare. This chapter provides an overview of how GBMs, when used as electrical transducers, are enabling very attractive functionalities towards new-age bioelectronics solutions. In doing so, this a special focus is given to GBMs produced via chemical routes for realization of devices, surface functionalization, and related transduction approaches. Finally, the chapter evaluates their role in bioelectronics based on relevant material properties, current impact and critical challenges blocking their way towards real healthcare applications.

Keywords

Two-Dimensional Materials, Graphene, System-Integration, Surface Functionalization, Biosensors, Field-Effect Transistor, Electrochemical Detection

† Current Affiliation: Institute of Biotechnology, RWTH Aachen University, Worringer Weg 3, D-52074 Aachen, Germany

Contents

1.   Graphene based materials

Carbon, one of the most abundant element found in our universe, is known since antiquity. Carbon is a common element throughout all known life forms for its ability to form organic compounds and polymers, which serve as the basic building blocks for living organism on earth. Among its isotopes, ^{12}C is the most stable form with a $1s^2 2s^2 2p^2$ electronic configuration. Carbon is *tetravalent*, which means it forms four chemical bonds with other carbon atoms or elements via sp^3-, sp^2-, and sp- hybridization. Accordingly, carbon is found as different allotrope forms in nature and by far these are known as Diamond, Fullerene, carbon nanotubes (CNTs) and Graphite.[1] In diamond, sp^3 hybridized C-atoms form a tetragonal lattice giving it characteristic material properties. Other allotropes of C such as Fullerenes, CNTs and Graphite form sp^2 bonds; however, exhibit different crystal lattice forms.[2] Among these allotropes, Graphite attracted immense scientific interest in recent years for it being a natural source for the thinnest material known to the humankind i.e. atomic layers of C-atoms stacked together by van der Waals bonds. This two-dimensional (2D) lattice of C-atoms is more popularly known as *graphene*.[3, 4] Extraction of graphene from graphite and subsequent demonstration of electrical field-effect in such atomic layers is identified as a pivotal point in the development of next generation electronics beyond Moore's law.[4, 5] Graphene and its other analogues based on carbon such as graphene oxide (GO) or graphite oxides (GrO) are grouped under the family of Graphene based materials (GBMs), which display distinct material properties depending on their source materials, extraction routes, elemental composition, sizes and number of the atomic layers.

Graphene, either directly extracted from graphite or being synthesized on atomically flat catalyst substrate using high temperature growth methods from molecular precursors; displays highest purity i.e. minimal doping, defects and low lattice heterogeneities. The triangular unit cells with 2 C-atoms in a 2D graphene lattice represents a very special electronic structure, where the energy dispersion is a function of the 2D wave-vector k in a hexagonal Brillouin zone with bands crossing near the k and k' points, where the electron energy is linearly dependent on the wave vector. The energy dispersion in graphene also have strong similarities to the Dirac spectrum of massless fermions and accordingly the charge-carriers in graphene can be described by a Dirac equation.[6] This interpretation is very unique arising from quantum-mechanical processes in the graphene lattice, where quasi-particles display a linear energy dispersion relation $E = \hbar k v_F$ ($\hbar$ is planck's constant, k is wave number, v_F is Fermi velocity) as if they were analogous to photons, however, they pertain particle-like behavior e.g. speed of transport, which is calculated as fermi-velocity, $v_F \sim c/300$ (c is speed of light in vacuum).[7] GBMs are therefore different from other metals or semiconductor materials, where charge-transport properties are described by Schrödinger's equation for non-relativistic quantum particles. Also, π and π^* bands in graphene (sp2 hybridization) are degenerated at the K-point of the reciprocal lattice called as Dirac-point, coinciding with the Fermi level and giving rise to the properties of graphene as a semi-metal. The valance band of graphene is completely filled at absolute zero temperature, while the conduction band is completely devoid of any states. By applying an electrical field, the Fermi-level can be moved with respect to the Dirac-point enabling a modulation of the charge-carrier flow in the graphene lattice, as firstly demonstrated by Geim and Novoselov in 2004.[3, 8, 9] Because of the unique crystal structure, and related electronic properties, graphene displays very high charge-carrier mobilities in the order of 10,000 cm^2/V.S (values up to 200,000 cm^2/V.S were reported for suspended graphene).[3, 10, 11] High charge-carrier mobilities and optical transparency of graphene make it an ideal candidate for high-speed and radio-frequency devices, nanoelectronic mechanical systems, photovoltaics and optoelectronics.[12-17]

Other than the electrical nature discussed above, GBMs also exhibit surface characteristic, which are very unique in the nanoscale world. Atomically flat C-C lattice with high mechanical strength offers many advantages for using graphene as a substrate material including the study of single molecules using different electrical, optical, microscopy techniques. In fact, graphite and highly ordered pyrolytic graphene (HOPG) have been the gold standard in scanning probe microscopy techniques for imaging biomolecules.[18] Consequently, surface modification of graphitic surfaces and immobilization of biomolecules is one of the most studied topic, and there exist advanced surface chemistry approaches to carry out biofunctionalization. Biomodification approaches on GBMs will be discussed in this chapter later on by giving a brief account of recent progress made in this field. GBM characteristics such as thermodynamically stable atomic lattice, mechanical strength, highest surface-to-volume (S/V) ratio, high charge-carrier mobilities and astute surface chemistry put together a unique combination of critical merits to realize an ultimate transducer. The estimated specific surface area of single layer graphene is 2630 m^2/g (SWCNTs or graphene on substrate – 1315 m^2/g).[19] Single and few-layer GBMs virtually sporting all of their conducting matter on their surface accessible for outer environment makes them an ideal choice for sensing molecular analytes from gaseous or liquid media. Naturally, GBMs have attracted a lot of attention in recent years for the realization of high-performance biosensor platforms showing very low limits of detection.

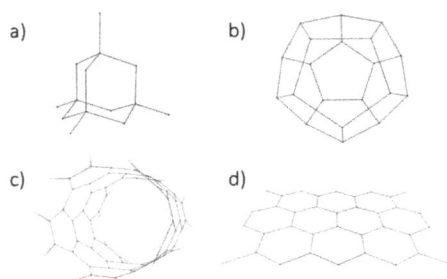

Figure 1: Sketches showing different carbon allotropes found in nature. a) Diamond, where C-atoms are bonded together in a tetragonal lattice formed by sp^3 hybridization, b) Fullerenes, which are cage-like lattices of C-atoms bonded together by sp^2 hybridization, c) CNTs are one-dimensional tubes formed by C-atoms bonded together by sp^2 hybridization. CNTs display different chirality in their lattice, also determining their material properties, CNTs can be made of more than one concentric tube of C-atoms and are therefore distinguished as single-walled and multi-walled CNTs. d) two-dimensional lattice of C-atoms bonded together by sp^2 hybridization is called Graphene and is the basic building block of all GBMs.

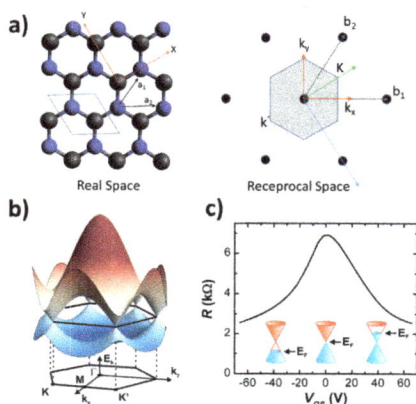

Figure 2: Two-dimensional atomic lattice of graphene and electronic properties. a) The lattice structure and Brillouin zone of graphene with its unit cell consisting of the two inequivalent carbon atoms depicted inside the rhombus (black and grey circles). a_1, a_2 and b_1, b_2 are the unit vectors (in real and reciprocal space, respectively), b) electron band structure in graphene where the dispersion relation near K and K' is linear, c) fermi level shift and ambipolar carrier transport.

The evolution of modern bioelectronics can be traced back to the first realization of a pH sensitive electrode and later modifications to render them specificity towards specific biomolecules such as enzyme and nucleic acids.[20-22] Major improvements in the field of bioelectronics came with the development of new lithography techniques for the realization of high surface area electrodes in microscale dimensions. Electrochemical sensors based on high surface-to-volume (S/V) ratio microelectrode arrays (MEAs) and nanoelectrode arrays (NEAs) have already flourished in the assembly of biosensor and bioelectronics platforms for highly-sensitive detection of biomolecules and for bioassays with living cells.[23] At the same time, microfabrication of silicon (Si) based ion-sensitive field-effect transistors (ISFETs) enabled a new approach towards surface-charge based detection of biomolecules. Complementary metal oxide semiconductor (CMOS) process compatibility of Si ISFETs makes them an ideal choice for the realization of biosensor platforms with multiplex readout capabilities and high-level integration, which are critical for clinical applications at point-of-care (PoC).[24, 25] Use of label-free nanoscale transduction approaches for clinical applications is, however, very challenging. The requirements of detecting analytes of very low concentrations from biological samples pose multitudes of technical and fundamental challenges for specific signal transduction. Biological samples such as blood or serum are usually very complex in their composition with different proteins, lipoproteins, lipids, hormones, nucleic acids, organelles and other biological constituents suspended in very high ionic strength aqueous media. Direct detection of low concentration analytes such as DNA, RNA and signaling proteins as biomarkers without pre-amplification steps require highly specific biochemical interactions between the nanoscale transducer and target analytes.

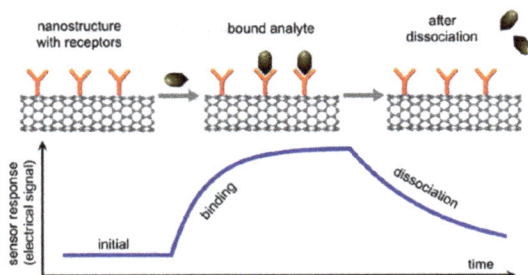

Figure 3: A label-free electrical biosensor, receptor-analyte interaction at the transducer surface and characteristic sensor response.

Figure 3 depicts such an electrical transducer based on nanomaterials. The nanoscale transducer which in this case is a GBM, is surface modified with analyte-specific receptor molecules. Electrical output of this device, which is usually the change in drain current (I_d) or equivalent gate voltage (V) are taken as the sensor signal, which are measured overtime. Placing a sample with analyte molecules onto this transducer surface eventually results in a specific interaction between analyte and receptor molecules when the analyte is bound to the transducer surface. This specific binding of the analyte to the transducer surface via specific receptor molecules brings a change in the electrical properties of the transducer material, and can be observed in the change of the signal output of the transducer. The change in the output signal of the transducer overtime is termed as sensor response, which typically follows a binding curve as shown in the schematic in figure 3. The signal saturates after a while when the rate of binding of analyte to the transducer surface and dissociation of the analytes off the transducer surface reach an equilibrium. The sensor signal of a high performance transducer showing specific molecular binding starts coming back to their original values when the equilibrium is shifted towards dissociation. A typical dissociation curve such as shown here signifies an efficient and reversible molecular binding between target and analyte, which is a critical feature of biosensor operation for real applications. The lowest detectable concentration of analyte is called as limit-of-detection (LoD) of the biosensor platform. As can be inferred from the above discussion, LoD values may be greatly influenced by non-specific molecular interactions on the transducer surface, which is a common problem when carrying out measurements in real biological samples. It is also important to notice the saturation point of the sensor signal for a biosensor, which signifies the maximum detectable concentration of an analyte. This analyte detection window between the LoD and the saturation point is called as the sensor dynamic range for the biosensor platform. Deploying a biosensor for clinical application requires an efficient and robust detection of target analytes below and above their clinically relevant concentration ranges in the biological samples. Therefore, other than the high-sensitivity of the transducer platform, characteristics such as surface properties and optimal biofunctionalization approaches to reduce non-specific binding of molecules and longer shelf life determine the successful practical implementation of a biosensor.

GBM, in this respect show tremendous potential not only for their high sensitivity because of high S/V ratio and unique electrical conduction, but also due to the rich surface chemistries exhibited by them for the construction of versatile biofunctional interfaces. So far, GBMs have been demonstrated as label-free electrical detectors for glucose, creatinine, DNAs, RNAs, antibodies, signaling proteins, hormones, glycoproteins, lipids and many other biomarker analytes that are relevant for clinical analysis and medical diagnostics.[26] In some cases, GBMs have also been deployed as an efficient platform for following biological processes such as the activity of enzymatic biochemical reactions. Other than this, GBMs are also deployed as a screening platform and realization of single cell-based assays and drug discovery applications. In all above examples, the chemical nature of GBMs is exploited to constitute efficient biofunctional layers demonstrating versatility of GBMs. It is therefore important to understand the origins, synthesis, and chemical behavior of different GBMs as enabling factors for their successful implementation in different healthcare fields. Keeping in mind the realization of biosensor platforms based on GBMs, we will give a brief overview of the relevant syntheses and fabrication processes in the following sections.

2.   Syntheses of graphene based materials

Graphene: In 2004, a group of scientists at Manchester, UK first reported the preparation of graphene via the micro-mechanical exfoliation of graphite.[27] Repeated tape pasting approach separated the graphite crystals into thinner layers. The optically translucent flakes were suspended in acetone solvent, followed by their transfer onto a silicon substrate. The obtained graphene displayed minimal defects and high electron mobility. However, the micromechanical exfoliation approach to obtain graphene is rather random and not suitable for either mass production of graphene layers or scalable transfer onto the other substrates. Graphene layers can also be produced using epitaxial growth techniques that involve heating of silicon carbide (SiC) substrates to very high temperatures in very low-pressure conditions in order to deplete silicon atoms and stimulate the formation of a graphene lattice.[28] The lateral dimensions of the graphene grown in epitaxy processes are therefore related to the size of SiC substrates and provide high quality graphene for scalable integration.[29] Graphene can also be grown on metal catalysts using epitaxial growth approaches.[30, 31] The quality of graphene layers and homogeneity is, however, influenced heavily by the properties of the substrate.[32] The epitaxial technique produces graphene with weak anti-localizations, which is not present in peeled/scotch-taped graphene. Epitaxy growth of graphene on metal substrates utilizes the origin and atomic framework of a metal substrate (ruthenium, iridium, nickel, copper, etc.) as a seed for the growth. Graphene on ruthenium is generally of non-uniform thickness while grown on iridium is highly organized, has uniform thickness, and is easy to be peeled off.[31] Graphene growth on iridium is, however, rippled as compared to other substrates. Availability of these long-range ripples contributes in formation of small gaps in the electronic band-structure (Dirac cone) of graphene. In another approach, graphene is grown using metal-carbon melts where a metal is melted in the presence of a carbon source. This source could be in the form of graphite powder, chunk or crucible, which is merely kept in contact with the molten metal. The technique employs a specific temperature to dissolve C-atoms inside a transition metal melt and precipitates single layer graphene at further lower temperatures.[33] CNTs can also be used as a source for production of graphene nanoribbons (GNRs). In this method, CNTs are unzipped longitudinally by treating with strong acids such as sulphuric acid (H_2SO_4) and potassium permanganate ($KMnO_4$) or metal-initiated catalysis.[34] For example, Mn^+ ions acts as a catalyst favoring the breaking of C-C bonds in in longitudinal direction.[35] Overall, vapor phase methods for the production of graphene have evolved significantly over the years where high-quality graphene is grown on different substrates. Controlling the atomic lattice, doping and underlying growth mechanisms, however, remain to be studied in detail.[36]

Oxides of GBMs: Oxides of GBMs such as Graphene oxide (GO), Graphite oxide (GrO), Graphene oxide quantum dots (GOQDs), Graphite platelets (GrPs) are mono-, few- or multi-layered carbon nanomaterial with a high oxygen content obtained by oxidation and subsequent exfoliation of graphite.[37, 38] GO typically exhibit C/O ratio of less than 3. Unlike pristine graphene, GO contains numerous polar groups, which makes it easily processable in aqueous solutions. In addition, carboxyl (-COOH), epoxy(-C-O-C-), hydroxyl (-OH) moieties of GO can be used for functionalization of the sheets with various molecules, nanoparticles (NPs), QDs, etc.[39] These chemical groups also play important role for the assembly of the GO flakes onto various substrates of interest such as glass, polymers etc. The presence of oxygen containing groups, however, disturbs the electronic structure of GO and makes it insulating. For electronic applications, GO is chemically or thermally treated to reduce the O content and to recover the ordered C-C lattice with delocalized electron density. GO subjected to such reduction procedures is therefore called reduced graphene oxide (rGO) and with the emergence of high efficiency reduction methods, provides a suitable alternative to graphene produced via gaseous growth methods.

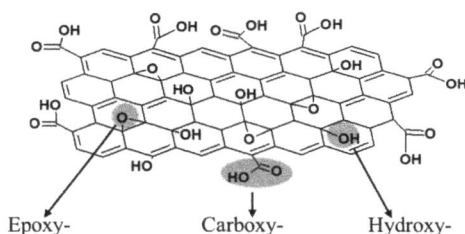

Epoxy-   Carboxy-   Hydroxy-

Figure 4: Chemical makeup of oxides of GBMs showing several oxygen-containing functional groups. The presence of such surface functional groups render them hydrophilic nature and access to wide range of chemical functionalization routes.

Owing to its defective structure and the remaining functional groups, rGO is extremely advantageous for pH sensing.[40, 41] This pH dependency is caused by two factors: the interaction of functional groups at the rGO surface, such as –OH and –COOH groups, with H^+ ions of the electrolytes, giving rise to a change in the surface charge density, and the change of the Gouy-Chapman diffuse electric double layer causing electrostatic gating effects. This intrinsic property of rGO can easily be exploited for biosensing purposes, but only few studies focused on this approach. In a later section, build-up of the solid-liquid interface at a typical GBM FET biosensors will be discussed in detail.

Starting material for production of GO is graphite which is a naturally occurring mineral. GO was first synthesized by Schafhaeutl, Brodie and later by Saudenmeier (combination of potassium chlorate ($KClO_3$) with nitric acid (HNO_3).[42] Among many oxidative routes for the production of GBM oxides, the protocol developed by Hummers and Offeman was commonly used.[43] In this approach, graphite is oxidized by potassium permanganate ($KMnO_4$) in the mixture of nitric and sulfuric acids. To ensure safety of the process, oxidation is performed in the ice bath and the oxidizing agent is added very slowly to avoid overheating of the solution. The reaction mixture slowly turns into brownish paste after taken off the ice bath. Addition of water to it results in rapid increase of temperature to 98°C. At this temperature, the oxidation of graphite progresses further. Oxidation is stopped with hydrogen peroxide, which reduces the remaining $KMnO_4$ and dimanganese heptaoxide (Mn_2O_7), which is a very strong oxidizing agent. After this, the suspension is filtered, and the obtained pellet is washed with large volumes of water to get rid of the chemical components of the oxidizing mixture. The resulting yellowish brown filter pellet comprises of GrO. In the last step for the preparation of GO flakes, the pellet is suspended again in water followed with ultrasonication and/or shaking procedures for efficient exfoliation. Usually, several centrifugation/resuspension cycles are performed to further purify the GO suspension from remaining salts.

There are numerous modifications of the above detailed Hummer's protocol aiming to increase efficiency of the production, have a better control over the oxidation of graphite, increase flake size, decrease amount of contaminants, shortening of the reaction time and improve environmental safety of the protocol.[42, 44, 45] Mercano et al. were able to minimize exothermic effects in Hummers approach and increase the density of oxidative functional groups by including phosphoric acid in the reaction mixture, and yet preserving the aromatic structure of the exfoliated graphitic lattice.[46] The resulting GBM products provide better charge-carrier conductivity for the assembly of field-effect based devices. More recently, Lu et al. made significant additions to the exfoliation and purification procedures for the production of GO flakes with high density of oxidative functional groups in a *low temperature exfoliation followed with desalination* (LTEDS) protocol, which were later on used for the realization of GO thin films over silicon and glass wafers.[47]

Starting material's type and quality, duration of the every step of the exfoliation procedure, temperature variations of the reaction mixture throughout the reaction course, addition or replacement of chemicals needed for reaction, thoroughness of washing, additional desalination steps and exfoliation strategy result in GO with different properties. Carbon to oxygen ratios, distribution of flake size, and concentration of GO and purity of the solution are among other parameters that may differ from batch to batch.[42, 48] Other strategies towards production of GBM oxides include ferro-induced oxidation, microwave assisted exfoliation, electrochemical and exfoliation of graphite and using expanded graphite as a source material.[44, 49-51] Electrochemical exfoliation of graphite for the production of graphene sheets with preferred lattice arrangements such as AB-stacked bilayers with a large lateral size (several to several tens of micrometers) have been reported.[52] Such electrochemically exfoliated GBMs are generally produced in organic solvents or in the presence of ionic components to provide a conducting media in an electrochemical cell and hence exhibit some level of chemical doping. Doping of GBMs may result in affecting the electronic and optical properties of the material produced and determine their applications. While EC exfoliated GBMs have been used to make transparent thin-films for photovoltaic applications, the usage for biosensor applications maybe limited for the lack of biocompatible characteristics.

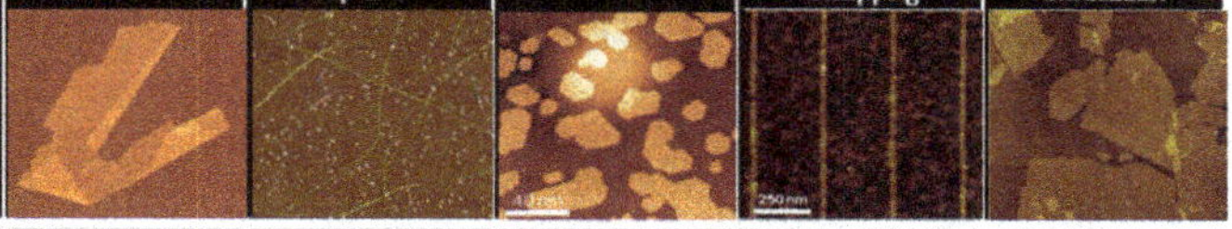

Synthesis method / Merits	Micromechanical exfoliation	Chemical vapour deposition	Epitaxial methods	CNT Unzipping	Chemical exfoliation
Scalability	- - -	+ + +	+ +	-	+ + +
Yield	-	-	- -	+ +	+ + +
Quality	+	+ +	+ + +	+ +	+
Cost	+ + +	- -	- - -	-	+ + +
Purity	+	+ +	+ + +	+ +	-

Figure 5: Comparison of main syntheses pathways for GBMs. Upper row shows AFM scan images of typical GBMs produced in the corresponding method. All the GBMs are then compared on terms-of-merits; scalability of the synthesis method, yield of the material with specific and reproducible material properties, ensample quality of the product for electronic applications, costs incurred for the synthesis method and the purity of the material such as minimal doping, defects and presence of residues specific to the synthesis method, respectively.

Main methods for the synthesis of GBMs are shown in figure 5 on the right side. Merit points for the synthesis procedures are given depending on key quality parameters such as large volume production of the material, efficiency of the synthesis methods, quality of the product, and cost-factors involved in the production. As it can be seen, CVD and epitaxy approaches scale high on quality and purity of the GBMs grown, but are costly maneuvers compared to other methods. Scalability especially is a big concern for electronic applications where GBMs from CVD, Epitaxy and chemical exfoliation are most popular methods.

3.   Surface (bio)functionalization of GBMs

The usage of GBMs for most of the applications such as sensors, catalysis, microelectronics, optoelectronics and molecular filters etc. require modification of their surface. Most of the modifications are targeted to enhance material properties (such as

Organic Bioelectronics for Life Science and Healthcare   Materials Research Forum LLC
Materials Research Foundations **56** (2019) 185-242   doi: https://doi.org/10.21741/9781644900376-6

electrical or optical properties) or to improve their biocompatibility, stability, solubility or selectivity towards preferred molecules or analytes.[89, 90] The process of modifying the surface in order to obtain certain functionality or immobilize desired biomolecules is termed as chemical functionalization or biofunctionalization. In case of biofunctionalization, the layer of biomolecules on the surface can be called as biofunctional layer. In the case of biosensing applications, functionalization of GBMs is important for rendering surface physicochemical properties that are required for high-performance operation and detection of target biomolecule in a complex liquid media. Bio-specificity of the transducer surface is usually achieved by coating or immobilizing analyte-specific receptor biomolecules to their surface. These receptors are specially designed with suitable functional groups that can adhere to the transducer surface either by formation of covalent or non-covalent bonds.[53, 54]

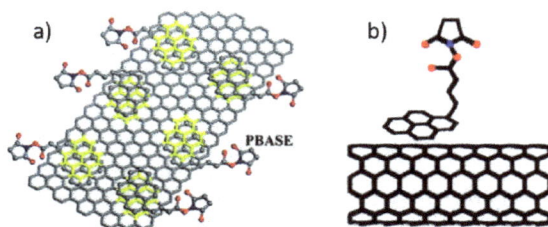

Figure 6: Non-covalent adsorption of molecules onto a graphitic lattice is exploited for the immobilization of biomolecules for biosensor applications.

Non-covalent functionalization: Immobilization of biomolecules to the GBMs without the formation of covalent linkage may be realized just by different kinds of physical interactions such π-π stacking. This is possible due to the extreme hydrophobic nature of the graphitic lattice, which encourages adsorption of molecules with aromatic residues. Therefore, receptor biomolecules such as nucleic acids or some hydrophobic proteins, which have aromatic residues, easily adsorb on to carbon nanostructures and GBMs.[19, 53, 55-57] Molecules with pyrene- or pyrene-like rings and other aromatic compounds are used as anchors for the attachment of receptor molecules such as shown in figure 6. Similar strategies for the adsorption of biomolecules on to the graphitic lattice have been adopted to create novel interfaces for biosensor applications. Macromolecules displaying negatively or positively charged regions can be used to form self-limiting assemblies on graphitic surfaces, which were modified with differently charged ammonium and sulfonate substituted pyrenes.[58] Also, biomolecules with amphiphilic nature (containing hydrophobic and hydrophilic regions) are adsorbed onto the graphitic lattice

Organic Bioelectronics for Life Science and Healthcare   Materials Research Forum LLC
Materials Research Foundations **56** (2019) 185-242   doi: https://doi.org/10.21741/9781644900376-6

which may stabilize the aqueous dispersions of GBMs.[59] Noncovalent approaches for the decoration of DNA molecules onto GO and rGO have been realized via self-assembly and also by integrating other nanomaterials into GO-DNA conjugates.[60] Other than the stacking orders and surface charge interactions, the defects in GBMs have been found to affect the wetting properties of the surface which may influence the adsorption of the biomolecules.[61] Non-covalent functionalization being chemically mild offer interesting routes to alter the surface properties of GBMs, are suitable for making transducers in a bio-friendly approach. Although, noncovalent functionalization processes require more focus into understanding of the underlying mechanisms, in order to develop robust and more controllable surface architectures deployable for intensive biochemical processes. Noncovalent approaches for realization of high performance biofunctional layers are witnessing major improvements such as development of receptor biomolecules with compatible physicochemical properties and high affinity towards target analytes and transducer surface. Formation of stable self-assembly biofunctional layers display immense potential to develop capacitance-sensitive readout approaches (impedimetric bioassays) for the detection of large biomolecules such as proteins where field-effect based detection pose fundamental limitations. For example, a graphene-coated nanoscale interdigitated electrode arrays (nanoIDE-arrays) sensor platform with non-covalent functionalization of single chain fragment variant (scFv) antibodies as receptor molecules was demonstrated recently.[62] The graphene layers suspended over nanoIDE-arrays provided an ideal surface for self-assembly of scFv receptor (figure 7). The assembled platform exhibited sensor dynamic ranges including clinically relevant concentrations of myeloperoxidase (MPO) and fatty-acid binding proteins (FABP).[62] Interestingly, these studies were carried out in physiological saline, showing their potential for the further development of high performance biosensor platforms for large biomolecules which do not suffer the solid-liquid interface issues such as Debye-screening of biomolecular charges often encountered for field-effect based biosensors. Graphene-IDE platforms with noncovalent surface modifications also displayed very low sensor-to-sensor variations and maybe expanded further into developing assays for clinical studies and work in real biological samples after more optimizations of such biofunctional layers.[63]

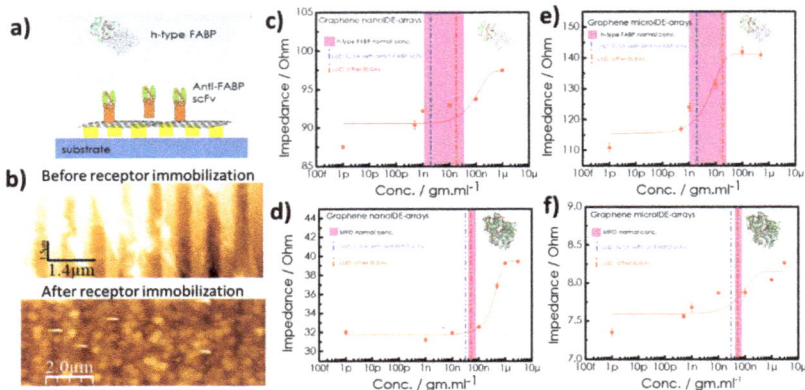

Figure 7: Non-covalently functionalized graphene coated IDE-arrays for label-free sensing of cardiac disease biomarkers. (a) the construction of biosensor platform in scheme, (b) AFM scan images of the graphene-coated IDE-arrays before and after the functionalization and self-assembly of anti-FABP scFv antibodies, (c-f) dose response curves of FABP and MPO biomarkers in physiological saline displayed in the graph where impedance of the biosensor platform at 30Hz is measured against analyte concentration.

Covalent functionalization: The presence of -COOH, -C-O-C- and –OH oxidative functional groups in GBMs provide a wide range of options for attachment of biomolecules using covalent chemistry approaches. Covalent bond formation is fundamental in organic chemistry for the synthesis of new compounds.[64] Covalent chemistry of carboxylic acids, epoxides and hydroxyl has been studied since the beginning of modern organic chemistry and in-depth studies have been carried out in c-based systems. Before GBMs, covalent chemistry was studied on amorphous carbon, graphite, fullerenes and carbon nanotubes for a range of applications. Use of covalent approaches to link biomolecular receptors to GBMs was therefore straightforward and has been widely used in the last decade.[65]

Covalent bonds are strongest of the molecular bonds in nature and therefor provide a robust linkage. In addition, covalent bonds also allow delocalization of electrons and are therefore conductive in nature. For example, a nucleotide sequence having an amino (-NH_2) functional group would be an ideal choice for the formation of a carbodiimide (-$CO-NH_2-$) link to the graphic lattice by using its -COOH group. In past, formation of carbodiimides has been studied in detail and implemented for biofunctionalization of graphene and oxides of GBMs.[66] As for the electronic nature of the covalent linkage, any changes in the distribution of the surface charges on a biomolecules are transmitted to the graphitic lattice and vice-versa. Therefore, covalent linkage are also an ideal option for transduction approaches where charge-transfer through the biofunctional layer is deemed necessary. Use of ion field-effect based (ISFET) devices for direct detection of charged biomolecules and associated system advantages popularized the covalent linkage chemistry for biosensor systems.

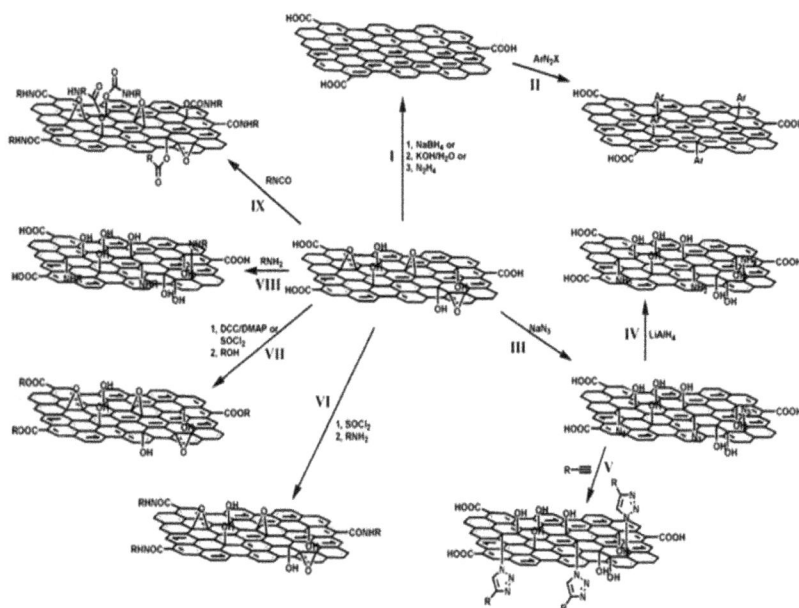

Figure 8. Surface functionalization of GBMs using covalent bond formation. I: reduction of GO II: covalent surface functionalization of reduced graphene via diazonium reaction. III: functionalization of GO by the reaction between GO and sodium azide. IV: reduction azide-GO with LiAlH₄ resulting in the amino-functionalized GO. V: functionalization of azide-GO through click chemistry. VI: modification of GO with long alkyl chains by the acylation reaction between the carboxyl acid groups of GO and alkylamine. VII: esterification of GO by dicyclohexylcarbodiimide chemistry or the acylation reaction between the carboxyl acid groups of GO and ROH alkylamine.VIII: nucleophilic ring-opening reaction between the epoxy groups of GO and the amine groups of an amine-terminated organic molecules (RNH₂). IX: the treatment of GO with organic isocyanates leading to the derivation of both the edge carboxyl and surface hydroxyl functional groups via formation of amides or carbamate esters (RNCO).

Organic Bioelectronics for Life Science and Healthcare   Materials Research Forum LLC
Materials Research Foundations **56** (2019) 185-242   doi: https://doi.org/10.21741/9781644900376-6

Formation of covalent bonds require higher activation energies, which can be sourced chemically or physically depending on the material system and efficiency. For example, in one of the most commonly used biofunctionalization approaches, thermally activated reactions are first carried out for the oxidation of GBMs. It results in the formation of the reactive carboxyl, epoxide or hydroxyl groups.[66] These oxygen-containing activated functional groups act as catalytic sites for the subsequent chemical reactions for the attachment of a receptor with the help of various chemical reactions. Now a biomolecule (receptor), of interest can be coupled to carboxyl groups via esterification or amidification reactions.[54, 65-69] An example of chemical reactions that make use of the oxygen-containing groups in oxidized graphene for covalent bond formations are shown in figure 8.[57, 70] In other works, products of the addition reactions are often used for subsequent attachment of the receptor molecules. For instance, fluorine of fluorinated GBMs can be replaced by nucleophilic substitution using alcohols, amines, Grignard reagents or alkyl lithium compounds.[57, 68, 70] Photochemical activation is also used for the modification of graphitic surfaces. Similarly, under intense UV-irradiation, graphene can react with benzoyl peroxides.[71] Conjugated polymeric materials such as thiophenes are well known with their potential applications in optoelectronics due to high charge mobility arising from delocalized π electrons along their molecular chains. Amine-terminated oligothiophenes have been functionalized onto GO flakes through formation of amide bonds.[72] Other than this, electrochemical functionalization also provide a favorable route for the attachment of biomolecules on to carbon surfaces in a very versatile manner in a covalent or non-covalent fashion.[73-75] For performing electrochemical reactions, an electrochemical cell is used. A typical electrochemical cell consists of a working electrode (WE), a reference electrode (RE) and a counter electrode (CE). Graphitic surface that is to be modified forms the WE for electrochemical reaction. While oxidation of an electrochemically active component in the solution can be achieved by applying a positive overpotential, a reverse situation of reduction process can be induced by applying a negative overpotential. In such electrochemical processes, RE is used to set the reference potential and CE is used for monitoring the electrical current generated in the cell during the electrochemical process. By careful selection of an electrochemical processes, covalent and non-covalent modification both can be realized, for example, by using receptor molecules containing aminophenyl groups.[76] If the GBM is used as a WE, application of positive voltages would lead to the polymerization of electroactive molecule onto the GBM surface forming a functional layer. For inducing a covalent functionalization in this case, aminophenyls are converted into their diazonium salts, which are highly reactive. Electrochemical reduction causes the formation of aryl radicals, which can be coupled to the GBM surface.

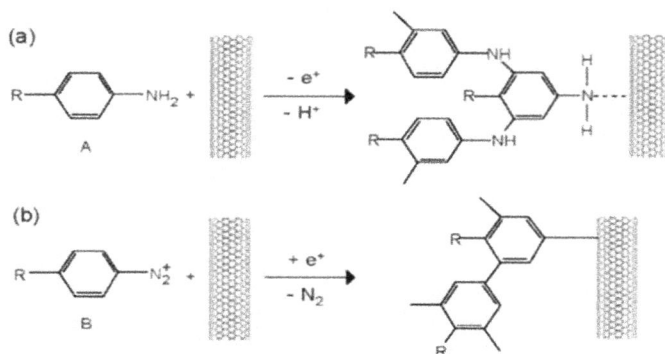

Figure 9. Schematic showing (a) oxidative electrochemical modification of C-lattice with 4-aminobenzyl-R and (b) reductive electrochemical modification with an aryl diazonium salt. In the first case, the C-lattice is covered by a polymeric layer of aminobenzyl groups without the formation of a chemical bond, whereas in the second case, polyphenyl groups covalently attach to the C-lattice.

Other than acting as an 'anchor' for the biomolecules to attach on the transducer element, surface functionalization serves other important roles for the realization of a high performance biosensor platform. Surface modification is often used in order to reduce non-specific binding of the various biomolecules to the surface of the transducing component. For example, non-specific binding of the oligonucleotides onto a graphitic surface can be minimized by non-covalently attached polyaminobenzoic acids, which can in turn also be used subsequently for the immobilization of receptor DNA strands. Surface functionalization of GBMs with conducting polymers is also used to enhance their charge-carrying capacity as well as field-effect mobilities, which in turn enhances the sensor performance. In several studies, surface functionalization approaches have been used to minimize the effect of lattice defects on the electrical properties of GBMs. Surface functionalization strategies can also be used to selectively tune the work function of GBMs.[77-79] In addition, surface modification by specialized organic or polymeric compounds is also used to provide mechanical strength to the graphitic lattice for NEMS applications. Because of the great importance attached to the surface modification of GBMs for biosensing applications, modification approaches deserve special attention. From the chemical point of view, GBMs and CNT surfaces are essentially the same so many of the surface modification chemistries worked out for CNTs can be applied to the GBM surfaces as well. The degree of functionalization, however, varies significantly for different carbon nanomaterials and their different areas. The edges and defect sites of

GBMs are generally more reactive.[57, 80, 81] Preferable reactive site in GBMs are usually geometrically strained regions [96] as against smaller diameter CNTs due to their increased curvature.[56, 82]

The alteration of surfaces and tuning of physicochemical character of GBMs for various applications is gaining interest and expanding rapidly. Over the last years, there have been reports where high-end surface modification strategies have been use to systematically tune the surface properties of GBMs and standardize their response towards analytes in liquids.[75, 83] It is widely known that the surface properties of GBMs vastly vary depending on their synthesis, processing and other parameters including the underlying substrate and device configurations. In this case, standardization of electrochemical properties of transducer surface is important in order to realize sensor platforms for real applications.

4.   System integration of GBMs

Development of biosensor platforms for use in healthcare requires standardization of the device fabrication procedures at several levels. Unlike the current silicon based and other semiconductor technologies which have several decades of know-how and infrastructure developed that is able to reproduce material with standard properties and processes that render them precise device characteristics over wafer scales, graphene and other GBMs are still in their nascent stages of standardization for industry applications. At first, the sourcing of GBMs and techniques in which GBMs are synthesized are still under development. Very few facilities in the world are able to synthesize high quality monocrystalline graphene over wafer scales in a routine manner. The high-end cleanroom foundries with advanced equipment for the production of GBMs is largely driven by promising applications of graphene wafer-scale graphene in high-frequency electronics, optoelectronics and photovoltaics, which is a costly practice and remains in limited access in the foreseeable future. Towards the lower-end applications such as fabrication of gas sensors and electrochemical sensors, super-capacitors and other applications, which do not require ultra-high quality graphene, are largely being driven by smaller research labs so far, closer to the academic environment with moderate access to the cleanroom facilities equipped for growing, handling and processing two-dimensional (2D) materials. Therefore, the system integration approaches for GBMs are largely divided based on the growth techniques. Gaseous synthesis techniques such as CVD and epitaxy have evolved significantly over the past few years, yet there are serious limitations in getting high quality GBMs from these methods onto regular substrates used in concurrent MEMS and NEMS process lines. On the other hand, exfoliation of GBMs from graphite by micromechanical cleavage or chemical routes require elaborate

Organic Bioelectronics for Life Science and Healthcare   Materials Research Forum LLC

Materials Research Foundations **56** (2019) 185-242   doi: https://doi.org/10.21741/9781644900376-6

procedures and process steps that are difficult to standardize for scalable integration. Although, there have been important developments in this direction, which will be discussed later in this section. Fabrication of GBM based devices can be broadly put into two categories namely *bottom-up* and *top-down* approaches. The schematic shown in figure 10 sums up main pathways for the exfoliation/synthesis of GBMs and their manipulation for the fabrication of nanoscale devices.

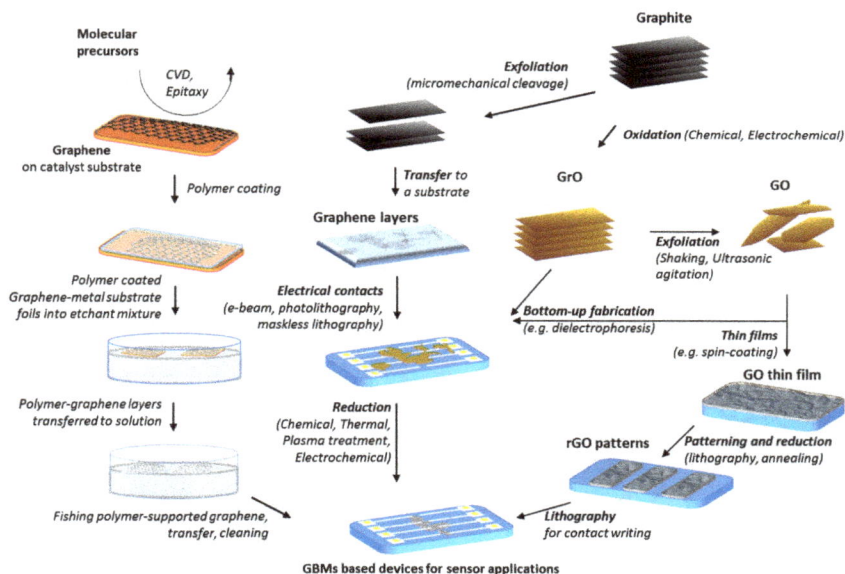

Figure 10: General fabrication routes for the realization of sensor platforms based on GBMs

Bottom-up fabrication: The device fabrication in a bottom-up approach usually means manipulation of individual flakes or layers for the realization of electrical devices. Micromechanical cleavage or chemical exfoliation are favored methods for the production of individual flakes that are then transferred onto a substrate of choice for further fabrication steps. The transfer methods may be solution based transfer or other physical manipulation techniques. After having the desired GBM layers on the substrate, electrical contacts are defined using e-beam lithography or massless lithography techniques. Contacting individual flakes in such massless process are easy to carry out, flexible and provide an efficient way to carry out fundamental studies at nanoscale. These lithography approaches, however, may be tedious depending on the quality of the material. One usually needs to locate the individual flakes on a marker substrate to assign coordinates for contact writing. Alternatively, the bottom-up approach also includes physical manipulation techniques such as dielectrophoresis (DEP). Using DEP, individual flakes of the GBMs are transferred onto microelectrode pairs working as source-drain electrodes. The electrodes are realized in standard photolithography process and therefore such chips can be produced in large quantities and are much faster to fabricate. DEP techniques have been demonstrated for scalable integration of nanomaterial onto desired locations and realization of biosensor platforms.[84-86] Approaches such as DEP are expected to serve a suitable alternative for fabrication processes that require heavy instrumentation and cleanroom facilities to operate.[87-89] Figure 11 on the right side shows typical devices fabricated by transferring GO flakes from their suspension onto the electrode pairs using DEP.[90, 91] While efforts continue to realize scalable approaches of nanomaterial manipulation, mass production of GBMs based device platforms using bottom-up fabrication remains very challenging. Alternatives towards solving compatibilities issues of GBMs with the standard top-down micro-/nano-fabrication processes are expected to progress towards industry requirements for integration of nanoelectronic platforms of sensors and other device applications.

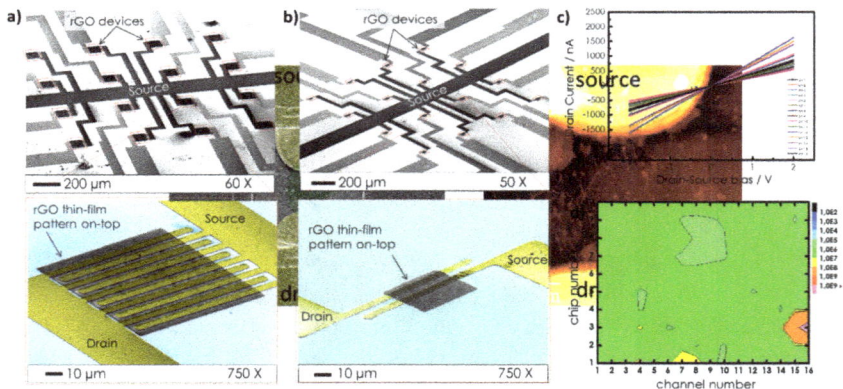

Figure 11: Bottom up fabricated devices. (a) SEM images of typical rGO chips where individual flake (shown in green) is placed onto the electrode-gap using dielectrophoresis technique, (b) AFM scan image of another similar device showing detailed surface characteristics.

Top-down fabrication: Top-down methods generally refer to fabrication approaches, where starting material is in the form of thin films over a wafer, which can be patterned using standard lithography processes. The lithography processes include photolithography, extreme UV lithography, nanoimprint lithography, which provide a high throughput and much faster alternatives for nanofabrication over a large area. Although in the very beginning stages, large area graphene grown in CVD and epitaxy have been used as a starting material for such top-down processes and carryout CMOS prototyping and other integrated circuits based on graphene for optoelectronic applications.[36, 92-94] Scalable approaches for integration of GBMs and other 2D materials into routine micro-/nanofabrication processes is crucial for the realization of 2D material based technologies for real industry applications.[20]

Figure 12: Wafer-scale nanofabrication of 2D devices using GO thin films. (a) Scanning electron micrograph showing 16 devices on a sensor chips that are produced top-down lithography process. GO thin-films are realized by spin-coating of GO solution onto silicon and glass wafers and subsequently reduced to form rGO. Photomask with desired electrode configurations are used to fabricate devices with interdigitated or (b) or single-pair electrodes, (c) IV characteristics of all the 16 devices from the sensor chip, (d) a heat-map showing resistance distribution of the devices on wafer-scale.

Thin films of GBMs from chemical exfoliation can also be used for top-down fabrication and realization of sensor platforms. Here, several approaches such as spin-coating, convective self-assembly (meniscus-dragging deposition), self-assembly, spray-coating, dip-coating, Langmuir–Blodgett (LB) assembly, layer-by-layer (LbL) assembly, and vacuum filtration have been used for realization of high quality thin-films.[95-103] Controlling the surface characteristics, material properties, homogeneity of these thin-films based on chemical approaches is, however, not as refined and may result in low performance of the sensor devices or variations in their sensor characteristics. Integration of chemically produced GBMs into nanoscale systems is a booming area in research and more advances in nanofabrication approaches are expected in near future. For example, Lu et al. recently demonstrated a more refined and controllable way to produce high quality GO layers by using a LTEDS approach.[47] These GO layers exhibit high-density oxidative functional groups

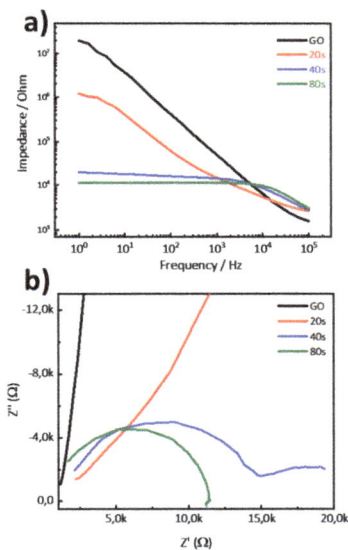

Figure 13: EIS characterization of thermal reduction/annealing of a GrO based device for biosensor applications

Organic Bioelectronics for Life Science and Healthcare   Materials Research Forum LLC
Materials Research Foundations **56** (2019) 185-242   doi: https://doi.org/10.21741/9781644900376-6

which facilitated their immobilization on surface modified silicon and glass wafers after spin-coating the GO layers. The spin-coating process was optimized on 4-inch wafers and provides highly homogeneous thin-films. The GO films can be converted to rGO by thermal annealing and deployed in a standard lithography process for patterning and writing electrical contacts. Figure 12 shows such rGO based nanoscale device coming out of this top-down fabrication approach. The methods enables fabrication of devices with electrode configurations, which can be decided arbitrarily such as show in the SEM images, and therefore provide a versatile platform for sensors integration. The devices show very low variations in the electrical properties and therefore are expected to be deployed for the development of high throughput electrical bioassays.

Reduction of GBM oxides: GO thin-films or devices prepared by one of the above mentioned methods may generally require improvement of their electrical characteristics before deploying them as electrical transducers. The high resistivity of GBM oxides can be reduced by restoration of the lattice defects, minimizing the density of the surface functional groups and improving the overlaps between individual layers towards enhancing the aromatic character of the graphitic lattice. The 'reduction' of GBMs is therefore an attempt to enhance the charge-carrier density in the graphitic lattice as well as improving the electrical interface to the metal contacts. In fact, GO reduction was the first method of graphene synthesis. A single layer-scaled synthesis of rGO was already stated by P. Boehm in 1962.[104] Detaching of Graphite oxide (GrO) can be achieved by regular heating of dispersed carbon powder with a small quantity of graphene flakes. Chemical exfoliation is achieved by using very strong oxidizing agents to produce GO from graphite. The GO is further chemically reduced by using hydrazine at high temperature annealing to form single layer graphene sheets. However, the quality of graphene produced by GrO reduction is inferior to graphene produced by the scotch-tape method due to abundance of several functional groups and defects. Thermal annealing of GBM oxides is one of the most popular approaches for device applications as it provides opportunity for the optimization of sensor characteristics in a controlled, minimally invasive and high throughput manner. Figure 13 shows such a thermal annealing effect in Ar atmosphere for GrO based devices converted to reduced graphite oxide (rGrO) and deployed later as glucose sensors. Graphs a and b shown in the figure represent Bode and Nyquist plots for the GO device and after annealing for 20, 40, and 80 s, respectively. Applying a 50mV sine wave with frequency ranging from 1 Hz to 100 kHz between the source and drain electrodes, EIS was recorded. In the successive plots, transition of very high resistivity of the GrO multilayers exhibiting a "capacitive" behavior to a typical resistive characteristics after the efficient reduction process and formation of rGrO devices can be seen.[91]

Thermal annealing of GBMs in specialized gaseous environments has also been used for doping.[105] Other than this, annealing methods can also be used to tune the work function of rGO thin-films.[79] In order to ascertain structure, surface and material properties, elaborate methods like microscopy and spectroscopy are routinely used for characterizing GBMs. GBMs, for their 2D material lattice show unique vibrational fingerprints in Raman spectroscopy, representative for their reduction state, material composition, crystallinity and orientation of atomic layers. Therefore, Raman analysis has become a characterization gold standard for GBMs. In addition to Raman, XPS and fluorescence spectroscopy techniques are regularly used for quality-control of GBMs during their production and processing.[106] Similarly, plasma induced reduction methods have also been described by Lee at al., who carried out a low temperature and normal pressure processes. A microplasma is created in the plasma chamber where Hydrogen has dissociated continuously and produces atomic hydrogen atoms. High conductivity flexible rGO thin-films were demonstrated by using the low temperature advantages of this method.[107]

In addition to thermal methods, electrochemical reduction of GBMs provide well-controlled processes and have been implemented by various research groups.[108-110] Electrochemical reduction of GO is carried out using an electrochemical cell where GO is deposited onto a conducting substrate to work as an electrode. Metal foams also have been used for the reduction of GO.[111] GBMs have also been reduced by photoactive reaction (photothermal and photochemical) where strong light sources such as lasers or arc discharges can be used to remove oxidative groups and repair the lattice structure.[112] In photochemical schemes, photogenerated electron-hole pairs enable different mechanisms for the removal of oxidative functional groups, leading to the restoration of sp^2 domains.[112, 113] Pei et al. provided a detailed overview of GO reduction methods compared the efficiencies and shortcomings of the above mentioned reduction approaches.[114]

5.　　Biosensor platforms of GBMs

GBMs as highly effective transducers, have found use in a variety of scientific and technological fields. Development of single component biological sensors using GBMs has been on the forefront. The use of GBMs in biosensors offers the development of promising devices with features like high sensitivity, label-free nature, real-time processing, multi-analyte sensing capability, miniaturization, and low power requirement. Electronic properties of 2D GBMs are greatly influenced by adsorption of molecules and atoms on their surface Such surface modifications at an electrically active surface may also serve as local 'doping sites'. The interaction of graphene with an analyte causes the introduction or withdrawal of electrons from the surface, which forms a basis for exploring the above mentioned nanomaterials for the development of highly sensitive sensors.

In the field of electrochemical biosensors, either graphene acts an electrode thus enhancing the electro-catalysis or provides high surface area for the immobilization of molecules. Most commonly, graphene and its composites are coated on the surface to enhance the peak current and decrease the redox potential of the target. The presence of functional groups at the edges of graphene controls the covalent immobilization of molecules and electron transfer rate. As an example, the graphene sheets can be used to realize a very useful design of direct charge transfer (DCT) electrodes. This design involves the fixing of graphene sheet as a channel between the source and drain, while it is coupled to the gate terminal by capacitance. Graphene also exhibits the chemical gating effect. In the case of graphene-based electrochemical bioassays, an event of target binding with the immobilized receptor causes the carrier concentrations to change, which alters the overall charge transport characteristics. GBMs have been deployed for detection of biomolecules using electrochemical impedance spectroscopy approaches.[115] Electrical impedance is a complex property, which involves resistive, capacitance and non-linear elements. Biomolecular interactions at the GBM surfaces influence the impedance characteristics of the device in specific ways and can be used for realization of bioassays.

Due to the high electron mobility and high surface-to-volume ratio, graphene-based immunosensors provide ultra-sensitive detection. As another brilliant feature, the graphene-based electrochemical sensors eliminate the problem of electrical noise. Graphene-based sensors show their potential to detect analytes even in a small volume of the sample with minimum false negatives or false positive results. The behavior of these devices is extremely sensitive to any surface adsorption/perturbation and is proportional to the analyte concentration. Electrochemical detection of glucose, dopamine, proteins, etc. has been demonstrated using graphene-modified electrodes.[77-81][116]

ISFETs based on GBMs: An ISFET is a voltage-controlled device that is able to alter current through a semiconducting channel by the usage of an electric field. In a GBM-ISFET, the GBM layer acts as 2D semiconductor channel between the source and drain electrodes. As charged biological molecules bind to the GBM surface, a proportional change of charge-carriers occur following the principles of electrochemical gating. Here, it is important to review the electrochemical gating at the interface between the GBM flakes and the electrolyte. A model for rGrO-electrolyte interface (rGrO as GBM) with negatively charged functional layer, originating from the oxidative functional groups, is shown in figure 14 a and b.[115] Field-effect response of the device while varying the

Organic Bioelectronics for Life Science and Healthcare   Materials Research Forum LLC
Materials Research Foundations **56** (2019) 185-242   doi: https://doi.org/10.21741/9781644900376-6

solution pH from 5 to 8 shows a consistent shift of Dirac-point towards positive gate voltages.[112-114] The model considers a negatively charged functional layer (FL) on top of rGrO, which attracts positively charged ions on the Outer Helmholtz Plane (OHP) and forms the electrical double layer (EDL). The red coloured line in figure 14a shows the electrostatic potential profile across this interface. Applying Gouy-Chapman theory for solid-liquid interfaces gives us the charge distribution at the OHP of the interface, which can be shown as-

$$\sigma_{OHP}(\psi_{OHP}) = -[(2k\,\varepsilon_0\varepsilon_r)/l_d\beta e]\sinh(\beta e\psi_{OHP}/2), \text{ Eq. 1}$$

here, ψ_{OHP} is the potential at OHP, $\beta = 1/k_b T$, and l_d is the Debye screening length, which can be given by $l_d = \frac{2\sqrt{\varepsilon_0\varepsilon_r k_b T}}{2N_a e^2 I}$. I in this equation is the total ionic strength of the solution, while ε_0 and ε_r are the permittivity of free space and relative permittivity, respectively, and e is the electronic charge. The electrostatic potential at the rGrO-electrolyte interface (ψ_0) can be given as-

$$\psi_0(\psi_{OHP}) = \psi_{OHP} - \sigma_{OHP}(\psi_{OHP})/C_{Stern}, \text{ Eq. 2}$$

where C_{Stern} is the specific capacitance of the Stern layer of thickness t_{Stern} given by $\varepsilon_0\varepsilon_{rStern}/t_{Stern}$.

The acid-base dissociation constants (pK_a/pK_b) of the ionizable functional groups in the FL determine the sign of the charge distribution (σ_0) and the potential (ψ_0) at the FL-Stern interface. Langmuir-Freundlich theory can be now be applied to calculate the net charge density by assuming highest surface charge densities at maximum protonation and dissociation of ionizable functional groups.[117] The charge-carrier density in the rGrO upon applying a particular voltage at the RE can now be given as-

$$\sigma_{rGrO}(\psi_{OHP}) = \psi_{rGrO} - \psi_0(\psi_{OHP}) \times C_{FL}, \text{ Eq. 3}$$

where ψ_{rGrO} is the electrostatic potential of rGrO. C_{FL} can be given as $C_{FL} = \varepsilon_0\varepsilon_{rFL}/t_{FL}$ with ε_{rFL} being the relative permittivity of the functional layer and t_{FL} the thickness of the functional layer.[75]

This model framework explains the effect of the electrolyte pH and the ionic strength on the field-effect characteristics of the rGrO devices. The framework also can be used for explaining the biosensor operation later on, such as charge detection and position of the ambipolar field-effect curves of a device deployed for DNA sensing. The model also includes the composition (such as permittivity and thickness) of the biofunctional layers on GBM-ISFETs. Based on this model, the effect of electrolyte pH and ionic strength on the rGrO devices can be evaluated and related to the biosensor operation for DNA sensing experiments.[85, 118, 119]

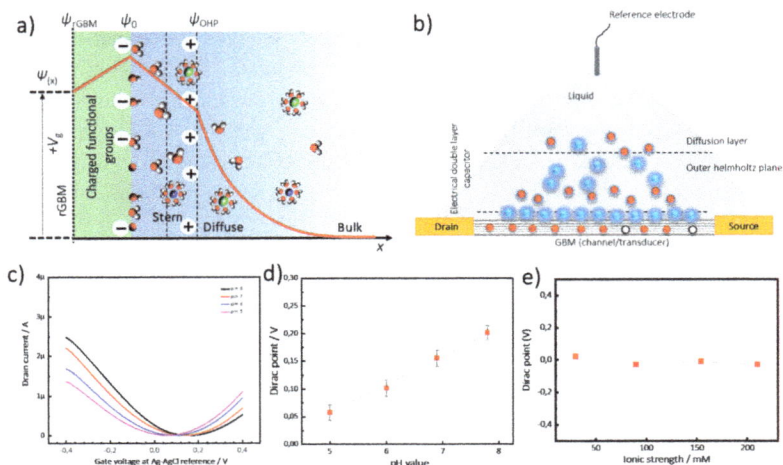

Figure 14: Operation of GBM ISFET biosensors. (a) Representation of the GBM-electrolyte electrochemical interface, containing the negatively charged functional layer as well as the Stern and the diffuse layers. The red line denotes the electrostatic potential profile across the solid-liquid interface, (b) ISFET configuration and build-up of solid-liquid interface for GBMs, (c) typical field-effect characteristics of a GBM-ISFET upon changing the solution pH. (c) field-effect characteristics of a typical device when deployed in liquid buffer solutions of different pH values, (d) typical pH sensor response, and (e) field-effect behavior of the devices shown for buffer solutions with different ionic strengths.

6. Biosensor platforms of GBMs: Applications in healthcare

The health-sector plays a critical role in the life of humans and animals. It is also the backbone of the world economy and therefore, monitoring of health is of foremost importance. The conventional techniques for health and safety monitoring are based on lab-based bio-chemical assays, immunoassays, chromatography, etc. Laboratory-based techniques have high sensitivity, but on the downside, they are time-consuming and expensive. The major disadvantage of conventional techniques is their lack of on-site testing capability. Thus, there is an urgent need to develop simpler, rapid, multiplex, selective, sensitive, and field-deployable devices for health diagnostics. Biosensors can potentially provide test results in minutes and help to make the decision on the spot. Incorporation of novel nanomaterials benefits in overcoming boundaries of conventional analytical methods and boosts the biosensing capability.[120] Nanomaterials-based bioassays have the potential to improve the analytical accuracy needed to detect the abnormal level of markers in complex blood and serum. One of the widely used nanomaterials in this regard is graphene.

In the following subsections, we discuss the application of graphene in biosensing assays for allergens, toxins, pesticides, and pathogens. The table given towards the end of this chapter will summarize different biosensor platforms based on GBMs that have been developed in recent years towards meeting the diagnostics challenges in healthcare.

Biosensors for allergens: The presence of allergens in processed agro and food products is associated with high risk to the consumer. Accidental exposure to even a trace amount of allergen may result in life-threatening reactions. Detection of allergens in food products is challenging, as they are often present in ultra-low concentration and due to the interference of complex food matrix. Allergies from nuts were reported to affect more than 2% of the US population and are more prevalent among children.[121, 122] Amongst the eight common allergens (milk, eggs, peanuts, nuts, fish, soy, wheat, and crustaceans), wheat and peanut allergies are most widespread. The allergic response appears to be primarily due to an immunological hypersensitivity mediated by allergen-specific immunoglobulin E.[122] Consumption of allergens often leads to digestive disorders, respiratory problems, edema, hypotension, urticaria, atopic dermatitis, and IgE-mediated anaphylactic shock. Conventional enzyme-linked immunosorbent assay and chemical analysis are usually performed in a centralized lab, requiring considerable analysis time and cost. Consequently, there is a need for the development of a sensitive, accurate and facile biosensor to detect these allergens in food products.

Recently, Chekin et al. developed a sensitive label-free voltammetric immunosensor for the detection of gluten.[123] In this sensor, porous rGO was covalently functionalized with anti-gliadin antibodies using pyrenecarboxylic acid as a linker molecule. The porous rGO was formed by etching pristine rGO using hydrogen peroxide. Porous rGO benefits in the increased active area and better mass transport from bulk to the electrode surface. This sensor achieved a an LoD of 1.2 ng/mL over a detection range of 1.2–34 ng/mL. Similarly, Eissa et al. reported a graphene based aptameric electrochemical biosensor for the detection of milk allergen.[124] The graphene-modified carbon electrodes were integrated with an aptamer targeting β-lactoglobulin (milk allergen). The adsorption of aptamer on graphene "turned-off" the signal from a redox couple present in the buffer. Upon binding of the allergen with the aptamer, negatively charged aptamer-protein complex desorbed from graphene and "turned-on" the signal. The LoD for β-lactoglobulin was calculated to be 20 pg/mL. In accordance with the growing trend, graphene nano-composites have also been used to detect food allergens. A graphene-gold nano-composite modified carbon electrode for the detection of peanut allergen was proposed by Sun et al.[125]. Gold NPs effectively prevented the aggregation of graphene sheets and enhanced the electron conductivity. A stem-loop DNA complementary to the Ara h1 gene (peanut allergen) was immobilized on the nano-composite. For the analysis, the Ara h1 gene was extracted from the commercial peanut milk and validated on the biosensor. The authors reason that heat treatment during commercial production often denatures Ara h1 proteins and hampers detection. However, the DNA remains intact and therefore can be used to verify the presence of peanut allergen. The biosensor showed high selectivity and reached a low LoD of 0.041 fM for the Ara h1 gene.

Zhang et al. reported a DNA aptamer based fluorescent assay for the shrimp allergen.[126]. In this work, the interaction of aptamer with tropomyosin (shrimp allergen) prevented its adsorption on the GO surface. The aptamer-tropomyosin complex was then dyed with OliGreen, which produced a positive fluorescent signal. The unreacted aptamers were adsorbed on the GO and could not be dyed. This tropomyosin assay had a LoD of 4.2 nM and a detection range of 0.5–50 µg/mL.

Biosensors for Bisphenol-A: Bisphenol-A (BPA), chemically known as 4,4`-(propane-2,2-diyl)diphenol, is a highly studied food contaminant. BPA is a precursor monomer applied during the synthesis of polyepoxides and thermoplastics. The majority of packaging materials and food containers, such as bottles, cans, tableware, and ovenware, are made using polycarbonate. Polyepoxides are internally coated on the cans and bottles for storing processed foods. BPA gradually leaches into the stored food products and its toxicity is widely reported in the literature.[127] BPA is a potent endocrine-disrupting compound (EDC). The chemical structure of BPA is similar to estradiol and diethylstilbestrol (endocrine hormones), so it has an affinity towards estrogen receptors. Low doses of BPA, even at sub-ng levels (0.23 ng/L), are found to be toxic to human health.[128, 129]. Consumption of BPA-contaminated food is known to affect the brain functions, endocrine gland, and reproductive organs[130]. Health departments throughout the world are noticing the detrimental effect of BPA on health and very recently BPA-free food containers are being promoted. Therefore, a number of publications have discussed the use of biosensors to detect the presence of BPA in food products.

Graphene and graphene based nano-composites are popular for detecting BPA in food samples.[131-133] In the majority of the biosensors, graphene nano-composites are employed for one-step direct electro-oxidation of BPA and measurement of the resultant current signal. A graphene-modified glassy carbon electrode was developed by Ntsendwana et al. to detect BPA in bottled water. Nano-dimensional graphene showed excellent electrocatalytic behavior and resulted in a detection limit of 0.469 µM with a detection range of 0.5–1 µM.[134] Similarly, Ndlovu et al. applied exfoliated graphene functionalized electrode to the detection of BPA.[135] The fabricated exfoliated graphene coated electrode eliminated the negative effect of phenol fouling during the determination of BPA. The device showed a detection range of 1.56–50 μM and a calculated detection limit of 0.76 μM. A graphene-CNT nano-composite synthesized by a green and facile route was loaded with platinum NPs and applied to the electrochemical detection of BPA by Zheng et al.[136]. The platinum-graphene-CNT composite benefited in large surface area and highly efficient accumulation ability. The authors reported a LoD of 0.42 μM and demonstrated the potential application for detecting BPA in thermal printing papers.

Graphene also increased the sensitivity of BPA detection while its other properties, such as fluorescence quenching, were explored. Zhou et al. developed a photoelectrochemical aptasensor for BPA.[137] In this paper, a nano-composite composed of titanium dioxide NPs and nitrogen-doped graphene was synthesized by a simple one-pot thermal treatment method. Compared with pristine titanium dioxide NPs, the nano-composite displayed enhanced performances, attributed to the presence of nitrogen-doped graphene. The nano-composite efficiently confined the recombination of photoinduced electron-hole pairs, increased charge transfer, and extended photoresponse. Aptasensor was successfully able to detect BPA and displayed a wide linear range from 1 fM-10 nM and low LoD of 0.3 fM.

Recently, Hu et al. proposed a "turn-on" fluorescent method for the detection of BPA.[138] The detection was based on Förster resonance energy transfer (FRET) between fluorescein-labeled BPA aptamer and magnetic GO. The interaction of BPA with aptamer adsorbed GO "turned-on" the fluorescence. The LoD of 0.071 ng/mL was obtained with the linear range of 0.2–10 ng/mL. In another work, Mitra et al. proposed a "turn-on" fluorescent assay for BPA using a dextran-fluorescein coated rGO.[139] In this approach, due to competitive interaction with BPA, the fluorescent probe detached itself from the graphene surface and regained its fluorescence. Recovered fluorescence was measured as the positive signal for the quantification of BPA.

Biosensor for Cancer Markers: Conventional instruments for cancer diagnosis including magnetic resonance imaging (MRI), computed tomography scan, are expensive and require long-waiting time, whilst the outcomes have not approached to the successful early-stage diagnosis yet. Due to the special properties of GBMs, e.g., good electrical and thermal conductivity, luminescence, and mechanic flexibility, these ultra-thin 2D nanostructures have been extensively used as platforms for detecting biomolecules and cells.

Carcinoembryonic antigen (CEA) is a protein that can be measured in the blood of a cancer patient. Recently, a label-free immunosensor based on the antibody-modified graphene FET was reported.[140] The surface modification is applied via a non-covalent functionalization and π-stacking using a pyrene and a reactive succinimide ester group to interact with graphene. The GFET biosensor shows the specific monitoring of the CEA protein in real-time with high sensitivity of <100 pg/mL.

An electrochemical immunosensor for carcinoembryonic antigen (CEA) using nanosilver-coated magnetic beads and gold–graphene nanolabels was developed.[141] The redox-active magnetic nanostructures, based on the combination of a magnetic nanocore, a layer of electroactive poly-o-phenylenediamine (PPD), and a silver metallic shell, displayed good adsorption properties for the attachment of capture antibodies and were integrated into a magnetic carbon paste electrode (CPE). Moreover, these magnetic nanocomposites can efficiently encapsulate the redox electroactive species avoiding the contamination of the electron mediators in the detection solution. AuNP-GO conjugated with Horseradish peroxidase (HRP) labeled detector antibodies were also used as labels in this sandwich-type immunoassay protocol. The doped PPD acted as cross-linkage and mediator for the electron transfer of HRP reduction of H_2O_2 monitored by differential pulse voltammetry (DPV). This nanoparticle-based immunocomposites allowed the detection of CEA at a concentration as low as 1.0 pg/mL and was successfully applied for its detection in clinical serum specimens.

A disposable electrochemical immunosensor for simultaneous assay of a panel of breast cancer tumor markers was constructed using an AuNPs–Gr-modified carbon electrode array and alkaline phosphatase (ALP)-labeled detector antibody (Ab2) functionalized Au cluster (AuCs)–Gr as nanocarriers.[142] The AgNPs deposition, catalyzed by both ALP and AuCs–Gr after the hydrolysis of 3-indoxyl phosphate was measured by anodic linear sweep voltammetric stripping analysis. This ultrasensitive cancer antigen (CA) multianalyte immunosensor, with LoDs of 1.5×10^{-3} U/mL (CA153), 3.4×10^{-4} U/mL (CA125), and 1.2×10^{-3} ng/mL (CEA), allowed a simple detection method for fast measurement of various tumor markers. The sensor platform also and avoided cross-talk, which represented significant clinical value for application in cancer screening, providing great potential for convenient point-of-care testing and commercial application.

An electrochemical immunosensor for CEA was developed by immobilizing an HRP-labeled anti-CEA antibody on an AuNPs–Gr composite modified glassy carbon electrode (GCE).[143] HRP was casted on the modified electrode surface not only as a blocker for non-specific sites but also as a catalyzer for redox reaction of hydroquinone (HQ) and H_2O_2. The biospecific interaction between CEA and the antibodies assembled on electrode surface resulted in the notably decreased cathodic peak current response of HQ due to the enhanced steric hindrance. Under the optimized conditions, the peak current change derived from the HQ DPV measurements was proportional to the CEA concentration from 0.10 to 80 ng/mL with a LOD of 0.04 ng/mL. In addition, this new immunosensing platform was applied to the determination of clinical serum specimens, with results in agreement with those derived from an enzyme linked immunosorbent assay (ELISA).

Biosensor for Cardiac Markers: Cardiac troponin-I (cTnI) is a biomarker of acute myocardial infarction. A sandwich-type cTnI biosensor was constructed by Liu et al. was based on GO-Ph-AuNP modified GCE as sensing platform and GO tailored with ferrocene (FcGO) as signal reporter labels.[144] cTnI capture antibodies were immobilized on both the WE surface and FcGO labels using aryldiazonium salt coupling chemistry.

cTnI has been label-free detected by using a one-step microwave-assisted unscrolling of CNTs to form functionalized rebar graphene (f-RGr).[145] Tuteja et al. made a functionalized rebar graphene (fRG) through the microwave-assisted unscrolling of CTs. They assembled the fRG in a FET configuration and this electric chip showed enhanced electronic properties. Then via modifying anti-cardiac marker troponin I (cTnI) antibodies, they constructed an immunosensor for the detection of cTnI and exhibited high sensitivity towards the cTnI. Well-characterized f-RGr functionalized with specific anti-cTnI antibodies via available functional carboxylic groups on f-RGr was immobilized on interdigitated electrodes, and then used as a biochip device for the detection of the cardiac marker. In the RGr nanocomposite, the CNTs form networks that serve as bridges across graphene boundaries and provide greater surface area resulting in improved electronic properties as compared to MWCNTs. The developed biochip was demonstrated to offer higher selectivity and sensitivity in comparison to other conventional methods, which was mainly attributed to the electron mobility of the f-RGr channel with highly specific antibodies. In the f-RGr based FET sensor, RGr acted as the channel where resistance was monitored with respect to voltage in a concentration dependent manner of the target antigen cTnI. A change in resistance was observed due to the charges induced in the f-RGr channel with the formation of the antibodyantigen immunocomplex, the conductance decreasing gradually with the increase in concentration of highly charged cTnI protein.

Recently developed approach, Kumar et al., 2015 reported the development of aptamer functionalized rGO/CNT nanostructured electrodes for the label free electrochemical sensing of cardiac marker myoglobin (Mb).[146] Tuteja et al. reported a method of producing graphene films through carboxylic acid by lithium ion intercalation mediated exfoliation process.[147] After that, the graphene sheets were used to construct an electrochemical conductance mode micro-device in the drain-source configuration for the labelfree detection of cTnI. The experimental results showed that detection limit is as low as 0.1 pg/mL. Singal et al. demonstrated the covalently linking of PtNPs to GR-CNT surface, after which, the resulting Pt/GR-CNT nanocomposite was further employed for the construction of a sensitive impedimetric cTnI sensor.[148] It was concluded that the superior sensitivity of the sensor was mainly ascribed to the combination of large surface area GR-CNT and ultra-high density of active graphene edge planes together with high surface to volume ratio of electroactive PtNPs.

Tuteja et al. employed electrochemical functionalization of graphene with 2-Aminobenzyl amine (2-ABA) as a derivative of aniline possessing an extra aliphatic amine group and carried out immunosensing of cardiac troponin I (cTnI).[149] Here, Graphene dispersion was drop casted onto IDEs, and modified electrochemically with 2-ABA. The cTnI antibodies attached the 2-ABA functionalized graphene (f-GN) sensor via Schiff reaction-based chemistry. The proposed method showed the linear range of antigen detection between (0.01-1 ng/mL) and the limit of detection was achieved in 0.01 ng/mL.

Recently, a label-free immunosensor for cardiac biomarker myoglobin (cMyo) detection using a GQDs-modified electrode, with a detection limit of 0.01 ng/mL was demonstrated.[150] To achieve a higher sensitivity, specific anti-myoglobin antibodies were selected and conjugated with the electrode. The results suggested that the values of charge transfer resistance (R_{ct}) of the immunosensor were found to increase linearly (from 0.20 to 0.31 kΩ) with the increasing cMyo concentration (from 0.01 to 100 ng/mL), which was due to the effective immobilization of antibodies and excellent electron mobility of the sensor. The above conclusions have shown that GQDs will become increasingly widely used as biosensors for detection and tracking in the medical diagnostics area.

Biosensors for influenza: The RNA viruses, Orthomyxoviruses family, cause pandemic influenza, resulting in high morbidity and mortality every year. Influenza virus A, B, and C viruses belonging to this family infects birds, swine, cattle, and humans as well.[151] Influenza outbreaks result in socioeconomic losses around the world. A recent report estimated a loss of $309.9 million in poultry production and related businesses in greater Minnesota alone.[152] Hemagglutinin (HA) and neuraminidase (NA) are the two glycoproteins located on the virus surface and act as antigens. Compared to the conventional methods like ELISA and nucleic acid amplification, biosensors are rapid, practical, and economical.

Recently, Singh et al. proposed a label-free detection of influenza viruses using rGO based electro-immunosensor.[153] The authors integrated a microfluidic platform with rGO based sensing layer. The working electrode was functionalized with rGO followed

Organic Bioelectronics for Life Science and Healthcare   Materials Research Forum LLC
Materials Research Foundations **56** (2019) 185-242   doi: https://doi.org/10.21741/9781644900376-6

by anti-H1N1 antibodies through carbodiimide linking. Chronoamperometric analysis showed an enhanced LoD of 0.5 PFU/mL and a detection range of 1–104 PFU/mL.

Huang et al. developed a dual graphene nano-composite based electro-immunoassay for the detection of avian influenza virus H7.[154] In this sandwich-type immunoassay, the graphene-gold nano-composite modified electrode was functionalized with anti-H7 capture antibodies. Additionally, anti-H7 detecting antibodies immobilized on graphene-silver nano-composite was used as labels. Silver NPs produce very distinct signals enabling electrochemical analysis. The performed electro-immunoassay showed high signal amplification and a detection range of 1.6 mg/mL-16 ng/mL with a LoD of 1.6 pg/mL. A microfluidic integrated GO ISFET for for viral genosensing via a flow-through strategy and detection of H5N1 gene was reported by Chan et al.[155] In this work, different DNA probe immobilization approaches were compared and π-π stacking was found to be optimal. The electrical detection of viral DNA achieved a LoD of 5 pM.

Biosensors for microbial pathogens: Conventional methods employed for pathogen and endotoxin detection are tedious and require a few days for the results. Electrical biosensing technologies for microbial pathogens detection are therefore attractive alternative. In recent years, graphene has significantly boosted the performances of biosensors for pathogen detection. Pandey et al. developed a graphene-interfaced capacitive biosensor for E. coli detection.[156] The graphene coated capacitor IDEs were functionalized with anti-E. coli O157:H7 antibodies using pyrenebutanoic acid succinimidyl ester as a linker. One-step direct capturing of E. coli changed the capacitance of sensor and resulted in a sensitivity as low as 10–100 cells/mL. The chip also incorporated packed silica microbeads zones to filter and enrich the norovirus-infected samples. The nano-composite contributed in a reliable surface for aptamer immobilization and facilitated in signal enhancement. Electrochemical studies revealed LoD of 100 pM for norovirus, with a total detection time of 35 min. In another work, Chan et al. fabricated a rGO based electrochemical biosensor for the detection of botulinum toxin, which is a neurotoxic protein produced by the *Clostridium Botulinum*.[157] Here, a gold electrode, first modified with reduced GO, was functionalized with SNAP-25-GFP peptide via pyrenebutyric acid linker. The introduction of botulinum toxin specifically cleaved SNAP-25-GFP and decreased the surface passivity of graphene. This enzymatic activity of botulinum was detected by monitoring the enhancement in the electrochemical signal of redox couple present in the buffer. A linear detection range of 1 pg/mL–1 ng/mL and LoD of 8.6 pg/mL for botulinum was achieved.

Zuo et al. reported a polydimethylsiloxane/paper/glass hybrid microfluidic chip for multiplexed detection of the pathogen using aptamer adsorbed GO[158]. A one-step "turn-on" homogenous assay was performed to detect *L. Acidophilus*, *S. Aureus*, and *S. Enterica*. The Cy3 labeled aptamer adsorbed onto GO interacted with the pathogen, resulting in fluorescence recovery. The concentration of pathogens was determined by analysis of the fluorescence intensity. Spiked samples of *L. Acidophilus* were detected using this platform with LoD of 11 CFU/mL. Recently, Shi et al. developed a FRET based genosensor using graphene QDs and gold NPs for the detection of Staphylococcus aureus [159]. This genosensor was realized by immobilizing capture probes on graphene QDs and reporter probes on gold NPs. The capture probes and reporter probes co-hybridized upon addition of target DNA (from *S. Aureus*), which brought QDs and gold NPs into close proximity. The excitation of QDs produced a fluorescence response that was quenched by a proximal gold NPs. The change in fluorescence signal was used to quantify and a LoD of 1 nM for bacterial DNA was obtained.

Biosensors for mycotoxin: Mycotoxins are a group of toxic secondary metabolites of fungi such as Aspergillus, Fusarium, and Penicillium, and have different organic structures.[160] Deoxynivalenol, zearalenone, ochratoxin, fumonisins, and aflatoxins are common mycotoxins that are often found in food. Mycotoxins can cause acute and chronic illnesses, like cancer, and impair vital organs such as the liver, kidney, and brain. Since fungi produce these toxins, mycotoxins are associated with diseased or moldy crops and rotten foods. As per a report from the United Nations, 25% of agricultural products throughout the world is remarkably contaminated with mycotoxins.[161] Recently, Srivastava et al. reported a GO based impedimetric biosensor for the detection of aflatoxin.[162] In this study, GO was coated on the gold WE, followed by functionalization with aflatoxin antibodies. The antibodies were attached on the edges of GO using carbodiimide crosslinking. This biosensor showed a detection range of 0.5–5 ng/mL, 0.23 ng/mL detection limit, and a stability of 5 weeks. Several graphene and NPs based composites have become a popular substrate in many biosensors.[163] Lu et al. proposed a label-free electrochemical immunosensor for the detection of two mycotoxins, fumonisin B1 (FB1) and deoxynivalenol (DON) using graphene-gold nano-composite.[164] In this work, the carbon WE was modified by polypyrrole, electrochemically rGO and gold NPs. The nano-composite was then functionalized with an anti-toxin antibody. The gold NPs attached on rGO sheet resulted in an improved electrocatalytic activity and loading of the antibodies due to the increased surface area. This sensor exhibited an enhanced sensitivity and wide linear detection range of 4.2 ppb and 0.2–4.5 ppm for FB1 and 8.6 ppb and 0.05–1 ppm for DON, respectively. Similarly, Gan et al. developed a GO and magnetic NPs composite for the electrochemiluminescent detection of aflatoxin.[165] In this assay, graphene nano-composite was employed as an adsorbent to extract aflatoxin from food samples. Additionally, a cadmium telluride QDs and CNT composites were synthesized for the labeling of antibody. The presence of QDs in the immunocomplex produced an electrochemiluminescent signal, which was used for quantification. A detection range from 1 pg/mL–10 µg/mL, with a very low LoD of 0.3 pg/mL was obtained.

Apart from electrochemical biosensors, graphene has also been used in fluorescent biosensors. In the majority of the biosensors, graphene acts as a fluorescence quencher and therefore works as a "turn-on/turn-off" biosensor.[166] Here, Zhang et al. reported a fluorescent sensor for aflatoxin using GO. Carboxyl-X-rhodamine labeled aptamers were adsorbed on GO for target detection. Adsorption of aptamer "turned-off" the fluorescent signal and protected it from nuclease cleavage. The addition of aflatoxin in the reaction mixture desorbed the aptamer and a "turned-on" signal. The assay was highly selective against other common molecules found in foods and showed a detection range of 1.0–100 ng/mL for aflatoxin.[167] Recently, Goud et al. compared the quenching properties of graphite, GO and carboxy GO for aptamer based sensing of zearalenone (ZEN).[168] Carboxy GO was found to be the most efficient quencher and the assay was able to detect ZEN in the concentration range of 0.5–64 ng/mL with a limit of detection of 0.5 ng/mL. This aptamer based bioassay was effectively applied to the investigation of ZEN in alcoholic beverages.

Detection of pesticides: The extensive use of pesticides in agriculture leads to their inclusion in the food chain questioning the quality and safety of food products.[169] The pesticide residues in agro products adversely affect human and animal health because of their high biological activity and toxicity. The WHO under the Pest Control Products Act regulates the use of pesticides and has set a maximum residue limit for each pesticide in food. For these reasons, the development of rapid, selective and sensitive sensing technologies that allow determination of pesticides remains a challenge for the scientific community. In this regard, graphene has emerged as an interesting material for the development of pesticides-sensing devices. Graphene has been used as a substrate, signal amplifier, as well as nano-catalyst. A number of biosensors for pesticides are based on investigating the enzymatic activity of acetylcholinesterase (AChE).[170-172] AChE catalyzes the breakdown of acetylcholine, a neurotransmitter, into choline. Pesticides inhibit the activity of AChE and therefore a reduction in enzymatic activity is used a measure to quantify pesticides. Li et al. reported a porous-rGO modified glassy carbon electrode for the detection of carbaryl.[173] Carbaryl inhibited the activity of immobilized AChE leading to a decrease in oxidation current of enzyme product. Herein, the porous-rGO not only provided increased surface area but also facilitated the diffusion and mass transport of reactants. It was found that the inhibition activity of carbaryl is proportional to its concentration ranging from 0.001–0.05 µg/mL with a LoD of 0.5

ng/mL. Zhang et al. reported a similar biosensor for paraoxon.[174] Here, a convenient methodology to synthesize functionalized GO was worked out, which had an affinity for histidine-tagged AChE. According to the inhibition of AChE activity by paraoxon, a LoD of 0.65 nM was obtained. Oliveira et al. reported a bi-enzymatic biosensor for carbamates based on graphene doped carbon paste electrode.[175] The electrode was further modified with a mixture of polyphenoloxidases (laccase and tyrosinase), gold NPs, and chitosan. The polyphenoloxidases provided high and selective catalytic activity towards phenolic compounds, whereas chitosan, gold NPs, and graphene aided in the high superficial area, conductivity, and electrocatalytic activity. The biosensor was successfully applied to the detection of multi-analytes, including formetanate hydrochloride, carbaryl, propoxur, and ziram in citrus samples.

Immunosensors for pesticides detection using biofunctionalized graphene were reported by Mehta et al. and Sharma et al.[176, 177] In these works, the authors functionalized carboxyl GO with antibodies for electrochemical detection of parathion and diuron, respectively. A novel "on-off-on" switch based photoelectrochemical aptasensor for acetamiprid was recently reported by Yan et al.[178] The authors presented a thiophene-sulfur-doped graphene and zinc oxide nanocomposite for achieving sensitive photo-electrochemical aptasensors. Here, doped graphene improved the photoactivity and photostability of zinc oxide by enhancing the interfacial charge transfer and decreasing the band gap. The thiophene-sulfur-doped graphene exhibited higher photocurrent sensitivity of about 2.6 times than that of pristine graphene nano-composite. Desorption of aptamer "switched-on" the biosensor and the response for acetamiprid was obtained in the range of 1–1000 ng/mL with a LoD of 0.33 ng/mL. Liang et al. proposed a sensitive electrochemiluminescent assay integrating graphene nanosheets, cadmium telluride QDs, and AChE.[171, 178, 179] The bio-nano composite yielded a cathodic electrochemiluminescent emitter for organophosphate sensing. The AChE reactions consumed dissolved oxygen and inhibited electrochemiluminescence. The addition of organophosphates inhibited enzymatic activity and thus promoted electrochemiluminescence. Under the optimized experimental conditions, the detection limit for methyl parathion was found to be as low as 0.06 ng/mL. Similarly, a fluorescent biosensor using graphene QDs was reported by Caballero-Díaz et al.[180] Graphene QDs are a graphene sheet with a radial thickness of few nanometers and lateral size of less than 100 nm. They have unique properties due to their exceptional quantum confinement and edge effects. This paper combined the use of nitrogen-doped graphene QDs and AChE as biorecognition element for fenoxycarb determination. The enzymatic products quenched the native fluorescence of graphene QDs. Introduction of fenoxycarb inhibited the enzyme activity and recovered the fluorescence. The assay resulted in sensitive detection of pesticide at concentrations ranging from 6–70 μM and has a detection limit of 3.15 μM.

In the field of biodetection researches have probed tremendous variations of possible applications of GO and reduced GO in unmodified or functionalized forms as well as their composites with diverse polymers and inorganic materials.[181]

Biosensors for progesterone: Progesterone is a sex hormone that controls the menstrual cycle, pregnancy, animal growth, and development in females of various species [182]. Quantifying the levels of progesterone in animal milk helps predict their reproductive status. Additionally, consumption of elevated levels of progesterone in human is reported to cause breast tenderness, stomach upset, vaginal discharge, breast and lung cancers.[182] The cost of erroneous estrus detection in USA is calculated up to $600 million annually. Precise detection of progesterone is essential to avoid economic losses and long calving intervals. Unfortunately, there are no reliable PoC tools available to test the concentrations of progesterone in milk. Graphene based biosensors can greatly assist farmers to have a portable and easy-to-use biodetector for measurement of progesterone levels.

Recently, Arvand et al. developed an electroanalytical platform for the simultaneous determination of progesterone and estradiol using graphene QDs.[183] For this purpose, graphene QDs doped poly(sulfosalicylic acid) was immobilized onto the WE of the sensor. The analytes were detected based on their inherent electro-catalytic signal. The graphene QDs complexes exhibited a distinct response towards the hormones. Under the optimal conditions, the sensor showed a detection limit of 0.23 nM with a detection range of 0.001–6.0 μM for estradiol and a LoD of 0.31 nM for progesterone. Dong et al. reported a thionine-doped GO for electrochemical immunosensing of progesterone.[184] The GO was then functionalized with P4coating antigen. The analysis was based on the competitive interaction of surface coated P4 antigen and free progesterone from the sample with biotinylated anti-progesterone antibodies in the buffer. The proposed immunosensor showed a detection limit of 0.0063 ng/mL for progesterone in milk samples.

Single cell assays: Graphitic materials show high biocompatibility for interfacing with living matter.[185] GBMs, therefore exhibit ideal properties to use as transducer alements for developing platforms for cell-on-chip and organs-on-chip platforms. Given their high carrier mobilities, low noise, assembly of high performance screening platforms employing electrical impedance spectroscopy and field-effect has been possible and display excellent resolutions and stability.[86] The advantages of using a planar and transparent material is evident in the combination of existing optical approaches and design of hybrid platforms for parallel screening techniques. With the advent of scalable approaches for fabrication of GBM based platform, arrays of such devices are fabricated on a chip, integrated with microfluidics solutions and deployed for high through-put screening.[186] Graphene coated electrochemical chips have been used for screening of cancer cells where specific recognition of cells was performed by surface modification of graphene by AS1411 aptamers.[187] Other than impedance spectroscopy and other AC readout approaches, graphene-FETs have been deployed for detection of cells and proteins for screening purposes.[85-87]

7.   Challenges and opportunities

Graphene based materials, have already been used as active components in a variety of sensor platforms that can be deployed in PoC systems, in wide application areas such as cancer-screening, immunoassays, detection of hormone levels and to monitor outer environments closely related to personal healthcare. The wide range of physicochemical characteristics available with GBMs make them an ideal candidate for sensor integration at nanoscale.[93] The next challenges and opportunity for the use of GBMs are truly in the efforts to standardize the production of GBMs, stratification of application specific properties and large-scale integration for the assembly of nanoscale sensor platforms. Cost-reduction by integration of chemically produced GBMs remains a unique opportunity for further successes in biosensor integration. In addition, GBMs also give unique opportunities to integrate parallel readout technologies using optical and electromechanical signal transduction, which are truly beneficial for development of stable and reliable bioassays. Clinical validation of bioassays require high throughput analyses, a topic which can only be addressed by solving scalability and performance issues. Further, realization of GBM biosensors with added functionalities such as flexible and transparent polymer substrates will allow their use in wearable for medical-diagnostics and health monitoring.[188] On the fundamental side, new strategies for the realization of smart biofunctional layers with highly specific receptor-analyte interaction, pH and temperature tolerance of the sensor signals, and minimal non-specific biomolecular adsorption on the surfaces are expected to solve major performance issues for GBM bioelectronics.[46] Wetting at 2D interfaces is also a novel topic with a lot of resonance in designing new solid-liquid interfaces, which may provide possibilities to tune biomolecular interactions at surfaces and address some of the fundamental challenges currently experienced with the implementation of nanomaterials for future healthcare applications.[189-193]

Table 1. Graphene based biosensors for Healthcare

Analyte	Transducer	Technique	LoD	Ref.
IgG/Anti IgG antibody	Multilayer TRGO/Au NPs, annealed in Ar, 200°C	ISFET	2 ng/ml – 20 mg/ml	[194]
DNA/PNA (peptide nucleic acid)	Chemical rGO, Ar, 150 °C	ISFET	100 fM	[195]
Escherichia coli/ anti-E. coli antibody	Thermal rGO on Au IDEs	ISFET	10 – 10000 cfu mL^{-1}	[196]
DNA/DNA	Pt NPs-GO films, Ar/H$_2$,1000°C, 2h	ISFET	2.4 nM	[197]
Matrilysin activity using polypeptide (JR2EC)	drop casted rGO	EIS/ISFET	10 ng/mL, (buffer), 40 ng/mL (Human plasma)	[198]
brain natriuretic peptide (BNP) /anti-BNP Ab	Pt NPs coated rGO	ISFET	100 fM (filtered blood)	[199]
pH sensing, cell adhesion detection	DEP trapped GO, 300 °C in Ar	EIS, ISFET	Single cell resolution	[86]
Urea/urease	Drop-casting, GO, hydrazine at 80 °C overnight	ISFET	1–1000 µm with a LOD of 1 µm	[40]
Ebola Virus Glycoprotein (EGP) /Ebola virus antibody	GO, 400°C for 10 min, Au NPs sputtered on RGO	ISFET	1 ng/ml (diluted serum/plasma)	[200]
H5N1 avian infuenza virus DNA sequence	spin-coated GO, hydrazine, 85 °C	ISFET	5 pM, assay time 1 h	[155]
Pesticide				
Organophosphate	Graphene, polyaniline	Electrochemical	20 ng/mL	[38]
Malation, methidathion, chlorpyrifos ethyl	Nafion, Ag NPs, Amine-rGO	Electrochemical	4.5 ng/mL, 9.5 ng/mL, 14 ng/mL	[40]
Carbaryl	Porous-rGO	Electrochemical	0.5 ng/mL	[41]
Paraoxon	Functionalized-GO	Electrochemical	0.65 nM	[42]
Carbamates	Graphene doped carbon paste	Electrochemical	1.68 nM	[43]
Parathion	Graphene	Electrochemical	52 pg/L	[99]
Diuron	Functionalized graphene–GO	Electrochemical	0.01 pg/mL	[22]
Acetamiprid	Thiophene doped grapheme, ZnO NPs	Photoelectrochemical	0.33 ng/mL	[45]
Methyl parathion	Graphene nanosheets, CdTe Qds	Electrochemiluminescence	0.06 ng/mL	[39]
Fenoxycarb	Graphene QDs	Fluorescence	3.15 µM	[46]

Analyte	Material	Method	LOD	Ref
Mycotoxin				
Aflatoxin	GO	Electrochemical	0.23 ng/mL	[49]
Fumonisin B1, Deoxynivalenol	Polypyrrole, rGO, Au NPs	Electrochemical	4.2 ppb, 8.6 ppb	[51]
Aflatoxin	GO, magnetic NPs	Electrochemical	0.3 pg/mL	[52]
Ochratoxin A	GO	Electrochemiluminescence	18.7 nM	[53]
Aflatoxin	GO	Fluorescence	0.35 ng/mL	[54]
Zearalenone	Carboxy GO	Fluorescence	0.5 ng/mL	[55]
Allergens				
Gluten	Porous-rGO	Electrochemical	1.2 ng/mL	[58]
Milk	Graphene	Electrochemical	20 pg/mL	[59]
Peanut	Graphene, Au NPs	Electrochemical	0.041 fM	[60]
Shrimp	GO	Fluorescence	4.2 nM	[61]
Bisphenol-A				
BPA	AuPd NPs, graphene nanosheets	Electrochemical	8 nM	[66]
BPA	Polypyrrole, graphene QDs	Electrochemical	40 nM	[67]
BPA	GO, hydroxyapatite	Electrochemical	60 pM	[68]
BPA	Nano-dimensional graphene	Electrochemical	0.469 μM	[69]
BPA	Exfoliated graphene	Electrochemical	0.76 μM	[70]
BPA	Graphene, CNT, Pt NPs	Electrochemical	0.42 μM	[71]
BPA	Nitrogen-doped graphene, TiO_2 NPs	Photoelectrochemical	0.3 fM	[72]
BPA	Magnetic GO	Fluorescence	71 pg/mL	[73]
BPA	Dextran-fluorescein coated rGO	Fluorescence	8 pM	[74]
Pathogens				
E. coli	Graphene	Capacitance	10 cells/mL	[75]
Botulinum toxin	rGO	Electrochemical	8.6 pg/mL	[76]
L. acidophilus	GO	Fluorescence	11 CFU/mL	[77]
S. aureus	Graphene QDs, Au NPs	FRET	1 nM	[78]
Cancer Markers				
carcinoembryonic antigen	Graphene	FET	100 pg/ml	[79]
carcinoembryonic antigen	gold–graphene nanolabels	Electrochemical	1.0 pg mL^{-1}	[80]
(CA153), (CA125), 1(CEA)	gold–graphene nanolabels	Electrochemical	1.5×10−3 U mL−1 (CA153), 3.4×10−4 U mL−1 (CA125), and 1.2×10−3 ng mL−1 (CEA),	[81]
CEA	gold nanoparticle–graphene composite	Electrochemical	0.04 ng mL^{-1}	[82]
Breast cancer (BRCA 1 gene)	Graphene-Gold NPs	Electrochemical	5.896 femtogram/ml	[100]

Analyte	Material	Method	LOD	Ref
overexpressed nucleolin (Any type of cancer)	nanocomposite Graphene	Electrochemical	794 cells/mL	[101]
Anti-CRP Ab	Graphene	CRET	0.93 ng/mL	[102]
Pyrophosphate (PPi) Melanoma cancer	GO	Fluorescence	0.60×10^{-7} M	[103]
Cardiac Markers				
cTnI	Nanocomposites based on GO and AuNPs	Electrochemical	0.05 ng mL-1	[83]
cTnI	functionalized rebar graphene (frG)	FET	1 pg/mL	[84]
Myoglobin	functionalized rGO/CNT nanostructures	Electrochemical	0.34 ng/mL	[85]
cTnI	Graphene	FET	0.1 pg/mL	[86]
cTnI	Pt/GR-CNT nanocomposite	Electrochemical	1.0 pg mL−1	[87]
cTnI	Graphene nanocomposite	FET	0.01 ng/mL	[88]
Myoglobin	Graphene QDs	Electrochemical	0.01 ng mL−1	[89]
Progesterone				
Progesterone, estradiol	Graphene QDs	Electrochemical	0.23 nmol/L, 0.31 nmol/L	[92]
Progesterone	Thionine doped GO	Electrochemical	0.0063 ng/mL	[93]
Influenza				
H1N1	rGO	Electrochemical	0.5 PFU/mL	[96]
Influenza virus H7	Graphene, Au NPs, Graphene, Ag NPs	Electrochemical	1.6 pg/mL	[97]
H5N1	GO	Transistor	5 pM	[98]

References

[1]   Jariwala, D., et al., *Carbon nanomaterials for electronics, optoelectronics, photovoltaics, and sensing.* Chem. Soc. Rev., 2013. **42**(7): p. 2824-2860. https://doi.org/10.1039/C2CS35335K

[2]   Ajayan, P.M., *Nanotubes from carbon.* Chemical Reviews, 1999. **99**: p. 1787–1799. https://doi.org/10.1021/cr970102g

[3]   Das Sarma, S., et al., *Electronic transport in two-dimensional graphene.* Reviews of Modern Physics, 2011. **83**(2): p. 407-470. https://doi.org/10.1103/RevModPhys.83.407

[4]   Novoselov, K.S., et al., *Two-dimensional atomic crystals.* Proc Natl Acad Sci U S A, 2005. **102**(30): p. 10451-3. https://doi.org/10.1073/pnas.0502848102

[5]   Novoselov, K.S., et al., *A roadmap for graphene.* Nature, 2012. **490**(7419): p. 192-200. https://doi.org/10.1038/nature11458

[6]   Horng, J., et al., *Drude conductivity of Dirac fermions in graphene.* Physical Review B, 2011. **83**(16). https://doi.org/10.1103/PhysRevB.83.165113

[7]   Hwang, E.H., S. Adam, and S.D. Sarma, *Carrier transport in two-dimensional graphene layers.* Phys Rev Lett, 2007. **98**(18): p. 186806. https://doi.org/10.1103/PhysRevLett.98.186806

[8]   K. S. Novoselov, et al., Electrical field effet in atomically thin carbon films graphene. Science, 2004. **306**: p. 666-669. https://doi.org/10.1126/science.1102896

[9]   Craciun, M.F., et al., *Tuneable electronic properties in graphene.* Nano Today, 2011. **6**(1): p. 42-60. https://doi.org/10.1016/j.nantod.2010.12.001

[10]   Jens Baringhaus, et al., Exceptional ballistic transport in epitaxial graphene nanoribbons. Nature, 2014. **506**: p. 349-354. https://doi.org/10.1038/nature12952

[11]   Du, X., et al., *Approaching ballistic transport in suspended graphene.* Nat Nanotechnol, 2008. **3**(8): p. 491-5. https://doi.org/10.1038/nnano.2008.199

[12]   F. Bonaccorso, Z.S., T. Hasan and A. C. Ferrari, *Graphene photonics and optoelectronics.* Nature Photonics, 2010. **4**: p. 611-622. https://doi.org/10.1038/nphoton.2010.186

[13]   Garaj, S., et al., *Graphene as a subnanometre trans-electrode membrane.* Nature, 2010. **467**(7312): p. 190-3. https://doi.org/10.1038/nature09379

[14]   Y.-M. Lin, et al., 100-GHz Transistors from Wafer-Scale Epitaxial Graphene. Science, 2010. **327**: p. 662. https://doi.org/10.1126/science.1184289

[15]   Barton, R.A., et al., High, size-dependent quality factor in an array of graphene mechanical resonators. Nano Lett, 2011. **11**(3): p. 1232-6. https://doi.org/10.1021/nl1042227

[16]   Goki Eda and Stefan A. Maier, *Two-Dimensional Crystals: Managing Light for Optoelectronics.* ACS Nano, 2013. **7**(7): p. 5660–5665. https://doi.org/10.1021/nn403159y

[17]   Yanqing Wu, et al., *State-of-the-Art Graphene High-Frequency Electronics.* Nano Letters, 2012. **12**: p. 3062−3067. https://doi.org/10.1021/nl300904k

[18]   Keller, D., C. Bustamante, and R.W. Keller, *Imaging of single uncoated DNA molecules by scanning tunneling microscopy.* Proc. Natl. Acad. Sci. U.S.A., 1989. **86**: p. 5356-5360. https://doi.org/10.1073/pnas.86.14.5356

[19]   Chun-Hua, L., et al., *A Graphene Platform for Sensing Biomolecules.* Angewandte Chemie International Edition, 2009. **48**(26): p. 4785-4787. https://doi.org/10.1002/anie.200901479

[20]   Bart H. van der Schoot and P. Bergveld, *ISFET Based Enzyme Sensors.* Biosensors, 1987. **3**(88): p. 161-186. https://doi.org/10.1016/0265-928X(87)80025-1

[21]   Bergveld, P., Development of an Ion-Sensitive Solid-State Device for Neurophysiological Measurements. IEEE Transactions on biomedical engineering, 1970: p. 70-71. https://doi.org/10.1109/TBME.1970.4502688

[22]   Bergveld, P., Thirty years of ISFETOLOGY - What happend in the past 30 years and what may happen in the next thirty years. Sens. Actuators, B, 2003. **88**: p. 1. https://doi.org/10.1016/S0925-4005(02)00301-5

[23]   Delle, L.E., et al., Scalable fabrication and application of nanoscale IDE-arrays as multi-electrode platform for label-free biosensing. Sensors and Actuators B: Chemical, 2018. **265**: p. 115-125. https://doi.org/10.1016/j.snb.2018.02.174

[24]   Pachauri, V. and S. Ingebrandt, *Biologically sensitive field-effect transistors: from ISFETs to NanoFETs.* Essays in Biochemistry, 2016. **60**(1): p. 81-90. https://doi.org/10.1042/EBC20150009

[25]   Achim, M., et al., Wafer-Scale Nanoimprint Lithography Process Towards Complementary Silicon Nanowire Field-Effect Transistors for Biosensor Applications. physica status solidi (a), 2018. **0**(0).

[26]   Janegitz, B.C., et al., *The application of graphene for in vitro and in vivo electrochemical biosensing.* Biosensors and Bioelectronics, 2017. **89**: p. 224-233. https://doi.org/10.1016/j.bios.2016.03.026

[27]   Bodenmann, A.K. and A.H. MacDonald, *Graphene: Exploring carbon flatland.* Physics Today, 2007. **60**(8): p. 35-41. https://doi.org/10.1063/1.2774096

[28]   Sutter, P., *Epitaxial graphene: How silicon leaves the scene.* Nature Materials, 2009. **8**: p. 171-172. https://doi.org/10.1038/nmat2392

[29]   Guanxiong Liu, et al., Epitaxial Graphene Nanoribbon Array Fabrication Using BCP-Assisted Nanolithography. ACS Nano, 2012. **6**(8): p. 6786–6792. https://doi.org/10.1021/nn301515a

[30]   Dabrowski, J., et al., Understanding the growth mechanism of graphene on Ge/Si(001) surfaces. Sci Rep, 2016. **6**: p. 31639. https://doi.org/10.1038/srep31639

[31]   Sutter, P.W., J.I. Flege, and E.A. Sutter, *Epitaxial graphene on ruthenium.* Nat Mater, 2008. **7**(5): p. 406-11. https://doi.org/10.1038/nmat2166

[32]   Tetlow, H., et al., *Growth of epitaxial graphene: Theory and experiment.* Physics Reports, 2014. **542**(3): p. 195-295. https://doi.org/10.1016/j.physrep.2014.03.003

[33]   Amini, S., et al., *Growth of large-area graphene films from metal-carbon melts.* Journal of Applied Physics, 2010. **108**(9). https://doi.org/10.1063/1.3498815

[34]   Kosynkin, D.V., et al., Longitudinal unzipping of carbon nanotubes to form graphene nanoribbons. Nature, 2009. **458**(7240): p. 872-6. https://doi.org/10.1038/nature07872

[35]   Wang, J., et al., Transition-metal-catalyzed unzipping of single-walled carbon nanotubes into narrow graphene nanoribbons at low temperature. Angew Chem Int Ed Engl, 2011. **50**(35): p. 8041-5. https://doi.org/10.1002/anie.201101022

[36]   Yuan Li and N. Chopra, Progress in large scale production of Graphene Vapor methods. JOM, 2015. **67**(1): p. 44-52. https://doi.org/10.1007/s11837-014-1237-z

[37]   Bianco, A., et al., All in the graphene family – A recommended nomenclature for two-dimensional carbon materials. Carbon, 2013. **65**: p. 1–6. https://doi.org/10.1016/j.carbon.2013.08.038

[38]   Wick, P., et al., *Classification framework for graphene-based materials.* Angew. Chem. Int. Ed. Engl., 2014. **53**(30): p. 7714-7718. https://doi.org/10.1002/anie.201403335

[39]   Lee, D.W., et al., The Structure of Graphite Oxide: Investigation of Its Surface Chemical Groups. J. Phys. Chem. B, 2010. **114**: p. 5723–5728. https://doi.org/10.1021/jp1002275

[40]   Piccinini, E., et al., Enzyme-polyelectrolyte multilayer assemblies on reduced graphene oxide field-effect transistors for biosensing applications. Biosensors & bioelectronics, 2017. **92**: p. 661–667. https://doi.org/10.1016/j.bios.2016.10.035

[41]   Reiner-Rozman, C., C. Kotlowski, and W. Knoll, *Electronic Biosensing with Functionalized rGO FETs.* Biosensors, 2016. **6**(2): p. 17. https://doi.org/10.3390/bios6020017

[42]   Talyzin, A.V., et al., Brodie vs Hummers graphite oxides for preparation of multi-layered materials. Carbon, 2017. **115**: p. 430-440. https://doi.org/10.1016/j.carbon.2016.12.097

[43]   Hummers, W.S. and R.E. Offeman, *Preparation of Graphitic oxide.* J. Am. Chem. Soc., 1958. **80**: p. 1339. https://doi.org/10.1021/ja01539a017

[44]   Cao, J., et al., Two-Step Electrochemical Intercalation and Oxidation of Graphite for the Mass Production of Graphene Oxide. Journal of the American Chemical Society, 2017. **139**(48): p. 17446–17456. https://doi.org/10.1021/jacs.7b08515

[45]   Siegfried Eigler, et al., *Wet Chemical Synthesis of Graphene.* Advanced Materials, 2013: p. 1-5.

[46]   Marcano, D.C., et al., *Improved Synthesis of Graphene Oxide.* ACS Nano, 2010. **4**(8): p. 4806-4814. https://doi.org/10.1021/nn1006368

[47]   Lu, X., et al., Front-End-of-Line Integration of Graphene Oxide for Graphene-Based Electrical Platforms. Advanced Materials Technologies, 2018: p. 1700318. https://doi.org/10.1002/admt.201700318

[48]   Poh, H.L., et al., Graphenes prepared by Staudenmaier, Hofmann and Hummers methods with consequent thermal exfoliation exhibit very different electrochemical properties. Nanoscale, 2012. **4**(11): p. 3515-3522. https://doi.org/10.1039/c2nr30490b

[49]   Yu, C., C.-F. Wang, and S. Chen, *Facile Access to Graphene Oxide from Ferro-Induced Oxidation.* Scientific reports, 2016. **6**: p. 17071. https://doi.org/10.1038/srep17071

[50]   Lu, J., et al., One-pot synthesis of fluorescent carbon nanoribbons, nanoparticles, and graphene by the exfoliation of graphite in ionic liquids. ACS nano, 2009. **3**(8): p. 2367–2375. https://doi.org/10.1021/nn900546b

[51]   Ahirwar, S., S. Mallick, and D. Bahadur, Electrochemical Method To Prepare Graphene Quantum Dots and Graphene Oxide Quantum Dots. ACS Omega, 2017. **2**(11): p. 8343–8353. https://doi.org/10.1021/acsomega.7b01539

[52]   Ching-Yuan Su, et al., High-Quality Thin Graphene Films from Fast Electrochemical Exfoliation. ACS Nano, 2011. **5**(3): p. 2332–2339. https://doi.org/10.1021/nn200025p

[53]   Tang, Q., Z. Zhou, and Z. Chen, *Graphene-related nanomaterials: tuning properties by functionalization.* Nanoscale, 2013. **5**(11): p. 4541-83. https://doi.org/10.1039/c3nr33218g

[54] Kuila, T., et al., *Chemical functionalization of graphene and its applications.* Progress in Materials Science, 2012. **57**(7): p. 1061-1105. https://doi.org/10.1016/j.pmatsci.2012.03.002

[55] Lu, C., et al., *A Graphene Platform for Sensing Biomolecules.* Angewandte Chemie, 2009. **121**(26): p. 4879-4881. https://doi.org/10.1002/ange.200901479

[56] Fabrice Balavoine, et al., Helical Crystallization of Proteins on Carbon Nanotubes: A First Step towards the Development of New Biosensors**. Angwandte Chemie Int. Edi., 1999. **38**(13/14): p. 1912-1915. https://doi.org/10.1002/(SICI)1521-3773(19990712)38:13/14<1912::AID-ANIE1912>3.0.CO;2-2

[57] Esther S. Jeng, A.E.M., Amanda C. Roy, Joseph B. Gastala, and Michael S. Strano, *Detection of DNA Hybridization Using the Near-Infrared Band-Gap Fluorescence of Single-Walled Carbon Nanotubes.* NanoLetters, 2006. **6**(3): p. 371-375. https://doi.org/10.1021/nl051829k

[58] Kian Ping Loh, et al., *The chemistry of graphene.* J. Mater. Chem., 2010. **20**: p. 2277–2289. https://doi.org/10.1039/b920539j

[59] Bosch-Navarro, C., et al., Non-covalent functionalization of graphene with a hydrophilic self-limiting monolayer for macro-molecule immobilization. FlatChem, 2017. **1**: p. 52-56. https://doi.org/10.1016/j.flatc.2016.11.001

[60] Georgakilas, V., et al., Noncovalent Functionalization of Graphene and Graphene Oxide for Energy Materials, Biosensing, Catalytic, and Biomedical Applications. Chem Rev, 2016. **116**(9): p. 5464-519. https://doi.org/10.1021/acs.chemrev.5b00620

[61] Zhou, H., et al., *Understanding Defect-Stabilized Noncovalent Functionalization of Graphene.* Advanced Materials Interfaces, 2015. **2**(17): p. 1500277. https://doi.org/10.1002/admi.201500277

[62] Delle, L.E., et al., ScFv-modified graphene-coated IDE-arrays for 'label-free' screening of cardiovascular disease biomarkers in physiological saline. Biosens Bioelectron, 2018. **102**: p. 574-581. https://doi.org/10.1016/j.bios.2017.12.005

[63] Delle, L., Advancing the performance of scalable nanoelectrochemical transducers: Nanoimprint fabrication, 2D material integration and biosensing optimization. 2017.

[64] Martin, G.a., Experimental Organic Chemistry.

[65] Akimitsu Narita, et al., *New advances in nanographene chemistry.* Chemical Society Reviews, 2015. **44**: p. 6616-6643. https://doi.org/10.1039/C5CS00183H

[66] Bonanni, A., A. Ambrosi, and M. Pumera, *Nucleic acid functionalized graphene for biosensing.* Chem. Eur. J., 2012. **18**(6): p. 1668-1673. https://doi.org/10.1002/chem.201102850

[67] Liu, Y., X. Dong, and P. Chen, *Biological and chemical sensors based on graphene materials.* Chem Soc Rev, 2012. **41**(6): p. 2283-307. https://doi.org/10.1039/C1CS15270J

[68] Georgakilas, V., et al., Functionalization of graphene: covalent and non-covalent approaches, derivatives and applications. Chem Rev, 2012. **112**(11): p. 6156-214. https://doi.org/10.1021/cr3000412

[69] Chen, D., H. Feng, and J. Li, Graphene oxide: preparation, functionalization, and electrochemical applications. Chem Rev, 2012. **112**(11): p. 6027-53. https://doi.org/10.1021/cr300115g

[70] Dreyer, D.R., et al., *The chemistry of graphene oxide.* Chem. Soc. Rev., 2010. **39**(1): p. 228-240. https://doi.org/10.1039/B917103G

[71] Steenackers, M., et al., *Polymer brushes on graphene.* J Am Chem Soc, 2011. **133**(27): p. 10490-8. https://doi.org/10.1021/ja201052q

[72] Liu, Y., et al., Synthesis, characterization and optical limiting property of covalently oligothiophene-functionalized graphene material. Carbon, 2009. **47**(13): p. 3113-3121. https://doi.org/10.1016/j.carbon.2009.07.027

[73] Balasubramanian, K. and M. Burghard, *Electrochemically functionalized carbon nanotubes for device applications.* Journal of Materials Chemistry, 2008. **18**(26): p. 3071. https://doi.org/10.1039/b718262g

[74] Balasubramanian, K. and K. Kern, 25th anniversary article: label-free electrical biodetection using carbon nanostructures. Adv. Mater., 2014. **26**(8): p. 1154-1175. https://doi.org/10.1002/adma.201304912

[75] Zuccaro, L., et al., Tuning the isoelectric point of graphene by electrochemical functionalization. Sci. Rep., 2015. **5**: p. 11794. https://doi.org/10.1038/srep11794

[76] Balasubramanian, K., et al., Electrical Transport and Confocal Raman Studies of Electrochemically Modified Individual Carbon Nanotubes. Advanced Materials, 2003. **15**(18): p. 1515-1518. https://doi.org/10.1002/adma.200305129

[77] Dreyer, D.R., A.D. Todd, and C.W. Bielawski, *Harnessing the chemistry of graphene oxide.* Chem. Soc. Rev., 2014. **43**(15): p. 5288-5301. https://doi.org/10.1039/C4CS00060A

[78] Garg, R., N.K. Dutta, and N.R. Choudhury, *Work Function Engineering of Graphene.* Nanomaterials (Basel), 2014. **4**(2): p. 267-300. https://doi.org/10.3390/nano4020267

[79] Sygellou, L., et al., *Work Function Tuning of Reduced Graphene Oxide Thin Films.* The Journal of Physical Chemistry C, 2016. **120**(1): p. 281-290. https://doi.org/10.1021/acs.jpcc.5b09234

[80]   Hirsch, A., *Functionalization of single walled carbon nanotubes.* Angwandte
      Chemie International Edition, 2002. **41**(11): p. 1853-1859.
      https://doi.org/10.1002/1521-3773(20020603)41:11<1853::AID-
      ANIE1853>3.0.CO;2-N

[81]   Bahr, J.L. and J.M. Tour, *Covalent chemistry of single-wall carbon nanotubes.*
      Journal of Materials Chemistry, 2002. **12**(7): p. 1952-1958.
      https://doi.org/10.1039/b201013p

[82]   Malig, J., et al., *Wet Chemistry of Graphene.* The Electrochemical Society
      Interface, 2011. **20**: p. 53-56. https://doi.org/10.1149/2.F06111if

[83]   Balasubramanian, K., L. Zuccaro, and K. Kern, *Tunable Enhancement of Raman
      Scattering in Graphene-Nanoparticle Hybrids.* Advanced Functional Materials, 2014.
      24(40): p. 6348-6358. https://doi.org/10.1002/adfm.201401796

[84]   Kurkina, T., et al., Self-Assembled Electrical Biodetector Based on Reduced
      Graphene Oxide. ACS Nano, 2012. **6**(6): p. 5514–5520.
      https://doi.org/10.1021/nn301429k

[85]   Lanche, R., et al., Graphite oxide electrical sensors are able to distinguish single
      nucleotide polymorphisms in physiological buffers. FlatChem, 2018. **7**: p. 1-9.
      https://doi.org/10.1016/j.flatc.2017.12.001

[86]   Lanche, R., et al., Reduced graphene oxide-based sensing platform for electric
      cell-substrate impedance sensing. Phys. Status Solidi A, 2014. **211**(6): p. 1404-1409.
      https://doi.org/10.1002/pssa.201330522

[87]   Monica, A.H., et al., Wafer-level assembly of carbon nanotube networks using
      dielectrophoresis. Nanotechnology, 2008. **19**(8): p. 085303.
      https://doi.org/10.1088/0957-4484/19/8/085303

[88]   Joung, D., et al., High yield fabrication of chemically reduced graphene oxide field
      effect transistors by dielectrophoresis. Nanotechnology, 2010. **21**(16): p. 165202-
      165207. https://doi.org/10.1088/0957-4484/21/16/165202

[89]   Collet, M., et al., Large-scale assembly of single nanowires through capillary-
      assisted dielectrophoresis. Adv. Mater., 2015. **27**(7): p. 1268-1273.
      https://doi.org/10.1002/adma.201403039

[90]   Lanche, R., et al., Routine fabrication of reduced graphene oxide microarray
      devices via all solution processing. physica status solidi (a), 2013. **210**(5): p. 968-974.
      https://doi.org/10.1002/pssa.201200910

[91]   Lanche, R., et al., Graphite oxide multilayers for device fabrication: Enzyme-
      based electrical sensing of glucose. physica status solidi (a), 2015. **212**(6): p. 1335-
      1341. https://doi.org/10.1002/pssa.201431936

[92]   Bianchi, M., et al., *Scaling of graphene integrated circuits.* Nanoscale, 2015. **7**(17): p. 8076-83. https://doi.org/10.1039/C5NR01126D

[93]   Banerjee, S.K., et al., *Graphene for CMOS and Beyond CMOS Applications.* Proceedings of the IEEE, 2011. **98**(12): p. 2032-2046. https://doi.org/10.1109/JPROC.2010.2064151

[94]   Nguyen, V.L. and Y.H. Lee, Towards Wafer-Scale Monocrystalline Graphene Growth and Characterization. Small, 2015. **11**(29): p. 3512-3528. https://doi.org/10.1002/smll.201500147

[95]   Zaharie-Butucel, D., et al., Flexible transparent sensors from reduced graphene oxide micro-stripes fabricated by convective self-assembly. Carbon, 2017. **113**: p. 361–370. https://doi.org/10.1016/j.carbon.2016.11.013

[96]   Cho, J., et al., Wafer-scale and environmentally-friendly deposition methodology for extremely uniform, high-performance transistor arrays with an ultra-low amount of polymer semiconductors. Journal of Materials Chemistry C, 2015. **3**(12): p. 2817–2822. https://doi.org/10.1039/C4TC02674H

[97]   Pham, V.H., et al., Fast and simple fabrication of a large transparent chemically-converted graphene film by spray-coating. Carbon, 2010. **48**(7): p. 1945–1951. https://doi.org/10.1016/j.carbon.2010.01.062

[98]   Gilje, S., et al., *A chemical route to graphene for device applications.* Nano letters, 2007. **7**(11): p. 3394–3398. https://doi.org/10.1021/nl0717715

[99]   Li, D., et al., *Processable aqueous dispersions of graphene nanosheets.* Nat Nanotechnol, 2008. **3**(2): p. 101-5. https://doi.org/10.1038/nnano.2007.451

[100]   Borini, S., et al., *Ultrafast graphene oxide humidity sensors.* ACS nano, 2013. **7**(12): p. 11166–11173. https://doi.org/10.1021/nn404889b

[101]   Yu, X., et al., Fabrication technologies and sensing applications of graphene-based composite films: Advances and challenges. Biosensors and Bioelectronics, 2017. **89**: p. 72-84. https://doi.org/10.1016/j.bios.2016.01.081

[102]   Kumar, R., et al., Direct laser writing of micro-supercapacitors on thick graphite oxide films and their electrochemical properties in different liquid inorganic electrolytes. Journal of Colloid and Interface Science, 2017. https://doi.org/10.1016/j.jcis.2017.08.005

[103]   Chen, X., et al., Controlling the Thickness of Thermally Expanded Films of Graphene Oxide. ACS Nano, 2017. **11**(1): p. 665-674. https://doi.org/10.1021/acsnano.6b06954

[104] Boehm, H.P., et al., *Das Adsorptionsverhalten sehr dunner Ko hlenstoff Folien.* Journal of Inorganic and General Chemistry, 1962. **316**(3-4): p. 119-127. https://doi.org/10.1002/zaac.19623160303

[105] Wang, Y., et al., Nitrogen-doped graphene and its application in electrochemical biosensing. ACS Nano, 2010. **4**(4): p. 1790-8. https://doi.org/10.1021/nn100315s

[106] Bonaccorso, F., et al., *Production and processing of graphene and 2d crystals.* Materials Today, 2012. **15**(12): p. 564-589. https://doi.org/10.1016/S1369-7021(13)70014-2

[107] Lee, S.W., et al., Plasma-Assisted Reduction of Graphene Oxide at Low Temperature and Atmospheric Pressure for Flexible Conductor Applications. J Phys Chem Lett, 2012. **3**(6): p. 772-7. https://doi.org/10.1021/jz300080p

[108] Zhijuan Wang, et al., Direct Electrochemical Reduction of Single-Layer Graphene Oxide and Subsequent Functionalization with Glucose Oxidase. JOurnal of Physical Chemistry C, 2009. **113**: p. 14071–14075. https://doi.org/10.1021/jp906348x

[109] Haque, A.M., et al., An electrochemically reduced graphene oxide-based electrochemical immunosensing platform for ultrasensitive antigen detection. Anal Chem, 2012. **84**(4): p. 1871-8. https://doi.org/10.1021/ac202562v

[110] Shao, Y., et al., Facile and controllable electrochemical reduction of graphene oxide and its applications. J. Mater. Chem., 2010. **20**(4): p. 743-748. https://doi.org/10.1039/B917975E

[111] Yang, J., et al., Direct Reduction of Graphene Oxide by Ni Foam as a High-Capacitance Supercapacitor Electrode. ACS Appl Mater Interfaces, 2016. **8**(3): p. 2297-305. https://doi.org/10.1021/acsami.5b11337

[112] Zhang, Y.-L., et al., *Photoreduction of Graphene Oxides: Methods, Properties, and Applications.* Advanced Optical Materials, 2014. **2**(1): p. 10-28. https://doi.org/10.1002/adom.201300317

[113] Yong-Lai, Z., et al., *Photoreduction of Graphene Oxides: Methods, Properties, and Applications.* Advanced Optical Materials, 2014. **2**(1): p. 10-28. https://doi.org/10.1002/adom.201300317

[114] Pei, S. and H.-M. Cheng, *The reduction of graphene oxide.* Carbon, 2012. **50**(9): p. 3210-3228. https://doi.org/10.1016/j.carbon.2011.11.010

[115] Bonanni, A., A.H. Loo, and M. Pumera, *Graphene for impedimetric biosensing.* TrAC Trends in Analytical Chemistry, 2012. **37**: p. 12-21. https://doi.org/10.1016/j.trac.2012.02.011

[116] Kim, J.E., et al., Highly sensitive graphene biosensor by monomolecular self-assembly of receptors on graphene surface. Applied Physics Letters, 2017. **110**(20): p. 203702. https://doi.org/10.1063/1.4983084

[117] Iddo Heller, S.C., Jaan Maennik, Marcel A. G. Zevenbergen, Cees Dekker, and Serge G. Lemay, *Influence of Electrolyte Composition on Liquid-Gated Carbon Nanotube and Graphene Transistors.* Journal of Americal Chemical Society, 2010. **132**: p. 17149–17156. https://doi.org/10.1021/ja104850n

[118] Giovanni, M., A. Bonanni, and M. Pumera, *Detection of DNA hybridization on chemically modified graphene platforms.* Analyst, 2012. **137**(3): p. 580-3. https://doi.org/10.1039/C1AN15910K

[119] Hwang, M.T., et al., Highly specific SNP detection using 2D graphene electronics and DNA strand displacement. Proc Natl Acad Sci U S A, 2016. **113**(26): p. 7088-93. https://doi.org/10.1073/pnas.1603753113

[120] Bellan, L.M., D. Wu, and R.S. Langer, *Current trends in nanobiosensor technology.* Wiley Interdiscip Rev Nanomed Nanobiotechnol, 2011. **3**(3): p. 229-246. https://doi.org/10.1002/wnan.136

[121] Karla Soares-Weiser, et al., *The diagnosis of food allergy: protocol for a systematic review.* Clinical and Translational Allergy, 2014. **3**(18): p. 76–86. https://doi.org/10.1186/2045-7022-3-18

[122] K., S.-W., et al., The diagnosis of food allergy: a systematic review and meta-analysis. Allergy, 2014. **69**(1): p. 76-86. https://doi.org/10.1111/all.12333

[123] Chekin, F., et al., Reduced Graphene Oxide Modified Electrodes for Sensitive Sensing of Gliadin in Food Samples. ACS Sensors, 2016. **1**(12): p. 1462-1470. https://doi.org/10.1021/acssensors.6b00608

[124] Eissa, S. and M. Zourob, In vitro selection of DNA aptamers targeting β-lactoglobulin and their integration in graphene-based biosensor for the detection of milk allergen. Biosensors and Bioelectronics, 2017. **91**: p. 169-174. https://doi.org/10.1016/j.bios.2016.12.020

[125] Sun, X., et al., Multilayer graphene–gold nanocomposite modified stem-loop DNA biosensor for peanut allergen-Ara h1 detection. Food Chemistry, 2015. **172**: p. 335-342. https://doi.org/10.1016/j.foodchem.2014.09.042

[126] Zhang, Y., et al., DNA aptamer for use in a fluorescent assay for the shrimp allergen tropomyosin. Microchimica Acta, 2017. **184**(2): p. 633-639. https://doi.org/10.1007/s00604-016-2042-x

[127] Chapin, R.E., et al., *NTP-CERHR expert panel report on the reproductive and developmental toxicity of bisphenol A.* Birth Defects Res B Dev Reprod Toxicol, 2008. **83**(3): p. 157-395. https://doi.org/10.1002/bdrb.20147

[128] Vandenberg, L.N., et al., Bisphenol-A and the great divide: a review of controversies in the field of endocrine disruption. Endocr Rev, 2009. **30**(1): p. 75-95. https://doi.org/10.1210/er.2008-0021

[129] vom Saal, F.S. and C. Hughes, An Extensive New Literature Concerning Low-Dose Effects of Bisphenol A Shows the Need for a New Risk Assessment. Environmental Health Perspectives, 2005. **113**(8): p. 926-933. https://doi.org/10.1289/ehp.7713

[130] Rubin, B.S., *Bisphenol A: An endocrine disruptor with widespread exposure and multiple effects.* The Journal of Steroid Biochemistry and Molecular Biology, 2011. **127**(1): p. 27-34. https://doi.org/10.1016/j.jsbmb.2011.05.002

[131] Su, B., et al., A sensitive bisphenol A voltammetric sensor relying on AuPd nanoparticles/graphene composites modified glassy carbon electrode. Talanta, 2017. **166**: p. 126-132. https://doi.org/10.1016/j.talanta.2017.01.049

[132] Tan, F., et al., An electrochemical sensor based on molecularly imprinted polypyrrole/graphene quantum dots composite for detection of bisphenol A in water samples. Sensors and Actuators B: Chemical, 2016. **233**: p. 599-606. https://doi.org/10.1016/j.snb.2016.04.146

[133] Alam, M.K., et al., Highly sensitive and selective detection of Bis-phenol A based on hydroxyapatite decorated reduced graphene oxide nanocomposites. Electrochimica Acta, 2017. **241**: p. 353-361. https://doi.org/10.1016/j.electacta.2017.04.135

[134] B. Ntsendwana, et al., Electrochemical Detection of Bisphenol A Using GrapheneModified Glassy Carbon Electrode. Int. J. Electrochem. Sci., 2012. 7 p. 3501-3512.

[135] Ndlovu, T., et al., *An Exfoliated Graphite-Based Bisphenol A Electrochemical Sensor.* Sensors, 2012. **12**(9): p. 11601-11611. https://doi.org/10.3390/s120911601

[136] Zheng, Z., et al., Pt/graphene–CNTs nanocomposite based electrochemical sensors for the determination of endocrine disruptor bisphenol A in thermal printing papers. Analyst, 2013. **138**(2): p. 693-701. https://doi.org/10.1039/C2AN36569C

[137] Zhou, L., et al., Femtomolar sensitivity of bisphenol A photoelectrochemical aptasensor induced by visible light-driven TiO2 nanoparticle-decorated nitrogen-doped graphene. Journal of Materials Chemistry B, 2016. **4**(37): p. 6249-6257. https://doi.org/10.1039/C6TB01414C

[138] Hu, L.-Y., et al., Magnetic separate "turn-on" fluorescent biosensor for Bisphenol A based on magnetic oxidation graphene. Talanta, 2017. **168**: p. 196-202. https://doi.org/10.1016/j.talanta.2017.03.055

[139] Mitra, R. and A. Saha, Reduced Graphene Oxide Based "Turn-On" Fluorescence Sensor for Highly Reproducible and Sensitive Detection of Small Organic Pollutants. ACS Sustainable Chemistry & Engineering, 2017. **5**(1): p. 604-615. https://doi.org/10.1021/acssuschemeng.6b01971

[140] Zhou, L., et al., Label-free graphene biosensor targeting cancer molecules based on non-covalent modification. Biosensors and Bioelectronics, 2017. **87**: p. 701-707. https://doi.org/10.1016/j.bios.2016.09.025

[141] Chen, H., et al., Electrochemical immunosensor for carcinoembryonic antigen based on nanosilver-coated magnetic beads and gold-graphene nanolabels. Talanta, 2012. **91**: p. 95-102. https://doi.org/10.1016/j.talanta.2012.01.025

[142] Ge, S., et al., Disposable electrochemical immunosensor for simultaneous assay of a panel of breast cancer tumor markers. Analyst, 2012. **137**(20): p. 4727-4733. https://doi.org/10.1039/c2an35967g

[143] Zhu, L., et al., Electrochemical immunoassay for carcinoembryonic antigen using gold nanoparticle–graphene composite modified glassy carbon electrode. Talanta, 2013. **116**: p. 809-815. https://doi.org/10.1016/j.talanta.2013.07.069

[144] Liu, G., et al., Nanocomposites of gold nanoparticles and graphene oxide towards an stable label-free electrochemical immunosensor for detection of cardiac marker troponin-I. Analytica Chimica Acta, 2016. **909**: p. 1-8. https://doi.org/10.1016/j.aca.2015.12.023

[145] Tuteja, S.K., et al., Biofunctionalized Rebar Graphene (f-RG) for Label-Free Detection of Cardiac Marker Troponin I. ACS Applied Materials & Interfaces, 2014. **6**(17): p. 14767-14771. https://doi.org/10.1021/am503524e

[146] Kumar, V., et al., Graphene-CNT nanohybrid aptasensor for label free detection of cardiac biomarker myoglobin. Biosens Bioelectron, 2015. **72**: p. 56-60. https://doi.org/10.1016/j.bios.2015.04.089

[147] Tuteja, S.K., et al., Graphene-gated biochip for the detection of cardiac marker Troponin I. Anal Chim Acta, 2014. **809**: p. 148-54. https://doi.org/10.1016/j.aca.2013.11.047

[148] Singal, S., et al., Immunoassay for troponin I using a glassy carbon electrode modified with a hybrid film consisting of graphene and multiwalled carbon nanotubes and decorated with platinum nanoparticles. Microchimica Acta, 2016. **183**(4): p. 1375-1384. https://doi.org/10.1007/s00604-016-1759-x

[149] Tuteja, S.K., et al., One step in-situ synthesis of amine functionalized graphene for immunosensing of cardiac marker cTnI. Biosens Bioelectron, 2015. **66**: p. 129-35. https://doi.org/10.1016/j.bios.2014.10.072

[150] Tuteja, S.K., et al., A label-free electrochemical immunosensor for the detection of cardiac marker using graphene quantum dots (GQDs). Biosens Bioelectron, 2016. **86**: p. 548-556. https://doi.org/10.1016/j.bios.2016.07.052

[151] Jeffery K. Taubenberger and D.M. Morens, *The Pathology of Influenza Virus Infections.* Annu Rev Pathol., 2008 **3**: p. 499–522. https://doi.org/10.1146/annurev.pathmechdis.3.121806.154316

[152] Peiris, J.S., M.D. de Jong, and Y. Guan, *Avian influenza virus (H5N1): a threat to human health.* Clin Microbiol Rev, 2007. **20**(2): p. 243-67. https://doi.org/10.1128/CMR.00037-06

[153] Singh, R., S. Hong, and J. Jang, Label-free Detection of Influenza Viruses using a Reduced Graphene Oxide-based Electrochemical Immunosensor Integrated with a Microfluidic Platform. Sci Rep, 2017. **7**: p. 42771. https://doi.org/10.1038/srep42771

[154] Huang, J., et al., Silver nanoparticles coated graphene electrochemical sensor for the ultrasensitive analysis of avian influenza virus H7. Analytica Chimica Acta, 2016. **913**: p. 121-127. https://doi.org/10.1016/j.aca.2016.01.050

[155] Chan, C., et al., A microfluidic flow-through chip integrated with reduced graphene oxide transistor for influenza virus gene detection. Sensors and Actuators B: Chemical, 2017. **251**: p. 927–933. https://doi.org/10.1016/j.snb.2017.05.147

[156] Pandey, A., et al., Graphene-interfaced electrical biosensor for label-free and sensitive detection of foodborne pathogenic E. coli O157:H7. Biosensors and Bioelectronics, 2017. **91**: p. 225-231. https://doi.org/10.1016/j.bios.2016.12.041

[157] Chan, C.-Y., et al., A reduced graphene oxide-Au based electrochemical biosensor for ultrasensitive detection of enzymatic activity of botulinum neurotoxin A. Sensors and Actuators B: Chemical, 2015. **220**: p. 131-137. https://doi.org/10.1016/j.snb.2015.05.052

[158] Zuo, P., et al., A PDMS/paper/glass hybrid microfluidic biochip integrated with aptamer-functionalized graphene oxide nano-biosensors for one-step multiplexed pathogen detection. Lab Chip, 2013. **13**(19): p. 3921-8. https://doi.org/10.1039/c3lc50654a

[159] Shi, J., et al., A fluorescence resonance energy transfer (FRET) biosensor based on graphene quantum dots (GQDs) and gold nanoparticles (AuNPs) for the detection of mecA gene sequence of Staphylococcus aureus. Biosens Bioelectron, 2015. **67**: p. 595-600. https://doi.org/10.1016/j.bios.2014.09.059

[160] Cheat, S., et al., The mycotoxins deoxynivalenol and nivalenol show in vivo synergism on jejunum enterocytes apoptosis. Food and Chemical Toxicology, 2016. **87**: p. 45-54. https://doi.org/10.1016/j.fct.2015.11.019

[161] Campagnollo, F.B., et al., The occurrence and effect of unit operations for dairy products processing on the fate of aflatoxin M1: A review. Food Control, 2016. **68**: p. 310-329. https://doi.org/10.1016/j.foodcont.2016.04.007

[162] Srivastava, S., et al., *Graphene Oxide-Based Biosensor for Food Toxin Detection.* Applied Biochemistry and Biotechnology, 2014. **174**(3): p. 960-970. https://doi.org/10.1007/s12010-014-0965-4

[163] Ahmed, S.R., et al., Size-controlled preparation of peroxidase-like graphene-gold nanoparticle hybrids for the visible detection of norovirus-like particles. Biosensors and Bioelectronics, 2017. **87**: p. 558-565. https://doi.org/10.1016/j.bios.2016.08.101

[164] Lu, L., et al., An Electrochemical Immunosensor for Rapid and Sensitive Detection of Mycotoxins Fumonisin B1 and Deoxynivalenol. Electrochimica Acta, 2016. **213**: p. 89-97. https://doi.org/10.1016/j.electacta.2016.07.096

[165] Gan, N., et al., An ultrasensitive electrochemiluminescent immunoassay for aflatoxin M1 in milk, based on extraction by magnetic graphene and detection by antibody-labeled CdTe quantumn dots-carbon nanotubes nanocomposite. Toxins (Basel), 2013. **5**(5): p. 865-83. https://doi.org/10.3390/toxins5050865

[166] Sheng, L., et al., PVP-coated graphene oxide for selective determination of ochratoxin A via quenching fluorescence of free aptamer. Biosens Bioelectron, 2011. **26**(8): p. 3494-9. https://doi.org/10.1016/j.bios.2011.01.032

[167] Zhang, J., et al., Size-dependent modulation of graphene oxide–aptamer interactions for an amplified fluorescence-based detection of aflatoxin B1 with a tunable dynamic range. Analyst, 2016. **141**(13): p. 4029-4034. https://doi.org/10.1039/C6AN00368K

[168] Yugender Goud, K., et al., Aptamer-based zearalenone assay based on the use of a fluorescein label and a functional graphene oxide as a quencher. Microchimica Acta, 2017. **184**(11): p. 4401-4408. https://doi.org/10.1007/s00604-017-2487-6

[169] Islam, M.N., S.F. Bint-E-Naser, and M.S. Khan, *Pesticide Food Laws and Regulations*, in *Pesticide Residue in Foods: Sources, Management, and Control*, M.S. Khan and M.S. Rahman, Editors. 2017, Springer International Publishing: Cham. p. 37-51. https://doi.org/10.1007/978-3-319-52683-6_3

[170] Li, Y., et al., An Acetylcholinesterase Biosensor Based on Graphene/Polyaniline Composite Film for Detection of Pesticides. Chinese Journal of Chemistry, 2016. **34**(1): p. 82-88. https://doi.org/10.1002/cjoc.201500747

[171] Liang, H., D. Song, and J. Gong, Signal-on electrochemiluminescence of biofunctional CdTe quantum dots for biosensing of organophosphate pesticides. Biosens Bioelectron, 2014. **53**: p. 363-9. https://doi.org/10.1016/j.bios.2013.10.011

[172] Guler, M., V. Turkoglu, and Z. Basi, Determination of malation, methidathion, and chlorpyrifos ethyl pesticides using acetylcholinesterase biosensor based on Nafion/Ag@rGO-NH2 nanocomposites. Electrochimica Acta, 2017. **240**: p. 129-135. https://doi.org/10.1016/j.electacta.2017.04.069

[173] Li, Y., et al., Porous-reduced graphene oxide for fabricating an amperometric acetylcholinesterase biosensor. Sensors and Actuators B: Chemical, 2013. **185**: p. 706-712. https://doi.org/10.1016/j.snb.2013.05.061

[174] Zhang, H., et al., Functionalized graphene oxide for the fabrication of paraoxon biosensors. Anal Chim Acta, 2014. **827**: p. 86-94. https://doi.org/10.1016/j.aca.2014.04.014

[175] Oliveira, T.M.B.F., et al., Sensitive bi-enzymatic biosensor based on polyphenoloxidases–gold nanoparticles–chitosan hybrid film–graphene doped carbon paste electrode for carbamates detection. Bioelectrochemistry, 2014. **98**: p. 20-29. https://doi.org/10.1016/j.bioelechem.2014.02.003

[176] Mehta, J., et al., Graphene modified screen printed immunosensor for highly sensitive detection of parathion. Biosens Bioelectron, 2016. **83**: p. 339-46. https://doi.org/10.1016/j.bios.2016.04.058

[177] Sharma, P., et al., Bio-functionalized graphene-graphene oxide nanocomposite based electrochemical immunosensing. Biosens Bioelectron, 2013. **39**(1): p. 99-105. https://doi.org/10.1016/j.bios.2012.06.061

[178] Yan, Y., et al., A facile strategy to construct pure thiophene-sulfur-doped graphene/ZnO nanoplates sensitized structure for fabricating a novel "on-off-on" switch photoelectrochemical aptasensor. Sensors and Actuators B: Chemical, 2017. **251**: p. 99-107. https://doi.org/10.1016/j.snb.2017.05.034

[179] Wang, X.-F., et al., Signal-On Electrochemiluminescence Biosensors Based on CdS-Carbon Nanotube Nanocomposite for the Sensitive Detection of Choline and Acetylcholine. Advanced Functional Materials, 2009. **19**(9): p. 1444-1450. https://doi.org/10.1002/adfm.200801313

[180] Caballero-Díaz, E., S. Benítez-Martínez, and M. Valcárcel, *Rapid and simple nanosensor by combination of graphene quantum dots and enzymatic inhibition mechanisms.* Sensors and Actuators B: Chemical, 2017. **240**: p. 90-99. https://doi.org/10.1016/j.snb.2016.08.153

[181] Park, J.W., et al., Polypyrrole nanotube embedded reduced graphene oxide transducer for field-effect transistor-type H2O2 biosensor. Analytical chemistry, 2014. **86**(3): p. 1822–1828. https://doi.org/10.1021/ac403770x

[182] Roney, J.R. and Z.L. Simmons, *Hormonal predictors of sexual motivation in natural menstrual cycles.* Hormones and Behavior, 2013. **63**(4): p. 636-645. https://doi.org/10.1016/j.yhbeh.2013.02.013

[183] Arvand, M. and S. Hemmati, Analytical methodology for the electro-catalytic determination of estradiol and progesterone based on graphene quantum dots and poly(sulfosalicylic acid) co-modified electrode. Talanta, 2017. **174**: p. 243-255. https://doi.org/10.1016/j.talanta.2017.05.083

[184] Dong, X.-X., et al., Development of a progesterone immunosensor based on thionine-graphene oxide composites platforms: Improvement by biotin-streptavidin-amplified system. Talanta, 2017. **170**: p. 502-508. https://doi.org/10.1016/j.talanta.2017.04.054

[185] Chang, Y., et al., *In vitro toxicity evaluation of graphene oxide on A549 cells.* Toxicol Lett, 2011. **200**(3): p. 201-10. https://doi.org/10.1016/j.toxlet.2010.11.016

[186] Ang, P.K., et al., Flow sensing of single cell by graphene transistor in a microfluidic channel. Nano Lett, 2011. **11**(12): p. 5240-6. https://doi.org/10.1021/nl202579k

[187] Feng, L., et al., A graphene functionalized electrochemical aptasensor for selective label-free detection of cancer cells. Biomaterials, 2011. **32**(11): p. 2930-7. https://doi.org/10.1016/j.biomaterials.2011.01.002

[188] Basiricò, L., et al., *Inkjet printing of transparent, flexible, organic transistors.* Thin Solid Films, 2011. **520**(4): p. 1291-1294. https://doi.org/10.1016/j.tsf.2011.04.188

[189] Akinwande, D., N. Petrone, and J. Hone, *Two-dimensional flexible nanoelectronics.* Nat Commun, 2014. **5**: p. 12. https://doi.org/10.1038/ncomms6678

[190] Das, S., et al., All two-dimensional, flexible, transparent, and thinnest thin film transistor. Nano Lett, 2014. **14**(5): p. 2861-6. https://doi.org/10.1021/nl5009037

[191] Fiori, G., et al., *Electronics based on two-dimensional materials.* Nat. Nanotechnol., 2014. **9**: p. 1-12. https://doi.org/10.1038/nnano.2014.283

[192] Nathan, A., et al., *Flexible Electronics: The Next Ubiquitous Platform.* Proceedings of the IEEE, 2012. **100**(Special Centennial Issue): p. 1486-1517. https://doi.org/10.1109/JPROC.2012.2190168

[193] Kim, D.H., et al., *Flexible and stretchable electronics for biointegrated devices.* Annu Rev Biomed Eng, 2012. **14**: p. 113-28. https://doi.org/10.1146/annurev-bioeng-071811-150018

[194] Mao, S., et al., Specific protein detection using thermally reduced graphene oxide sheet decorated with gold nanoparticle-antibody conjugates. Advanced materials (Deerfield Beach, Fla.), 2010. **22**(32): p. 3521–3526. https://doi.org/10.1002/adma.201000520

[195] Cai, B., et al., Ultrasensitive Label-Free Detection of PNA-DNA Hybridization by Reduced Graphene Oxide Field-Effect Transistor Biosensor. ACS Nano, 2014. **8**(3): p. 2632–2638. https://doi.org/10.1021/nn4063424

[196] Chang, J., et al., Ultrasonic-assisted self-assembly of monolayer graphene oxide for rapid detection of Escherichia coli bacteria. Nanoscale, 2013. **5**(9): p. 3620–3626. https://doi.org/10.1039/c3nr00141e

[197] Myung, S., et al., Label-free polypeptide-based enzyme detection using a graphene-nanoparticle hybrid sensor. Adv Mater, 2012. **24**(45): p. 6081-7. https://doi.org/10.1002/adma.201202961

[198] Chen, H., et al., Detection of Matrilysin Activity Using Polypeptide Functionalized Reduced Graphene Oxide Field-Effect Transistor Sensor. Analytical chemistry, 2016. **88**(6): p. 2994–2998. https://doi.org/10.1021/acs.analchem.5b04663

[199] Lei, Y.-M., et al., Detection of heart failure-related biomarker in whole blood with graphene field effect transistor biosensor. Biosensors & bioelectronics, 2017. **91**: p. 1–7. https://doi.org/10.1016/j.bios.2016.12.018

[200] Chen, Y., et al., Field-Effect Transistor Biosensor for Rapid Detection of Ebola Antigen. Scientific reports, 2017. **7**(1): p. 10974. https://doi.org/10.1038/s41598-017-11387-7

Organic Bioelectronics for Life Science and Healthcare   Materials Research Forum LLC
Materials Research Foundations **56** (2019) 213-262   doi: https://doi.org/10.21741/9781644900376-7

Chapter 7

Inkjet Printing for Biosensors and Bioelectronics

Nouran Adly[1], Philipp Rinklin[1], Bernhard Wolfrum[1*]

Neuroelectronics - Munich School of Bioengineering, Department of Electrical and Computer Engineering, Technical University of Munich, Boltzmannstraße 11, 85749, Garching, Germany

*bernhard.wolfrum@tum.de

Abstract

This chapter examines the role of inkjet printing for the fabrication of biosensors and bioelectronic devices. Inkjet printing is a promising technique that can be employed to directly fabricate sensors and electronics on a variety of substrates including flexible plastics and soft silicones. In contrast to state-of-the-art mask-based microfabrication it is capable of testing new materials and designs in a rapid prototyping approach. As such, it has the potential to play a powerful role in the development of customized devices. Here, we present an overview of recent research describing the application of inkjet technologies to print functional biosensors and bioelectronic devices. We discuss advantages and limitations of this approach and address possible routes to overcome current challenges for future developments in this field.

Keywords

Flexible Bioelectronics, Biosensors, Microelectrode Arrays, Inkjet Printing

Contents

1. Introduction

Inkjet printing has recently started to make significant contributions to the world of micro-manufacturing for developing affordable and large-area electronic devices. Inkjet printing is an additive manufacturing platform that has been successfully employed to fabricate flexible electronic devices and prototypes by many research labs. Devices produced using inkjet printing have heralded the next generation of flexible electronics capable of performing functions that were previously only accessible with costly components relying on state-of-the-art microfabrication technologies. Since inkjet printing utilizes digital information, it is capable of modifying the print pattern on a real-time basis. Consequently, design changes can be introduced without any additional cost allowing, for instance, to produce devices with individually unique features. This can prove helpful when exploring the use of different designs with the intention of optimizing performance, in particular in research and development settings. A further advantage of inkjet printing over other printing approaches (such as screen printing, flexography, and gravure) is that it employs a non-contact method of deposition, providing the possibility to achieve functionality on sensitive and fragile substrates that are incapable of withstanding the pressure associated with alternative printing platforms.

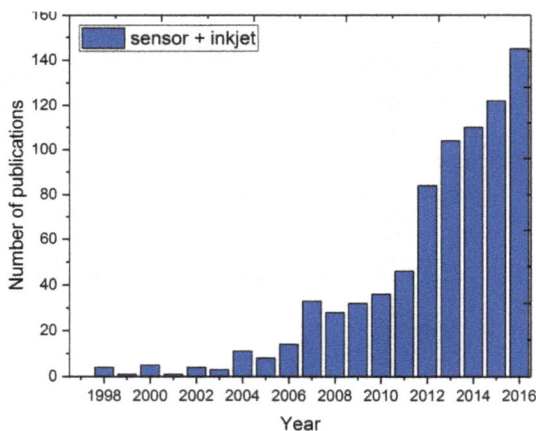

Figure 1: Number of publications per year on inkjet printed biosensors. The search was performed using "inkjet" and "sensor*" as search key terms on Web of Knowledge; December 2017. The general consensus is that inkjet printing will facilitate the production of flexible electronic devices in a cost-effective manner and enable a range of applications including wearable and disposable sensors. Currently, however, our understanding of the technology and its capabilities remains limited, and further expertise needs to be acquired to facilitate the printing of high-resolution devices for bioelectronic applications.*

This chapter will focus on recent progress in printed flexible electronics for healthcare applications. It is organized into two parts addressing biosensors and bioelectronic interfaces. More fundamental aspects on the basic principles of inkjet printing can be found in the literature [11].

A significant amount of reports have focused on tailoring functional nanomaterials into printable inks for inkjet printing to create devices that cannot be processed with conventional microfabrication [1,2]. For example, direct writing via inkjet printing on plastic substrates using functional inks such as carbon nanotubes [3,4], graphene [5,6], metallo-organic compounds [7,8], metal nanoparticles, organic semiconductors, and quantum dots [9,10] have all been reported.

2.   Biosensor

Over the past several decades, an enormous amount of physical, chemical and biological sensors have been reported using printing technologies. While some of these sensors could also be realized using existing silicon microfabrication approaches, the use of inkjet printing offers several advantages for the development of inexpensive sensors for which the sensor design can be modified rather effortlessly. As a consequence, growth in the field of inkjet printed sensors has been quite remarkable. In 2016, more than 140 papers were published on inkjet-printed sensors indicating a substantial growth of this field (Fig. 1). Recent reviews by Mattana *et al.* on inkjet-printed physical and chemical sensors [12], Moya *et al.* on inkjet-printed electrochemical sensors [13] and Salim et al. on inkjet-printed capacitive sensors [14] provide a good overview of the possible applications of inkjet printing and the feasibility of applying printed technology to this field.

In the current section, we focus on the application of inkjet printing in the field of electrochemical biosensors. Biosensors are analytical tools that combine the inherent specificity of a bioreceptor (recognition element) to bind to a biological target molecule with a physical, mechanical, or electrical transducer element. Potential applications of biosensors range from disease diagnosis to food safety and environmental monitoring [15–20]. Over the last decades, many efforts have been directed at integrating inkjet-printing tools with biosensor development, which offers several advantages over other printing techniques in terms of print quality, accuracy, and scaling capabilities. Perhaps the most straightforward example of a biosensor that can be simply developed by inkjet printing is the electrochemical glucose sensor. Glucose sensing based on a glucose oxidase (GOD) enzyme-modified electrode is one of the earliest and most well-known biosensors due to its importance for the diagnosis and management of diabetes mellitus, which is a global health problem [21,22]. Typically, the GOD enzyme serves as a catalyst for oxidizing glucose molecules in solution and producing hydrogen peroxide. The generation of hydrogen peroxide can be directly measured electrochemically or utilized as a substrate for oxidizing horseradish peroxidase (HRP) for signal enhancement. Oxidized HRP is further reduced by a reversible redox mediator such as ferrocenemethanol (FcMeOH), which is capable of transferring electrons from the working electrode to the redox center of the HRP [23–25].

Setti *et al.* fabricated an amperometric glucose biosensor by formulating an ink containing the GOD enzyme and printing it on an indium tin oxide (ITO) coated glass substrate [26], which demonstrated the possibility of inkjet-printing a glucose sensor. Yun *et al.* developed a glucose sensor by preparing a bioelectrical ink that consisted of a mixture of GOD, HRP, and poly(3,4-ethylene dioxythiophene):poly(styrenesulfonate) (PEDOT:PSS) [27]. The conducting polymer PEDOT:PSS was added to the ink mixture to serve as a matrix for encapsulating GOD and HRP. At the same time, it permits electron transfer between the enzymes and the working electrode. The sensor reached a detection limit of 270 μM and sensitivity of 205 A m^{-2} M^{-1}. Further recent examples are included in Table 1.

In addition to glucose sensors, inkjet printing has been adopted in the electrochemical detection of protein and DNA biomarkers. The former was reported in 2011 in a hallmark work by Jensen *et al.* [28] that demonstrated the feasibility of inkjet printing more than 50 gold electrode arrays in a single pass for immunosensor applications. This was made possible by the development of a stable gold nanoparticle ink. The objective of their work was to quantify protein biomarkers such as interleukin-6 (IL-6) in serum for cancer diagnosis in a rapid prototyping and low-cost approach. The first step in this immunosensor construction was the covalent immobilization of the antibody on a printed gold surface. To this end, a self-assembled monolayer (SAM) was deposited on the gold surface by thiol-gold interaction, followed by carbodiimide-based cross-linking of the antibodies. The printed arrays were applied to immunosensing of interleukin-6 in serum with a detection limit of 20 pg mL^{-1}.

Inkjet-printed DNA biosensors have also been explored for the diagnosis of diseases including human papillomavirus (HPV) [29] and human immunodeficiency virus (HIV) [30]. Similar to immunosensors, DNA biosensors combine bioreceptor layers, such as single-stranded DNA (ssDNA) or synthetic nucleic acid probes, with electrochemical transducers to produce a signal originating from the hybridization reaction. In general, the signal output due to the binding event is limited by the surface density of the hybridization probes, which often imposes a limit on the detection capability. As a consequence, there are several strategies to enhance the obtained signal for the selective detection of DNA, including the use of nanomaterial assemblies [31], the addition of nanotags, such as silver nanoparticles, to the DNA duplex [32], nuclease–assisted target recycling [33], polymerase-based target recycling amplification [34], and isothermal solid-phase amplification [35].

Table 1: *Summary of reports on recent inkjet-printed electrochemical biosensors*

Detection method	Detection Limit	Analyte	Substrate	Ink	Ref.
Differential pulse voltammetry	7.46×10^{-9} M	Dopamine	Indium tin oxide (ITO)	Gold and silver	[37]
Amperometry	n/a	H_2O_2	Paper	Graphene	[38]
Square-wave voltammetry	2.3×10^{-9} M	DNA	Paper	Graphene-polyaniline (G-PANI)	[29]
Amperometry	20 pg mL^{-1}	Interleukin-6	Kapton	Gold nano-particle (AuNP)	[28]
Potential step voltammetry	n/a	Dopamine	Paper	Carbon nanotube (CNT) ink	[39]
Amperometry	7.88 mg dL^{-1}	Triglyceride		Au/PEDOT-PSS Nano-composite	[40]
Amperometry	ATR =0.01 µg L^{-1} TSH =0.5 µIU mL^{-1}	Thyroid-stimulating hormone (TSH) Atrazine (ATR)	Polyethylene terephthalate (PET)	Ag and CNT inks	[41]
Amperometry	78 fg mL^{-1} for IL-6 19 fg mL^{-1} for IL-8	IL-6 and IL-8	Kapton	Gold nano-particle	[42]
Amperometry	0.3 mg dL^{-1}	Glucose	Polyimide (PI)	Graphene	[43]

Amperometry	Only two concentrations: $0,10^5$ and 10^6 CFU mL^{-1}	Salmonella	PET	Silver Gold	[44]
Amperometry	0.2×10^{-3} M	Cholesterol	Polyester film	Prussian blue nanoparticles (PBNPs).	[45]
Linear sweep voltammetry (LSV)	Only two concentrations 0.5×10^{-3} M and 1.0×10^{-3} M	Ascorbic acid	Kapton	Ag CNT	[46]

Another promising approach for signal amplification, which does not require enzymes or pre-labeling, is the use of redox-cycling in a dual-electrode configuration [36]. The redox-cycling concept is realized by placing two electrodes in direct proximity to one another. One of the electrodes is biased to the oxidizing potential of a redox-active probe, while the second electrode is set to its reducing potential. Consequently, the redox molecules in the solution (e.g. ferrocene dimethanol) can undergo repetitive oxidation and reduction events resulting in a current amplification. The appeal of this approach to DNA detection lies in the fact that no ssDNA labeling or specific enzymes are involved for the amplification to occur, making it a universal method for signal enhancement. However, one of the main challenges in constructing a redox-cycling sensor – in particular when it comes to inkjet printing – is to realize a sufficiently small gap between the two electrodes. Since the transfer of probe molecules between the two electrodes is driven by diffusion, gaps in the nano- or micrometer range are required to enable an efficient transport and current amplification.

A recent example of sensitive DNA detection was enabled by the printing of two closely spaced interdigitated electrodes, as depicted in Fig. 2 [30]. In this example, the anodic and cathodic electrodes of the sensor system were printed with a separation of 1 µm. Diffusion of redox-active molecules between the electrodes generated a redox cycling current. This strategy amplified the current signal by a factor of 4.8 compared to measurements of the same electrode configuration without redox cycling.

These printed microgap redox-cycling sensors were explored for DNA sensing by immobilizing the PNA probe on the surface of a carbon electrode as a capture probe via a pyrene group linked to the 3′ end. The pyrene rings subsequently formed a π–π interaction with the carbon surface [47,48]. A calibration curve for the detection of ssDNA with a sensitivity of 0.4 A M^{-1} and detection limit of 1 nM were reported. As mentioned earlier, the distance between the two electrodes has to be small in order to permit efficient diffusion of the redox molecules to increase the amplification. It is important to note here that the resolution of printed features, even though it has significantly improved over the last 10 years [49,49–53], is still limited by the size and size variations of the individual droplets. In this respect, inkjet printing has a significant disadvantage compared to optical or electron beam lithography, which can be used to reliably fabricate interdigitated arrays at sub-μm spacing [54].

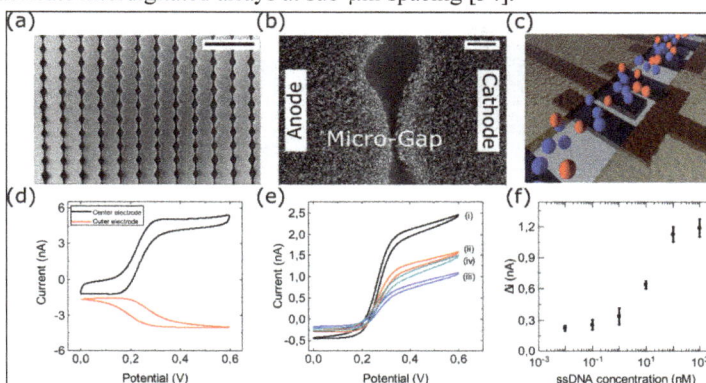

Figure 2: ssDNA detection in inkjet-printed microgap devices. (a) SEM image of high resolution inkjet-printed lines and (b) zoom-in on the gap between two printed lines, scale bars represent 100 and 5 μm, respectively. (c) Schematic overview of the redox cycling sensor comprising center and outer electrodes. (d) Typical cyclic voltammetry curves of the center and outer electrode during redox cycling mode using a 0.25 × 10⁻³ M ferrocene dimethanol solution acquired by scanning the center electrode (black line) from 0.0 to 0.6 V while fixing the potential of the outer electrode (red line) at 0.0 V. (e) Proof-of-principle of DNA detection using a printed microgap chip, current for (i) the bare carbon electrode; (ii) PNA immobilization on the carbon surface; (iii) addition of target ssDNA showing hybridization with the PNA probe; (iv) thermal denaturation of PNA–DNA duplex at 95 °C. (f) Calibration curve of redox-cycling sensors for ssDNA detection. (a), (b), and (d)-(f) modified from [30] with permission from John Wiley and Sons.

Another viable strategy to obtain an electrode separation down to the range of a few hundred nanometers relies on a vertical stacking of the two electrodes as shown in Fig. 3 [55]. In this approach, a cathodic bottom gold electrode and anodic top nanoporous carbon electrode were directly printed on a flexible polyethylene naphthalate (PEN) substrate. The two electrodes were separated by a nanoporous layer of polystyrene nanoparticles, which permits the diffusion of the redox molecule through the porous structure while keeping the carbon top electrode mechanically stable, as shown in Fig. 3. The three-dimensional device demonstrated a reliable electrochemical performance in phosphate buffered saline at pH 7.4. The printed device was capable of sensing ferrocene dimethanol as a redox-active probe. In principle, the approach can provide a low-cost and disposable sensitive electrochemical biosensor, eliminating the need for expensive fabrication and substrate pre-patterning.

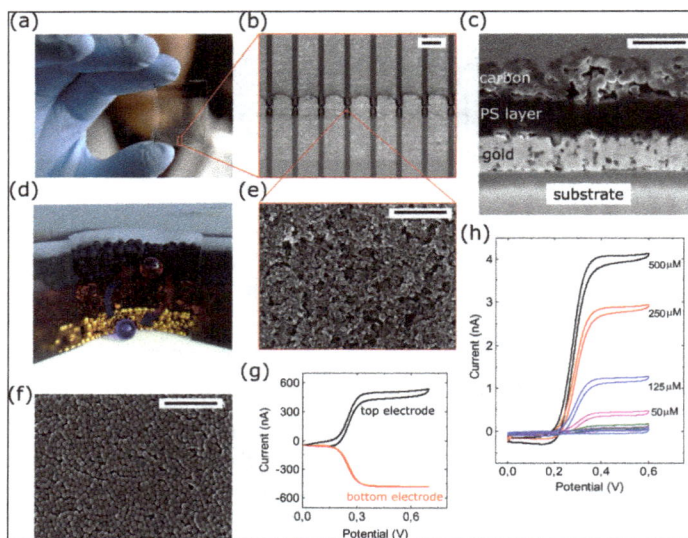

Figure 3: Vertically stacked redox cycling sensor. (a) Photograph of redox cycling device inkjet-printed onto a flexible PEN substrate. (b) Microscopic image of carbon microelectrodes covered with polyimide layer except for an electrode opening, the scale bar represents 200 µm (c) Cross-sectional SEM image of the printed device showing the vertical staking of the three printed layers, the scale bar represents 300 nm. (d) 3D illustration of the redox cycling device consisting of a porous carbon top, gold bottom electrode and a porous separation layer of printed polystyrene beads. (e) SEM image of the porous carbon layer and (f) polystyrene nanospheres, the scale bars in (e) and (f) represent 1 µm. (g) CV measurements of the carbon top electrode and the bottom gold electrode using 500 µM Fc(MeOH)₂. The potential of the bottom electrode is held at 0.0 V while the top electrode is swept from 0.0 V to 0.6 V at a scan rate of 20 mV s⁻¹. (h) The current measured during redox cycling mode for the top electrode at different concentrations of Fc(MeOH)₂. (a)-(c) and (e)-(f) modified from [55] – Published by The Royal Society of Chemistry. (d) Adapted with permission from [36]. Copyright 2016 American Chemical Society.

Organic Bioelectronics for Life Science and Healthcare   Materials Research Forum LLC
Materials Research Foundations **56** (2019) 213-262   doi: https://doi.org/10.21741/9781644900376-7

Overall, inkjet printing represents a cost-effective, practical, and reliable solution for the development of biosensors. It can be employed to fabricate different types of sensors, ranging from enzyme-based planar glucose biosensors to three-dimensional sensors with inherent redox cycling amplification. While inkjet printing has evolved in several aspects towards becoming a competitive sensor fabrication method, its full potential in this growing field has yet to be truly embraced. We believe future developments will critically depend on the availability of suitable ink formulations and the possibility to efficiently integrate fluid handling, readout electronics, and power management including batteries within the printed sensor concepts. While its full potential is waiting to be discovered, it is clear that inkjet printing has a very promising future when it comes to the development and rapid prototyping of sensitive low-cost biosensors.

3.   Bioelectronic Interfaces:

In the previous section, we have seen how inkjet printing can be utilized to fabricate sensors for the detection of specific molecules. In this part, we focus on how inkjet printing technology can be used to design and develop electronic devices suitable for interacting with biological systems. As reviewed recently by Khan *et al.*, new manufacturing approaches are being explored with the goal of demonstrating flexible and conformal sensing devices for health monitoring [56]. In this respect, inkjet printing on flexible substrates is emerging as an interesting alternative to conventional silicon microfabrication, which not only reduces the cost but also allows for printing larger areas of sensors since it is no longer limited by the wafer throughput. Moreover, inkjet printing is not restricted to certain substrate materials, unlike conventional silicon microfabrication. For instance, Tao *et al.* demonstrated the ability to print colorimetric bacterial sensors on surgical gloves and Petri dishes [57], substrates that are not easily functionalized using standard fabrication approaches. Furthermore, adding electronic functionality onto unconventional substrates such as sticking plaster for temperature sensing [58], conductive patterns on textile fabrics [59], and electrochemical biosensors on polydimethylsiloxane (PDMS) substrates have also been reported [60].

Very recently, there have been several reports on the development of inkjet printing processes for bioelectronic interfaces. Vuorinen and coworkers developed disposable conformable temperature sensors on the skin by inkjet printing graphene/PEDOT:PSS ink on screen-printed silver electrodes. The full sensor is developed on polyurethane plaster, which is a medical adhesive bandage that provides good adhesion to the skin [58]. In their study, a printed flexible thermistor was evaluated on human skin and achieved a sensitivity greater than 0.6 ppm $°C^{-1}$ under physiological conditions (35–45 °C). One of the earliest reports on incorporating inkjet printing for developing electrode arrays for health monitoring by interfacing with soft tissues *in vivo* was published in 2015 by Swisher *et al.* [61]. In their work, an impedimetric pressure sensor was fabricated and tested for *in vivo* measurements using a rat model for the early detection of tissue damage. The fabrication of this flexible electrode arrray used a combination of inkjet printing of a gold electrode array (55 electrodes of a diameter of 0.5 mm) and plasma etching of electrode vias. The electrode openings were subsequently covered by a hydrogel layer using a siloxane stencil, which has a defined opening and is pre-aligned on the electrode opening. The final device was placed on a wound region in rats and was able to record electrical impedance from each individual electrode for tissue mapping. In another study from the group of Ana Arias, a wearable sensor patch for electrocardiography (ECG) recording and temperature sensing on the skin was developed [62]. For ECG recordings, a pair of gold electrodes was patterned on a flexible Kapton substrate. The ECG electrodes were capable of recording a 1 mV peak-to-peak ECG signal using a 4.7 cm electrode spacing. Furthermore, a temperature-sensitive element composed of a nickel oxide (NiO) nanoparticle paste was placed between the two gold electrodes by stencil-printing for temperature measurements during rest and mild exercises along with the ECG recording. The thermistor demonstrated good sensitivity at physiological temperature characterized by a measured temperature coefficient, $\alpha \approx -5.8\% \, K^{-1}$, which represents the change in resistance with temperature. The sensor system was further integrated with a battery and an analog front end for amplifying and filtering the signal sent to a bluetooth system on chip. The designed system was capable of recording heartbeats and performing on-skin temperature measurements without mechanical flexural wear or tear-out. To demonstrate the flexibility of the printed gold electrodes, the authors carried out mechanical tests by bending the sample to radii of 10 and 5 mm and showing a minimum change in the electrical resistance. Another recent example demonstrated electromyography (EMG) and ECG recording using a flexible PEDOT:PSS multielectrode array for neuromuscular investigation as reported by Sanaur's group [63]. The electrode arrray was fabricated by inkjet printing using a silver nanoparticle ink that was subsequently covered by drop casting a PEDOT:PSS solution as a material for skin interfacing. Figure 4 shows the full design of such a multi-electrode array, incorporating the disposable connection pads, 300 μm wide tracks, and an array of four by seven electrodes. Using a healthy human volunteer, EMG and ECG signals were measured and compared to a commercial electrode system typically employed for surface electromyography.

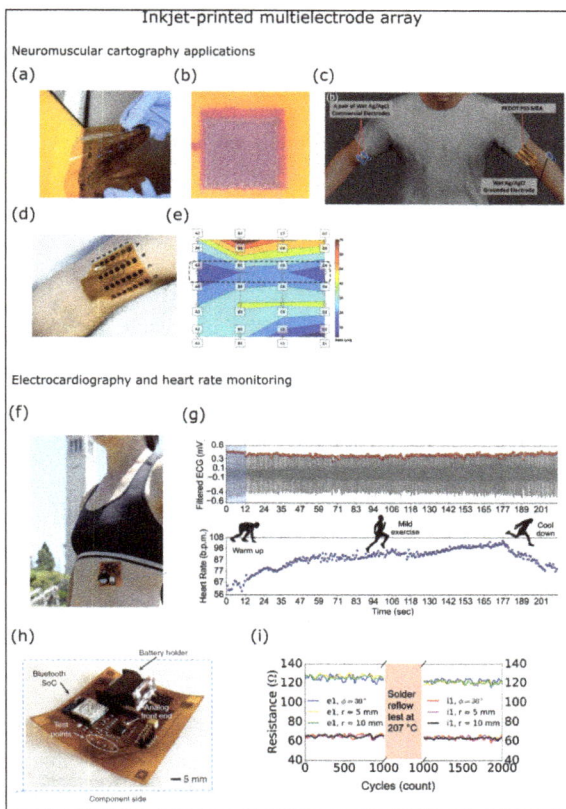

Figure 4: Inkjet printed electrodes for electrophysiological recordings (ECG and EMG). *The top panel depicts (a) photograph of printed silver layer on a flexible PI substrate, (b) printed graphics ink as insulator covering the silver layer and (c) illustration of the whole system used for recording surface electromyography (sEMG) on skin. (d)-(e) illustrate the recording of muscle activity using multi-electrode array system consisting of 28 silver electrodes coated with PEDOT:PSS and placed on the left wrist of human volunteer as shown in (d). (e) Cartography of the RMS values obtained using the sEMG recordings. The bottom panel shows (f) a photo of a wearable sensor patch, (g) filtered ECG and heart rate recordings from the patch, (h) a photograph of the component side of the wearable sensor patch, and (i) mechanical robustness test results. (a)-(e) modified from [63] with permission from Jon Wiley and Sons. (f)-(i) modified from [62] with permission from John Wiley and Sons.*

Organic Bioelectronics for Life Science and Healthcare   Materials Research Forum LLC
Materials Research Foundations **56** (2019) 213-262   doi: https://doi.org/10.21741/9781644900376-7

In order to get a deeper understanding of signal processing in cellular networks, it is desirable to gather information from individual cells, which is not possible with low-resolution ECG measurements. In this regard, microelectrode array (MEA) systems provide promising opportunities to study electrical signals in neuronal and cardiac cell networks. Furthermore, such devices can also be used to restore sensory function or treat disorders of the nervous system. However, action-potential recording and stimulation on a cellular level using printed devices is still rather limited. Currently, most extracellular recordings from electrogenic cells are obtained using devices based on standard microfabrication techniques. For instance, typical commercial clean-room processed planar MEAs for *in vitro* experiments consist of approximately 60–200 electrodes with electrode diameters ranging between 10 and 30 µm, while complementary metal–oxide–semiconductor (CMOS)-based MEAs feature up to several thousand electrodes [64,65]. Generally speaking, to achieve action potential recording on a cellular level, it is imperative that the MEA exhibits a high spatial resolution comparable to the size of individual cells. Given that typical printed structures do not fulfill this criterion, it is necessary to improve the printing resolution for the fabrication of MEAs via inkjet printing. Let us therefore consider existing methods for enhancing the printing resolution. As introduced earlier, efforts to push the printing resolution have been reported by several groups [66]. For example, high-resolution inkjet printing of an organic transistor with a channel length of 5 micrometers has been reported by Sirringhaus *et al.* It was made possible by fine-tuning the surface energy of the substrate prior to printing the organic transistor [52]. The source and gate electrodes were inkjet-printed onto a glass substrate. The channel distance between the source and the drain was defined by a narrow, repelling, hydrophobic barrier of polyimide. This barrier was used to define the critical device dimensions. However, depositing thin lines of polyimide was done by a photolithographic process, which required a shadow mask for defining the hydrophobic regions. After all, this approach reduced the number of the photolithographic steps that are typically employed for fabricating a device with equivalent performance. This helps to reduce the overall cost and complexity of fabricating organic transistors and is a first step towards the precise local control of deposited droplet volumes.

Later, the group of Sirringhaus presented a sub-hundred-nanometer channel length without any top-down patterning steps [51]. This small feature size was defined by a self-aligned gap. In their approach, the first electrode line was printed and then the surface energy of the line was selectively reduced by fluorination via plasma treatment using carbon tetrafluoride (CF4) prior to printing the second line. At the same time, the plasma exposure lead to an increase in the surface energy of the glass substrate due to the plasma etching. This treatment resulted in a high surface-energy contrast between the fluorinated

Organic Bioelectronics for Life Science and Healthcare   Materials Research Forum LLC
Materials Research Foundations **56** (2019) 213-262   doi: https://doi.org/10.21741/9781644900376-7

electrode surface and the substrate. Therefore, when printing the second line of the PEDOT:PSS ink, the droplets retract from the fluorinated PEDOT surface and impact the glass surface creating a thin void between the two printed lines.

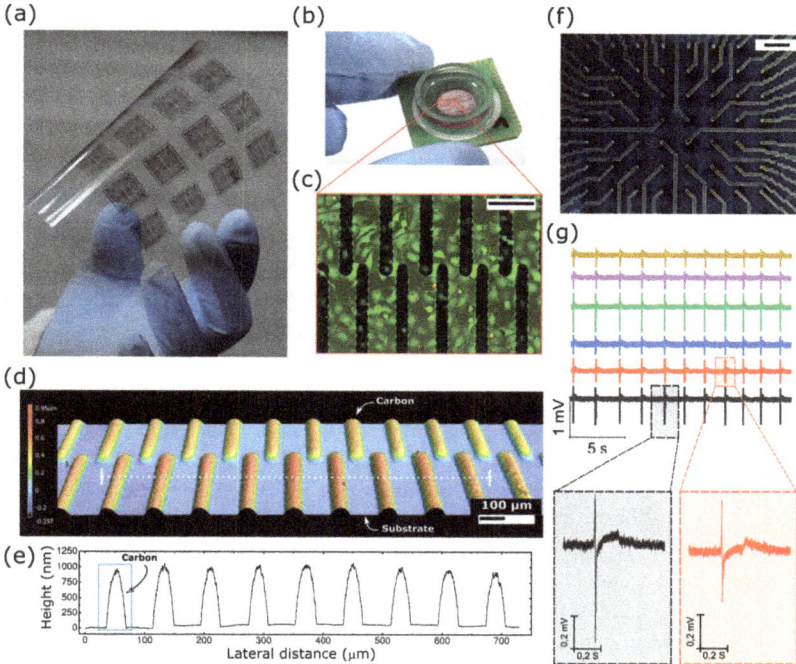

Figure 5: *Inkjet-printed microelectrodes for extracellular recording from cardiomyocyte-like HL-1 cells (a) Picture of large-area printed high-resolution MEA devices. (b) Photograph of an encapsulated MEA chip with cell culture media. (b) Live/dead staining of HL-1 cells cultured on printed MEA chip (green indicates living cells, red indicates dead cells, scale bar 100 µm). (d) 3D interferometer scan of carbon feedlines (dashed line indicates height profile in (b. (b) Height profile of printed feedlines indicated in (d) by dashed line. (f) Microscope image of the MEA printed with gold nanoparticle ink with an electrode pitch of 200 µm, scale bar 200 µm. (g) Time traces recording of extracellular action potentials from different channels using carbon MEA showing a maximum signal amplitude of ≈0.96 mV peak-to-peak, at a background noise of ≈50 µV peak-to-peak. Zoom into two different spikes with signal amplitude of ≈0.9 mV peak-to-peak (gray line) and ≈0.56 mV peak-to-peak (red line). (b)-(e) modified from [72] with permission from John Wiley and Sons.*

While these are very sophisticated techniques for increasing the spatial resolution of printed structures, the intriguing question remains how one can directly print mass-customized flexible MEA devices for measuring cellular signals. We have recently introduced high-resolution MEA devices printed directly on a flexible substrate for extracellular recording of action potentials [67]. The fabrication of these printed MEAs solely relied on inkjet printing without etching, lift-off processes, or prefabricated shadow masks. The success in developing an inkjet-printing process for a high-resolution MEA pattern was achieved by resolving commonly reported printing artifacts such as bulge formation, thin-line breakup and ink migration [68–72]. A series of high-resolution studies have been reported, providing a general printing methodology that can be applied to a wide range of substrates and electrode materials. For instance, the issue of thin-line instability has been resolved by tuning the drop spacing and the drying condition as has been demonstrated with gold, silver, and carbon inks. As another example, ink accumulation and fluid migration to the parent drop, which is crucial when producing individually addressable electrodes, has been resolved by controlling the printing velocity. However, substrate pre-patterning and ink modification approaches have been completely avoided in order to render the process of printing MEAs straightforward and applicable to a wide range of materials. As presented in Fig. 5, a sheet of 24 individual MEAs is printed with a writing speed of 84 mm s^{-1}. The individual MEA (comprising of 64 microelectrodes) is then cut using scissors and encapsulated for cell culture application. One of the advantages of using printing technology for cellular investigations is the possibility of depositing diverse materials that may be challenging to integrate in standard microfabrication techniques. However, depending on the ink, the individual printing paramters have to be adjusted. Here, we evaluated the printing process for the fabrication of microelectrode arrays composed of standard gold as well as nanoporous carbon electrodes. The gold MEAs were printed in two steps. First, a gold nanoparticle ink was used to print conductive tracks on a flexible polymer-based substrate. In a post-processing step, the printed gold was thermally sintered to achieve conducting feedlines. In a second step, the feedlines were passivated using a polyimide-based ink. The gold MEA layout consisted of 64 electrodes arranged in an eight-by-eight configuration, with a lateral spacing of 200 µm and an electrode diameter of ~30 µm. The carbon MEAs were printed in three steps. First, a silver nanoparticle ink was printed for the feedlines and bond pads. Second, a nanoporous carbon ink was printed as the active electrode material. Third, a polyimide ink was printed as an insulator. The carbon MEA was designed and printed in a line format with a feedline width of about 28 µm, where two adjacent carbon feedlines are separated by about 2-3 µm in distance. A dielectric polymer layer with a thickness of approximately 700 nm was printed on top of the silver and the highly porous carbon such that only the bond pads and microelectrodes (≈30 µm diameter) were exposed. HL-1 cells were grown on the printed carbon and gold MEAs and extracellular action potentials were recorded. An example of an extracellular voltage recording from cardiomyocyte-like HL-1 cells using such all-printed devices is shown in Fig. 5. Typically, these MEAs yielded relatively stable extracellular recordings with a signal-to-noise ratio comparable to standard microelectrodes. Examples of recorded signals are presented in Fig. 5 showing peak-to-peak amplitudes in the range of several hundred μV. Overall, both gold and carbon MEAs can be printed at high-resolution on flexible substrates. The printed MEAs provide good electrical and electrochemical characteristics suitable for cellular recording and possibly stimulation, which still needs to be demonstrated. Although the inkjet technology inherently limits the resolution of the structures, we believe that this approach can be used in the future to provide low-cost disposable alternatives to clean-room microfabricated electrode array devices.

Summary

Inkjet printing offers a range of distinct benefits that can facilitate the fabrication of cost-effective bioelectronics and biosensor devices. A variety of affinity-based electrochemical biosensors have already been demonstrated by printing either the biorecognition element, the transducer, or both directly on a suitable sensor substrate. Sensors can be enhanced by implementing signal amplification strategies, such as electrochemical redox cycling, directly in the design and fabrication process of the printed devices. The inherent limitation of the printing resolution that typically sets an upper boundary for the efficiency of in-plane redox cycling sensors, can be circumvented by printing critical features in a multi-layer approach.

We highlighted recent applications of inkjet-printed bioelectronics for wearables including electrocardiography and electromyography. Furthermore, we showed the potential of using inkjet printing to develop flexible microelectrode arrays for cell recordings. These devices are capable of monitoring single-cell action potentials and can be processed at a fraction of the cost typically required for clean-room fabricated devices. We believe that inkjet printing will further develop its potential and thereby become an increasingly appealing rapid prototyping tool for the purpose of developing new biosensors and bioelectronic devices for future applications.

References

[1]   B. Derby, *Annu. Rev. Mater. Res.* **2010**, *40*, 395. https://doi.org/10.1146/annurev-matsci-070909-104502

[2]   S. Magdassi, *The Chemistry of Inkjet Inks*, World Scientific Singapore, **2010**. https://doi.org/10.1142/6869

[3]   K. Kordás, T. Mustonen, G. Tóth, H. Jantunen, M. Lajunen, C. Soldano, S. Talapatra, S. Kar, R. Vajtai, P. M. Ajayan, *Small* **2006**, *2*, 1021. https://doi.org/10.1002/smll.200600061

[4]   P. Beecher, P. Servati, A. Rozhin, A. Colli, V. Scardaci, S. Pisana, T. Hasan, A. J. Flewitt, J. Robertson, G. W. Hsieh, F. M. Li, A. Nathan, A. C. Ferrari, W. I. Milne, *J. Appl. Phys.* **2007**, *102*, 043710. https://doi.org/10.1063/1.2770835

[5]   E. B. Secor, P. L. Prabhumirashi, K. Puntambekar, M. L. Geier, M. C. Hersam, *J. Phys. Chem. Lett.* **2013**, *4*, 1347. https://doi.org/10.1021/jz400644c

[6]   L. Huang, Y. Huang, J. Liang, X. Wan, Y. Chen, *Nano Res.* **2011**, *4*, 675. https://doi.org/10.1007/s12274-011-0123-z

[7]   A. Kamyshny, J. Steinke, S. Magdassi, *Open Appl. Phys. J.* **2011**, *4*.

[8]   A. L. Dearden, P. J. Smith, D.-Y. Shin, N. Reis, B. Derby, P. O'Brien, *Macromol. Rapid Commun.* **2005**, *26*, 315. https://doi.org/10.1002/marc.200400445

[9]   H. M. Haverinen, R. A. Myllylä, G. E. Jabbour, *Appl. Phys. Lett.* **2009**, *94*, 073108. https://doi.org/10.1063/1.3085771

[10]   H. M. Haverinen, R. A. Myllylä, G. E. Jabbour, *J. Disp. Technol.* **2010**, *6*, 87. https://doi.org/10.1109/JDT.2009.2039019

[11]   S. D. Hoath, Fundamentals of Inkjet Printing: The Science of Inkjet and Droplets, John Wiley & Sons, **2016**. https://doi.org/10.1002/9783527684724

[12]   G. Mattana, D. Briand, *Mater. Today* **2016**, *19*, 88. https://doi.org/10.1016/j.mattod.2015.08.001

[13]   A. Moya, G. Gabriel, R. Villa, F. Javier del Campo, *Curr. Opin. Electrochem.* **n.d.**, https://doi.org/10.1016/j.coelec.2017.05.003.

[14]   A. Salim, S. Lim, *Sensors* **2017**, *17*. https://doi.org/10.3390/s17112593

[15]   A. J. Baeumner, *Anal. Bioanal. Chem.* **2003**, *377*, 434. https://doi.org/10.1007/s00216-003-2158-9

[16]   S. Bobade, D. R. Kalorey, S. Warke, *Biosci. Biotechnol. Res. Commun.* **2016**, *9*, 132. https://doi.org/10.21786/bbrc/19.1/20

[17]   A. N. Sekretaryova, M. Eriksson, A. P. F. Turner, *Biotechnol. Adv.* **2016**, *34*, 177. https://doi.org/10.1016/j.biotechadv.2015.12.005

[18]   S. Mittal, H. Kaur, N. Gautam, A. K. Mantha, *Biosens. Bioelectron.* **2017**, *88*, 217. https://doi.org/10.1016/j.bios.2016.08.028

[19] A. P. F. Turner, *Chem. Soc. Rev.* **2013**, *42*, 3184.
https://doi.org/10.1039/c3cs35528d

[20] B. Bohunicky, S. A. Mousa, *Nanotechnol. Sci. Appl.* **2010**, *4*, 1.

[21] S. J. Updike, G. P. Hicks, *Nature* **1967**, *214*, 986.
https://doi.org/10.1038/214986a0

[22] G. Sánchez-Pomales, R. A. Zangmeister, "Recent Advances in Electrochemical Glycobiosensing," **2011**. https://doi.org/10.4061/2011/825790

[23] A. E. G. Cass, Graham. Davis, G. D. Francis, H. A. O. Hill, W. J. Aston, I. John. Higgins, E. V. Plotkin, L. D. L. Scott, A. P. F. Turner, *Anal. Chem.* **1984**, *56*, 667. https://doi.org/10.1021/ac00268a018

[24] S. Borgmann, A. Schulte, S. Neugebauer, W. Schuhmann, Adv. Electrochem. Sci. Eng. WILEY-VCH Verl. GmbH Co KGaA Weinh. Ger. **2011**.

[25] C. Chen, Q. Xie, D. Yang, H. Xiao, Y. Fu, Y. Tan, S. Yao, *RSC Adv.* **2013**, *3*, 4473. https://doi.org/10.1039/c2ra22351a

[26] L. Setti, A. Fraleoni-Morgera, B. Ballarin, A. Filippini, D. Frascaro, C. Piana, *Biosens. Bioelectron.* **2005**, *20*, 2019. https://doi.org/10.1016/j.bios.2004.09.022

[27] Y. H. Yun, B. K. Lee, J. S. Choi, S. Kim, B. Yoo, Y. S. Kim, K. Park, Y. W. Cho, *Anal. Sci. Int. J. Jpn. Soc. Anal. Chem.* **2011**, *27*, 375.
https://doi.org/10.2116/analsci.27.375

[28] G. C. Jensen, C. E. Krause, G. A. Sotzing, J. F. Rusling, *Phys. Chem. Chem. Phys.* **2011**, *13*, 4888. https://doi.org/10.1039/c0cp01755h

[29] P. Teengam, W. Siangproh, A. Tuantranont, C. S. Henry, T. Vilaivan, O. Chailapakul, *Anal. Chim. Acta* **2017**, *952*, 32.
https://doi.org/10.1016/j.aca.2016.11.071

[30] N. Adly, L. Feng, K. J. Krause, D. Mayer, A. Yakushenko, A. Offenhäusser, B. Wolfrum, *Adv. Biosyst.* **2017**, n/a.

[31] A. Miodek, N. Mejri, M. Gomgnimbou, C. Sola, H. Korri-Youssoufi, *Anal. Chem.* **2015**, *87*, 9257. https://doi.org/10.1021/acs.analchem.5b01761

[32] J. Zhuang, L. Fu, M. Xu, H. Yang, G. Chen, D. Tang, *Anal. Chim. Acta* **2013**, *783*, 17. https://doi.org/10.1016/j.aca.2013.04.049

[33] E. Xiong, X. Zhang, Y. Liu, J. Zhou, P. Yu, X. Li, J. Chen, *Anal. Chem.* **2015**, *87*, 7291. https://doi.org/10.1021/acs.analchem.5b01402

[34] L. Hu, T. Tan, G. Chen, K. Zhang, J.-J. Zhu, *Electrochem. Commun.* **2013**, *35*, 104. https://doi.org/10.1016/j.elecom.2013.08.004

Materials Research Forum LLC
doi: https://doi.org/10.21741/9781644900376-7

[35]   J. S. del Río, N. Yehia Adly, J. L. Acero-Sánchez, O. Y. F. Henry, C. K. O'Sullivan, *Biosens. Bioelectron.* **2014**, *54*, 674. https://doi.org/10.1016/j.bios.2013.11.035

[36]   B. Wolfrum, E. Kätelhön, A. Yakushenko, K. J. Krause, N. Adly, M. Hüske, P. Rinklin, *Acc. Chem. Res.* **2016**, *49*, 2031. https://doi.org/10.1021/acs.accounts.6b00333

[37]   J. Diao, A. Ding, Y. Liu, J. M. Razal, J. Chen, Z. Lu, B. Wang, *Adv. Mater. Interfaces* **2017**, *4*, n/a. https://doi.org/10.1002/admi.201700588

[38]   S. R. Das, Q. Nian, A. A. Cargill, J. A. Hondred, S. Ding, M. Saei, G. J. Cheng, J. C. Claussen, *Nanoscale* **2016**, *8*, 15870. https://doi.org/10.1039/C6NR04310K

[39]   T. H. da Costa, E. Song, R. P. Tortorich, J.-W. Choi, *ECS J. Solid State Sci. Technol.* **2015**, *4*, S3044. https://doi.org/10.1149/2.0121510jss

[40]   A. Phongphut, C. Sriprachuabwong, A. Wisitsoraat, A. Tuantranont, S. Prichanont, P. Sritongkham, *Sens. Actuators B Chem.* **2013**, *178*, 501. https://doi.org/10.1016/j.snb.2013.01.012

[41]   M. Jović, Y. Zhu, A. Lesch, A. Bondarenko, F. Cortés-Salazar, F. Gumy, H. H. Girault, *J. Electroanal. Chem.* **2017**, *786*, 69. https://doi.org/10.1016/j.jelechem.2016.12.051

[42]   C. E. Krause, B. A. Otieno, A. Latus, R. C. Faria, V. Patel, J. S. Gutkind, J. F. Rusling, *ChemistryOpen* **2013**, *2*, 141. https://doi.org/10.1002/open.201300018

[43]   Z. Pu, R. Wang, J. Wu, H. Yu, K. Xu, D. Li, *Sens. Actuators B Chem.* **2016**, *230*, 801. https://doi.org/10.1016/j.snb.2016.02.115

[44]   J. Chen, Y. Zhou, D. Wang, F. He, V. M. Rotello, K. R. Carter, J. J. Watkins, S. R. Nugen, *Lab. Chip* **2015**, *15*, 3086. https://doi.org/10.1039/C5LC00515A

[45]   S. Cinti, F. Arduini, D. Moscone, G. Palleschi, L. Gonzalez-Macia, A. J. Killard, *Sens. Actuators B Chem.* **2015**, *221*, 187. https://doi.org/10.1016/j.snb.2015.06.054

[46]   A. Lesch, F. Cortés-Salazar, M. Prudent, J. Delobel, S. Rastgar, N. Lion, J.-D. Tissot, P. Tacchini, H. H. Girault, *J. Electroanal. Chem.* **2014**, *717–718*, 61. https://doi.org/10.1016/j.jelechem.2013.12.027

[47]   Y. Zhang, C. Liu, W. Shi, Z. Wang, L. Dai, X. Zhang, *Langmuir* **2007**, *23*, 7911. https://doi.org/10.1021/la700876d

[48]   F. Liu, J. Y. Choi, T. S. Seo, *Chem. Commun.* **2010**, *46*, 2844. https://doi.org/10.1039/b923656b

[49] B. H. Kim, M. S. Onses, J. B. Lim, S. Nam, N. Oh, H. Kim, K. J. Yu, J. W. Lee, J.-H. Kim, S.-K. Kang, C. H. Lee, J. Lee, J. H. Shin, N. H. Kim, C. Leal, M. Shim, J. A. Rogers, *Nano Lett.* **2015**, *15*, 969. https://doi.org/10.1021/nl503779e

[50] S. H. Ko, H. Pan, C. P. Grigoropoulos, C. K. Luscombe, J. M. J. Fréchet, D. Poulikakos, *Nanotechnology* **2007**, *18*, 345202. https://doi.org/10.1088/0957-4484/18/34/345202

[51] C. W. Sele, T. von Werne, R. H. Friend, H. Sirringhaus, *Adv. Mater.* **2005**, *17*, 997. https://doi.org/10.1002/adma.200401285

[52] H. Sirringhaus, T. Kawase, R. H. Friend, T. Shimoda, M. Inbasekaran, W. Wu, E. P. Woo, *Science* **2000**, *290*, 2123. https://doi.org/10.1126/science.290.5499.2123

[53] L. Zhang, H. Liu, Y. Zhao, X. Sun, Y. Wen, Y. Guo, X. Gao, C. Di, G. Yu, Y. Liu, *Adv. Mater. Deerfield Beach Fla* **2012**, *24*, 436. https://doi.org/10.1002/adma.201103620

[54] E. D. Goluch, B. Wolfrum, P. S. Singh, M. A. G. Zevenbergen, S. G. Lemay, *Anal. Bioanal. Chem.* **2009**, *394*, 447. https://doi.org/10.1007/s00216-008-2575-x

[55] N. Y. Adly, B. Bachmann, K. J. Krause, A. Offenhäusser, B. Wolfrum, A. Yakushenko, **2017**, *7*, 5473. https://doi.org/10.1039/C6RA27170G

[56] Y. Khan, A. E. Ostfeld, C. M. Lochner, A. Pierre, A. C. Arias, *Adv. Mater.* **2016**, *28*, 4373. https://doi.org/10.1002/adma.201504366

[57] H. Tao, B. Marelli, M. Yang, B. An, M. S. Onses, J. A. Rogers, D. L. Kaplan, F. G. Omenetto, *Adv. Mater.* **2015**, *27*, 4273. https://doi.org/10.1002/adma.201501425

[58] T. Vuorinen, J. Niittynen, T. Kankkunen, T. M. Kraft, M. Mäntysalo, *Sci. Rep.* **2016**, *6*, 35289. https://doi.org/10.1038/srep35289

[59] P. Chen, H. Chen, J. Qiu, C. Zhou, *Nano Res.* **2010**, *3*, 594. https://doi.org/10.1007/s12274-010-0020-x

[60] J. Wu, R. Wang, H. Yu, G. Li, K. Xu, N. C. Tien, R. C. Roberts, D. Li, *Lab. Chip* **2015**, *15*, 690. https://doi.org/10.1039/C4LC01121J

[61] S. L. Swisher, M. C. Lin, A. Liao, E. J. Leeflang, Y. Khan, F. J. Pavinatto, K. Mann, A. Naujokas, D. Young, S. Roy, M. R. Harrison, A. C. Arias, V. Subramanian, M. M. Maharbiz, *Nat. Commun.* **2015**, *6*, 6575. https://doi.org/10.1038/ncomms7575

[62] Y. Khan, M. Garg, Q. Gui, M. Schadt, A. Gaikwad, D. Han, N. A. D. Yamamoto, P. Hart, R. Welte, W. Wilson, S. Czarnecki, M. Poliks, Z. Jin, K. Ghose, F. Egitto, J. Turner, A. C. Arias, *Adv. Funct. Mater.* **2016**, *26*, 8764. https://doi.org/10.1002/adfm.201603763

[63]   T. Roberts, J. B. De Graaf, C. Nicol, T. Hervé, M. Fiocchi, S. Sanaur, *Adv. Healthc. Mater.* **2016**, *5*, 1462. https://doi.org/10.1002/adhm.201600108

[64]   G. Bertotti, D. Velychko, N. Dodel, S. Keil, D. Wolansky, B. Tillak, M. Schreiter, A. Grall, P. Jesinger, S. Röhler, M. Eickenscheidt, A. Stett, A. Möller, K. H. Boven, G. Zeck, R. Thewes, in *2014 IEEE Biomed. Circuits Syst. Conf. BioCAS Proc.*, **2014**, pp. 304–307.

[65]   J. Müller, M. Ballini, P. Livi, Y. Chen, M. Radivojevic, A. Shadmani, V. Viswam, I. L. Jones, M. Fiscella, R. Diggelmann, A. Stettler, U. Frey, D. J. Bakkum, A. Hierlemann, *Lab. Chip* **2015**, *15*, 2767. https://doi.org/10.1039/C5LC00133A

[66]   T. Sekitani, Y. Noguchi, U. Zschieschang, H. Klauk, T. Someya, *Proc. Natl. Acad. Sci.* **2008**, *105*, 4976. https://doi.org/10.1073/pnas.0708340105

[67]   B. Bachmann, N. Adly, J. Schnitker, A. Yakushenko, P. Rinklin, A. Offenhaeusser, B. Wolfrum, *Flex. Print. Electron.* **2017**, DOI 10.1088/2058-8585/aa7928.

[68]   P. C. Duineveld, *J. Fluid Mech.* **2003**, *477*, 175. https://doi.org/10.1017/S0022112002003117

[69]   S. Dodds, M. S. Carvalho, S. Kumar, *J. Fluid Mech.* **2012**, *707*, 521. https://doi.org/10.1017/jfm.2012.296

[70]   S. Kumar, *Annu. Rev. Fluid Mech.* **2015**, *47*, 67. https://doi.org/10.1146/annurev-fluid-010814-014620

[71]   J. Stringer, B. Derby, *Langmuir* **2010**, *26*, 10365. https://doi.org/10.1021/la101296e

[72]   Schnitker Jan, Adly Nouran, Seyock Silke, Bachmann Bernd, Yakushenko Alexey, Wolfrum Bernhard, Offenhäusser Andreas, *Adv. Biosyst.* **2018**, *2*, 1700136. https://doi.org/10.1002/adbi.201700136

Organic Bioelectronics for Life Science and Healthcare   Materials Research Forum LLC
Materials Research Foundations **56** (2019) 263-314   doi: https://doi.org/10.21741/9781644900376-8

Chapter 8

Rapid Point-of-Care-Tests for Stroke Monitoring

Dorin Harpaz [1, 2, 3], Evgeni Eltzov [4], Raymond C.S. Seet [5], Robert S. Marks [1, 2, 6, 7, *] and Alfred I.Y. Tok [2, 3, *]

[1] Department of Biotechnology Engineering, Ben-Gurion University of the Negev, Beer-Sheva 84105, Israel

[2] School of Material Science & Engineering, Nanyang Technology University, 50 Nanyang Avenue, Singapore 639798

[3] Institute for Sports Research (ISR), Nanyang Technology University and Loughborough University, Nanyang Avenue, 639798 Singapore

[4] Agriculture Research Organization (ARO), Volcani Centre, Rishon LeTsiyon 15159, Israel

[5] MBBS, FRCP. Division of Neurology, Department of Medicine, Yong Loo Lin School of Medicine, National University of Singapore, NUHS Tower Block, 1E Kent Ridge Road, Singapore 119228

[6] The National Institute for Biotechnology in the Negev, Ben-Gurion University of the Negev, Beer-Sheva 84105, Israel

[7] The Ilse Katz Centre for Meso and Nanoscale Science and Technology, Ben-Gurion University of the Negev, Beer-Sheva 84105, Israel

* rsmarks@bgu.ac.il (R.S. Marks) and MIYTok@ntu.edu.sg (A.I.Y. Tok)

Abstract

Stroke contributes to at least 7 million deaths annually and is the second leading cause of death globally. Ischemic stroke is the most common type of stroke where an interruption of blood flow within brain tissues can lead to cessation of brain activities. Central to existing reperfusion treatment strategies is prompt restoration of blood supply to compromised brain tissues. Existing methods of monitoring the progress of stroke evolution are limited to clinical evaluation and neuroimaging techniques. With advances in biomarker research, there is now a growing interest in the use of stroke biomarkers to guide clinical decision-making, especially for on-site biomarker measurements, to make time-sensitive decisions on stroke prognosis. This chapter presents an overview on point-of-care-tests (POCT) used to guide stroke prognosis, with the aim of shortening time-to-treatment, classifying stroke subtypes and improving patient's outcome. First, the current stroke monitoring workflow is presented, which highlights gaps for an improved bedside biomarker assessment. Second, a detailed overview on the different POCTs used for stroke monitoring is specified. Last, a novel approach for the creation of a future successful stroke POCT is presented, which combines the use of a quantitative and multiplex POCT for the detection of stroke-specific biomarkers.

Keywords

Stroke, Prognosis, Point-of-Care-Test, Biomarkers, Time-Dependent Treatment, Multiplex and Quantitative Detection

Contents

1. Introduction

1.1 Stroke prognosis: what is missing?

Stroke, the world's second leading cause of death, results in 15 million stroke victims each year. More than 6 million die, while another 6 million suffer from permanent disability. The global death number is projected to rise to 7.8 million in 2030 [1]. Though the elderly are more affected by stroke, the younger population is also not spared [2]. Traditional risk factors include prior stroke, high blood pressure, heart disease, hypertension, dyslipidemia, diabetes mellitus, cigarette smoking, obesity and physical inactivity [3]. Stroke patient prognosis starts from symptoms onset and continue up to the recovery phase. It includes the following steps: (1) symptoms identification, (2) dispatch alert, (3) ambulance transportation to hospital, (4) admission to ED, (5) brain imaging, (6) nervous system evaluation, (7) subtype classification, (8) therapeutic treatment, (9) outcome improvement, and (10) secondary stroke prevention (***Figure 1***) [4]. The successful management of a stroke patient starts with recognizing the sudden onset of neurologic symptoms (e.g. facial asymmetry, arm and leg weakness and slurring of speech) [5]. In 70% of cases, the emergency medical service (EMS) is contacted for additional assessment upon symptom recognition [6-8]. This is followed by rapid transportation to the hospital, preferably with an urgent stroke unit or hospital pre-notification [9-11]. However, for more than 50% of cases, stroke clinical symptoms are not recognized by the EMS [7], thus significantly reducing the critical survival rate of the patient [12]. Pre-hospital identification of stroke patients can be improved by increasing access to EMS, improving EMS patient evaluation and transport to the hospital, integrating stroke specialists in pre-hospital stroke response teams [13], mobile telemedicine for remote clinical examination [14, 15], and integration of CT scanners and POCTs in ambulances [16]. These approaches will help to enable quicker admission as a stroke patient and shorten time to treatment [17, 18].

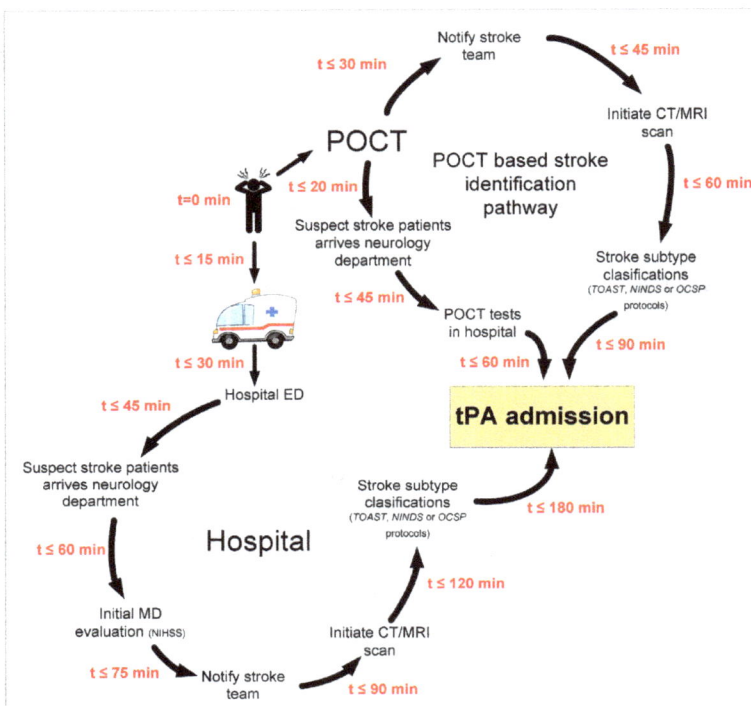

Figure 1. Stroke Patient Conventional Prognosis vs. Improved Using POCTs. *(1) Upon stroke symptoms onset, EMS dispatch is contacted (t = 0). (2) EMS follows 'case entry protocol' and evaluate the patient's medical status by specific parameters. (3) Patient is then transported to the hospital (t ≤ 15min). (4) Conventional prognosis: at best, stroke diagnosis has already been confirmed by the EMS, so he is immediately transported to a stroke specialized unit. Otherwise, the patient will go through the ED, like any other patient suffering from a range of medical problems (t ≤ 30min). The suspected stroke patient will then only be evaluated by a neurologist (t ≤ 60min).The most common neurologic examination used is NIHSS, and followed by a CT/MRI scan for initial stroke diagnosis (t ≤ 90min). Subsequently, stroke subtype classification (t≤120min) is conducted by either TOAST, NINDS or OCSP classification schemes. Admission of IV-tPA has to be within 3hr from stroke-symptoms-onset (t ≤ 180min). (5) Improved prognosis using POCTs: shorten the time from stroke-symptoms-onset to neurologist examination (t ≤ 20min). In addition, the time to CT/MRI scan can be reduced (t ≤ 45min) and, most importantly, the time to tPA admission can be reduced as well (60min ≤ t ≤ 90min) [4].*

Upon stroke patient admission to the ED, a neurological evaluation is completed typically by a scoring system, such as the National Institute of Health Stroke Scale (NIHSS) [19]. In addition, an evaluation of the nervous system is conducted using imaging techniques [3], such as computed tomography (CT) or magnetic resonance imaging (MRI), in order to differentiate ischemic from haemorrhagic stroke [20, 21]. Additional classification technologies that may be used include: blood tests (glucose, platelet count, prothrombin time (PT), partial thromboplastin time (PTT)), electrocardiogram, carotid ultrasound, carotid angiography, electrocardiogram and echocardiography [3, 22]. Most stroke events are classified as ischemic stroke (88%), which is caused by the blockage of a brain blood vessel due to clot formation and accumulation of fat in the vessel walls. In the case of ischemic stroke, recanalization of the occluded artery confers a high likelihood of clinical recovery. To date, intravenous recombinant tissue plasminogen activator (IV-tPA, Alteplase) is the only FDA-approved treatment for ischemic stroke, provided that it is administered within 4.5 hours following stroke onset [23-25]. More recently, randomized trial data support the use of endovascular treatment (through mechanical clot retrieval and aspiration systems) to facilitate arterial recanalization [26-28]. There are 5 ischemic stroke etiologies: (1) 20% atherothrombotic (large artery atherosclerosis, LAA), (2) 20% cardioembolic (CEI), (3) 25% small vessel disease (lacunar, LAC), (4) 5% other causes and (5) 30% unknown causes (cryptogenic strokes) [29-31]. The main value of subtyping ischemic stroke is in classifying patients for a better-targeted therapeutic decision-making process, in order to minimize time-to-treatment of thrombosis with IV-tPA administration. The gold standard method for ischemic stroke subtyping is the Trial of Org 10172 in Acute Stroke Treatment (TOAST) classification. Since its introduction in 1993, its reliability has been improved by the use of a computerized algorithm. However, the group of undetermined stroke cases is highly heterogeneous, and could flaw the medical decision-making process [32-36]. Additional methods used for ischemic stroke subtyping include the National Institute of Neurological Disorders and Stroke (NINDS) classification [37] and the Oxford Community Stroke Project (OCSP) classification [38, 39]. As previously mentioned, the less common stroke type is haemorragic stroke (12%), which is caused due to the rupturing of a weakened blood vessel which then bleeds into the surrounding brain [3, 5, 40]. Of the haemorrhagic strokes, 9% are due to an intra-cerebral haemorrhage (ICH), and 3% are due to a subarachnoid haemorrhage (SAH) [41]. Victims of haemorrhagic strokes are often younger and the fatality rate often higher than for ischemic stroke. Overall prognosis is also poorer for those who have haemorrhagic strokes. The symptoms of a haemorrhagic stroke usually appear suddenly and often include: very severe headaches, nausea and vomiting and partial or total loss of consciousness [3]. This distinction between haemorrhagic and ischemic stroke is thus critical for stroke management and treatment decisions [4].

Once therapeutic treatment is administered and their condition becomes stable, the patient can continue their recovery in a long-term care (LTC) facility. There are approximately 30 million stroke survivors globally and they make up approximately 25% of the residents in LTC facilities. According to the U.S. National Stroke Association, 10% of stroke survivors recover completely, 25% recover with minor impairments, 40% recover with moderate to severe impairments and require special care, while 10% require full care in a nursing home or other LTC facility. The remaining 15% of stroke victims die shortly after the stroke event [3]. The recovery process is personalized and can last between days to months. The main focus for stroke patient recovery is outcome improvement and secondary stroke prevention [42]. In stroke recovery, neuroimaging and neurophysiological assessments are used, such as: voluntary finger extension, shoulder abduction, diffusion-weighted imaging, fMRI activation pattern and MRI DTI sequence [43]. However, there are additional biological values that can add to a better planning of the patient's recovery, such as biomarkers measurement [4].

1.2   Time is Brain: need for rapid solutions

After stroke onset, brain cells die rapidly, which forms the 'time is brain' rationale. Unlike other diseases, stroke is unique because treatment delays can cause significant repercussions in terms of long-term disability as a minute's delay in delivering acute stroke treatment can result in 1.9 million loss of brain neurons. While the damage to the ischemic lesion core is irreversible, the perilesional penumbra still maintains metabolic and structural integrity so it can still be salvaged [16, 44]. There is a direct correlation between timely diagnosis and treatment administration to the patients' outcome [45-47]. As previously mentioned, treatment with IV-tPA significantly improves the outcomes of ischemic stroke patients [27, 44, 48, 49]. However, the treatment efficacy is time dependent, with only a minority of eligible ischemic stroke patients receiving recanalizing therapies [16, 50]. This is caused by pre-hospital and in-hospital time delays in stroke prognosis [51, 52]. Improved outcomes can be also achieved by monitoring post-hospital complications, such as hemorrhagic transformations or post stroke infections [53].

Despite recent advances, critical gaps still exist in the delivery of stroke care. Namely, insufficiently accurate ischemic stroke diagnosis, differentiation from stroke mimics (seizures, migraine, brain space-occupying lesions and complicated migraine), selection of patients for acute reperfusion treatment, better determination of stroke etiology and finally a critical triaging of patients for resource-intensive stroke units. Proposed pre-hospital strategies include public education on stroke symptoms awareness, prioritizing stroke by emergency medical services (EMS), increasing ease of access to medical

Organic Bioelectronics for Life Science and Healthcare   Materials Research Forum LLC
Materials Research Foundations **56** (2019) 263-314   doi: https://doi.org/10.21741/9781644900376-8

records, pre-hospital notification and mobile CT scanning. In-hospital strategies include a streamlined code stroke system, co-location of a CT scanner with ED, 24/7 availability of stroke physicians, use of POCT or laboratory testing and access to expert neuroimaging interpretation [52]. The use of alternative-biological information, such as blood biomarkers and genetic polymorphisms, has become an increasingly popular method for improving stroke outcome [52, 54, 55]. Several blood-based biomarkers are now available to guide such clinical decision-making in stroke monitoring to provide clinicians with an opportunity to confirm stroke risk and personalize stroke diagnosis [56].

Ideal brain biomarkers are usually proteins or small molecules, which can be detected in bio fluids, such as blood, saliva or urine. Brain biomarkers provide additional biological information on specific organs, mainly the brain and spinal cord. These biomarkers demonstrate: (1) specificity - uniquely present in the central nervous system (CNS) and reflect the extent of brain damage; (2) sensitivity - abundant and easily detected; and (3) selectivity - for example, reflecting therapeutic efficacy [57]. There is also a need for the biomarkers to be resistant to cytoplasmic and extracellular proteolytic activity and not be dependent on renal excretion [58]. Biomarkers can be classified into three main classes: (1) susceptibility - reveal subjects with genetically mediated predisposition to a specific condition; (2) effect - measure early biological effect (structural or functional changes in affected cells or tissues); and (3) exposure - measure chemicals or their metabolites to determine a patient's exposure to them [59]. The identification and use of such ideal brain biomarkers would be useful for identifying patients at risk of stroke, as well as in detecting and monitoring their treatment [58]. With the growing incidence of stroke, there is an emerging need for timely bedside assessment of blood-based biomarkers.

1.3   Rapid Point-of-Care-Tests

A POCT will enable a faster and better approach to caring for stroke patients in their most critical state, as it provides relevant information on-site to expedite specific diagnostic delays [60]. A POCT platform is a mobile laboratory device that is located at the site where the patient is treated, which can be used to perform the needed tests by the same personnel who is treating the patient, thus potentially reducing interface times and examination times. POCT holds many advantages, such as simple measurement procedures, use of low sample volumes, no sample processing, automated data processing and off-course mobility (*Figure 2*) [61]. The potential for POCT use in stroke care is therefore enormous. Conventional technologies are usually found in an analytical lab due to their more complicated set-up and operation procedure. Current conventional technologies do provide sufficient sensitivity and reliable results, however they are not always considered practical. An important factor of POCT development is that the test will be considered effective only if action is taken based on its results. Only when the POCT outcome is utilized on-site as a basis for patient treatment can POCT successfully and efficiently reduce hospital stay, improve adherence to treatment, and reduce complications [62]. The current POCT development and design research focuses on minimizing the device size, while still obtaining highly sensitive and accurate results [63]. POCT successes include the glucose biosensor strips [64] and lateral flow immunoassay strips with the most well-known being the pregnancy test. The conventional clinical technology is ELISA (enzyme-linked immunosorbent assay) [65] which can reach higher sensitivities than most POCTs available, but which require multiple steps and a complicated testing procedure. It is important to note that sensitivity is not always the most critical factor, especially in cases where there needs to be faster and more accessible treatments. POCTs may result in sufficient sensitivity and accuracy standards being obtained.

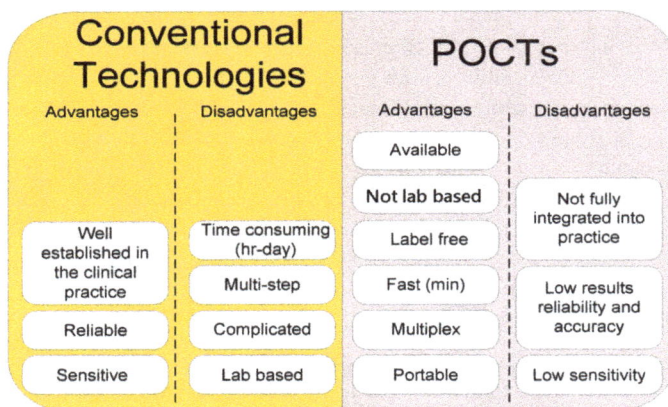

Figure 2. POCTs vs. Conventional Technologies. *The advantages of POCTs are: portability, a simple structure, easy to use, allows multiplex detection, gives results within minutes, and does not require labelling. However, a POCT demonstrates lower sensitivity and gives less reliable and accurate results. Conventional technologies are usually characterized with a more sensitive and reliable detection, but the limitations deny its usage as an on-site diagnostic tool. The disadvantages of conventional technologies are: lab facility requirement, results take hours or even days and complicated usage that requires professional personnel which results in a time-consuming process [4].*

However, not all POCT devices demonstrate the full features required of a diagnostic device. A POCT is not defined by any particular technology or method of use, in that it does not require reagent-free operation, battery-powered operation, or a specific degree of operator training [66]. However, a biosensor as a POCT will be a more accurate approach for a novel-stroke POCT. There is a variety of POCT-based sensor analytical formats, such as microfluidics [67], microarrays [68], paper-based immunoassays [69-71] and optical-based sensors [72-74]. A biosensor is usually described as a self-contained device, capable of providing selective and quantitative or semi-quantitative analytical information and which uses a biological recognition element and a transducer placed in intimate contact via some form of chemical immobilization [75-77]. A typical biosensor is designed based on three different aspects: the biospecific capture entity - biological detection of the target molecule; the chemical interface – controls the main function of the system; and transducer – signal detection and measurement (*Figure 3*). The biospecific capture entity (e.g., whole cells, enzymes, DNA or RNA strands, antibodies, antigens or biomimetic molecules) is chosen according to the target analyte, while the interfacial chemistry ensures that the biospecific capture entity molecule is immobilized upon the relevant transducer. A successful POCT sensor must fulfil at least three critical conditions after the immobilization steps: (1) maintain stability and activity of the biological part during and after functionalization; (2) maximize proximity of the biological layer to the transducer; (3) maintain sensitivity and specificity of the biological components to a target analyte [78]. Adsorption, cross-linking, covalent binding, entrapment, and less useful Langmuir–Blodgett (LMB) deposition, self-assembled monolayers and bulk modifications are the most reported methods used in biosensor functionalization applications [78, 79]. The transducer, which converts the molecular recognition event to a measurable signal, may be electrochemical, acoustic or of optical origin etc.[80]. An ongoing trend in biosensors research is the development of healthcare diagnostic tools [81].

2. POCTs Expedite Stroke Prognosis

Prior to the administration of IV-tPA treatment, laboratory results such as coagulation profile (international normalized ratio (INR) and activated partial thromboplastin time (APTT)), blood-count (platelet, leukocyte and erythrocyte count) and blood-chemistry (hemoglobin, glucose, c-glutamyltransferase and p-amylase test) were highly useful. However, in clinical practices, these valuable tests are not completed due to the time consuming diagnostic procedure [28, 82]. Most guidelines of stroke prognosis recommend that thrombolytic therapy should not be delayed to wait for these test results unless there is clinical suspicion of: bleeding abnormality, thrombocytopenia, and cases where either the patient has received anticoagulants (heparin or warfarin) or if the use of anticoagulants is not known [28]. This might lead to an increased risk of overlooking stroke mimics or patients with contraindications for thrombolysis treatment administration [82]. In the ED, 50% of non-strokes are misdiagnosed as stroke [83]. The current clinical standard is based solely on the clinician's assessment of symptoms and rudimentary stroke scale tools. There is no simple, immediate, and unbiased way to diagnose stroke [84]. For example, studies have showed that serum glucose could be safely obtained by paramedics and that INR POCT can reduce door-to-needle times [52, 85]. Part of these tests can be completed even before arriving to the hospital. Furthermore, a recent study [82] showed that the use of POCT instead of the central hospital laboratory could reduce the time-to-therapy from 84 ± 26 to 40 ± 24 min (p < 0.001) and that the results of most laboratory tests (except APTT and INR) revealed close agreement with those from a standard centralized hospital laboratory. Different POCTs have been tested for their usefulness in stroke monitoring (**Table 1**). However, their accuracy and effectiveness still have not been fully clinically tested [86].

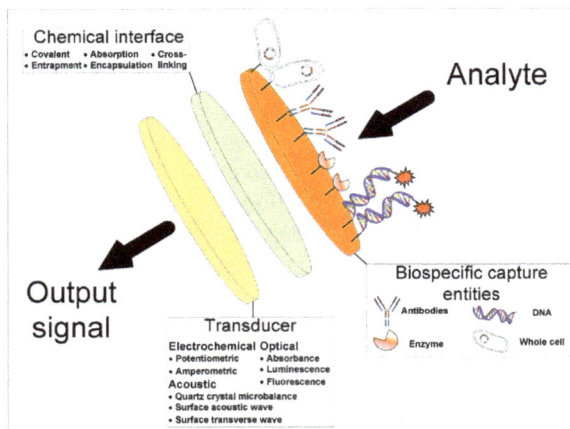

Figure 3. POCT Biosensor Structure. A POCT biosensor consists of three components: biospecific capture entity, chemical interface and transducer. A variety of biospecific capture entities are used for the bio-detection of the target analyte in biosensors, such as antibodies, enzymes, DNA and whole cells. The biosensor chemical interface can be based on either a covalent, entrapment, absorption, encapsulation or cross-linking binding. In addition, the transducer, which is used for the signal transmission and measurement, can be based on electrochemical signal (potentiometric or amperometric), optical signal (absorbance, luminescence or fluorescence) or acoustic signal (quartz crystal microbalance, surface acoustic wave or surface transverse wave) [4].

Table 1. POCTs for Stroke Monitoring [4].

Category	POCT Device	Description	Stroke Application	Ref.
Mobile Stroke Unit	Mobile Stroke Unit	Ambulance containing imaging tools (CT scanners), POCTs (coagulation profile, blood-count, and blood-chemistry), telemedicine connection to hospital and therapeutic treatments (thrombolysis for ischemic stroke and anticoagulant reversal for haemorrhagic stroke)	Diagnosis	[13, 88-93]
Imaging POCTs	CereTom® (Samsung/ Neurologica)	Portable CT scanner that produce images with sufficient quality, however doesn't allow the assessment of the neck vessels or the aortic arch	Ischemic vs. haemorrhagic stroke diagnosis	[87, 94]
	SOMATOM Scope (Siemens)	Portable CT scanner that allows higher resolution and assessment of neck vessels and the aortic arch, however it is bigger in size and require on-board power generation	Ischemic vs. haemorrhagic stroke diagnosis	[87, 95]
	Vivid q® (GE Healthcare)	Portable cardiac ultrasound device, show the condition of the carotid arteries in the neck and intracranial vessels	Ischemic vs. haemorrhagic stroke diagnosis	[96, 97]
Electro-chemical POCTs Assays	CoaguChek ® (Roche)	Amperometric determination of the PT time after activation of the coagulation with human recombinant thromboplastin, can determine the INR value from a drop of capillary whole blood, results within 1min	Diagnosis	[86, 93, 98-102]
	SMARTChip (Sarissa Biomedical)	3-layered micro-electrode consists of a biolayer, mediator and electrode, which measures purines from a drop of whole blood, results within 3-5 minutes	Diagnosis	[103-106]
	i-STAT® (Abbott)	Specific antibodies are located on an electrochemical sensor fabricated on a silicon chip, includes tests such as blood gases, electrolytes, metabolites, coagulation and cardiovascular biomarkers, results within 2 minutes	Diagnosis	[93, 107-111]
	pocH-100i (Sysmex)	Micro-fluidics. WBCs, RBCs and PLTs are counted using the direct current detection method with hydrodynamic focusing technology. Hemoglobin analysis is conducted using a non-cyanide method Provides a full blood count and a 3-part differential leukocyte count	Diagnosis	[82, 93, 112, 113]

Optical POCTs Assays	Hemochron ® Junior (ITC)	Whole blood micro-coagulation system , LED optical detectors detects the reduced sample flow once a clot is formed within the test channel	Diagnosis	[82, 114, 115]
	Reflotron® plus analyzer (Roche)	Blood-chemistry test (c-glutamyltransferase, p-amylase, and glucose), based on reflectance photometry, using LED as a light source and photodiodes as light detectors, results within 2-3 minutes	Diagnosis	[82, 116, 117]
	Cobas® h 232 (Roche)	Cardiac biomarkers (troponin T, NT-proBNP, D-dimer, myoglobin and CK-MB) immunoassay, based on gold-labelled monoclonal and biotinylated polyclonal antibodies, optical system converts the intensity of the signal line into a quantitative result	Prognosis (correlation to CEI) and stroke recovery (indication on second stroke reoccurrence)	[90, 118-147]
	Triage® BNP Test (Alere)	Fluorescence immunoassay for quantitative BNP measurement, the specimen reacts with fluorescent antibody conjugates, the concentration of BNP in the specimen is proportional to the fluorescence detected, results within 15 minutes	Prognosis (correlation to CEI) and stroke recovery (indication on second stroke reoccurrence)	[90, 118-146, 148]
	Cornell University	Luminescent assay of a biomimicry approach, based on enzymes tethered to nanoparticles, for the detection of NSE. Immobilization of pyruvate kinase (PK) and luciferase on silica NPs, results within 10 minutes	Diagnosis	[149-155]
	VerifyNow® (Accumetric s/Accriva Diagnostics)	Qualitative turbidimetric based optical detection system which measures platelet induced aggregation as an increase in light transmittance. As aggregation occurs, the system converts luminosity transmittance results into 'Aspirin Reaction Units', for the identification of aspirin non-responsive patients	Prevention of second stroke recurrence	[156-171, 173, 174]
Other POCTs	PFA-100®, (Dade/Siem ens)	Simulates under high shear stress, the interaction of platelets with an injured blood vessel, measures the time required for platelets to adhere and aggregate, used for identification of aspirin non-responsive patients	Prevention of second stroke recurrence	[156-173, 175]

Prediction Sciences LLC	Lateral flow based test for the measurement of the proteomic marker cellular fibronectin (c-Fn), elevated serum levels in stroke patients at IV-tPA admission can identify if the patient is at high or low risk for a subsequent hemorrhage, results within 10 minutes	Diagnosis	[176]
ReST™ (Valtari Bio™ Inc.)	Measurement of blood brain-specific biomarkers associated with immune responses, The degree and direction of the immune system activation allow the accurate identification of acute stroke from non-stroke, results within 10 min	Diagnosis	[83, 84, 177]

2.1   Mobile Stroke Unit

The mobile stroke unit (MSU) enables administration of treatment at the emergency site, which saves critical time and helps to expedite treatment with the time-dependent therapeutics. It was developed in 2003 and clinically tested in 2008. The MSU contains imaging tools, POCTs, telemedicine connection to hospital and therapeutic treatments (thrombolysis for ischemic stroke and anticoagulant reversal for haemorrhagic stroke). The size of the devices used has been minimized, in order to fit into the standard ambulance. Since then, non-contrast CT, CT angiography, and CT perfusion (excluding MRI) have been integrated into MSUs [87]. The integration of mobile CT scanner and POCTs in ambulances was first clinically tested in the MSU project of the University of Homburg, Saarland, Germany. These ambulances allow IV-tPA treatment on-site to test for ischemic stroke. After excluding the possibility of haemorrhagic stroke, the patient would then be transported to the hospital in a normal ambulance. The results of the controlled study showed a remarkable reduction of time from alarm to therapy decision (median of 35 minutes compared to 76 minutes in regular care) [88, 89]. However, scanning failures (mainly technical) were reported in a number of patients (12 of 53) [90]. Later, The Stroke Emergency Mobile (STEMO) project of the Charité in Berlin integrated new features into a fully equipped ambulance, enabling hyper-acute treatment and transport in the same vehicle [91, 92]. This pilot study showed encouraging results for the safety of treatment and in the number of prehospital IV-tPA applications (23 treatments within 52 days) with a mean call-to-needle time of 62 minutes compared to 98 minutes. The data suggests that prehospital stroke care in STEMO is feasible, without any safety concerns. A third clinical study of the mobile stroke treatment unit (MSTU) in the Cleveland Pre-Hospital Acute Stroke Treatment Study Group, also incorporated the presence of a registered nurse, paramedic, emergency medical technician, and a CT technologist. A cerebrovascular specialist evaluates the patient via telemedicine, and a neuroradiologist remotely evaluates images obtained by the portable CT scanner. In addition, a variety of POCTs were used, such as coagulation profile, complete blood-count, and blood-chemistry [93].

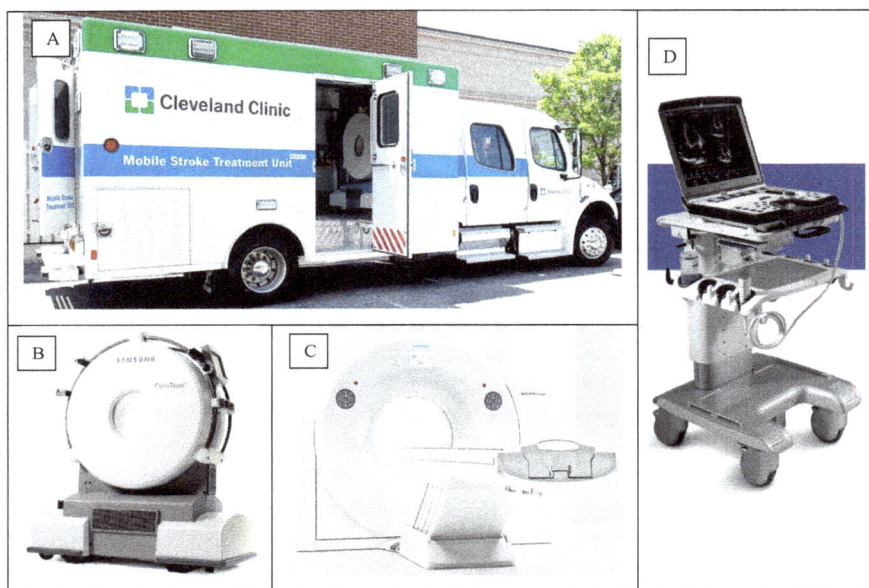

Figure 4. Mobile Stroke Unit and Imaging Tools*. A) Mobile Stroke Unit [278]. B) CereTom® (Samsung/ Neurologica). C) SOMATOM® (Siemens). D) Vivid q® (GE Healthcare) [97].*

2.2 Imaging POCTS

2.2.1 CereTom® (Samsung/Neurologica) & SOMATOM Scope (Siemens)

Most of the mobile brain imaging tools used in ambulances are portable, accumulator driven and radiation shielded, as they were originally designed for intensive care units. Two commonly used CT scanners were shown to be significantly useful when used in the MSUs: CereTom® (Samsung/Neurologica) and SOMATOM Scope (Siemens) [87]. CereTom® produces images with sufficient quality, but does not allow assessment of the neck vessels or the aortic arch. It is an 8-slice CT scanner that delivers non-contrast, angiography and contrast perfusion scans. It uses a radiolucent universal scan board made from carbon fiber material designed to fit any bed [94]. SOMATOM Scope, on the other hand, allows higher resolution and assessment of neck vessels and the aortic arch. However it is bigger in size and requires on-board power supply. It uses an ultra-fast ceramics (UFC) detector with adaptive signal boost that reduces image noise by up to 50%, and has a posterior fossa optimization filter that delivers reduced beam hardening artifacts in the base of the skull. Using Siemens' proprietary UFC, SOMATOM Scope provides excellent image detail even at low radiation doses. This industry-leading device achieves exceptional efficiency thanks to high X-ray absorption combined with extremely effective conversion of X-ray energy into visible light, reducing noise in the images [95].

2.2.2   Vivid q® (GE Healthcare)

Standard echocardiography (SE) is also used in ischemic stroke recovery, though it is less commonly utilized. In a recent study in the stroke unit of the University Hospital Würzburg, Germany, the feasibility and accuracy of the POCT echocardiography was compared with SE using Vivid q® (GE Healthcare), a portable cardiac ultrasound device [96]. Results showed that amongst the patients tested, there was good agreement for cardiac parameters, interrater agreement for left ventricular ejection fraction (LVEF), as well as good detection of systolic dysfunction. They concluded that the use of POCT echocardiography in a stroke unit setting is feasible, enabling reliable preliminary assessment of selected cardiac parameters. However, its potential clinical utility in triaging stroke patients needs to be further investigated in larger diagnostic studies. Vivid q® is a comprehensive, compact and portable ultrasound system, which produces raw data image quality with matrix array imaging which is one of the most advanced probe technologies available. The matrix probe technology allows electronic beam focusing in three dimensions, enabling higher image quality. With its 16 probes and compact size, the Vivid q® allows clinicians to share exceptional ultrasound imaging in any clinical environment [97].

2.3   Electrochemical POCTs Assays

2.3.1   CoaguChek® (Roche)

By using a coagumeter POCT, INR values can be measured immediately at the bedside [86]. Roche Company commercialized CoaguChek®, an electrochemical POCT, is a convenient, portable and user-friendly instrument for monitoring oral anticoagulation therapy and determining the INR value from a drop of capillary whole blood. This POCT was used for both pre-hospital stroke care [93, 98] and in-hospital stroke care [86, 99]. CoaguChek® test principle is based on amperometric (electrochemical) detection of the coagulation response of the platelet after activation with the human recombinant thromboplastin. After placing a blood drop on the test strip (8ul of fresh capillary whole blood), results are obtained within 1 min. The INR values can be measured from 0.8 to 8. The POCT can be directly connected using Bluetooth or USB to send the test results to a healthcare professional quickly and with ease [100]. The INR POCT results correlates well with laboratory values and can be used to shorten door-to-needle time [98, 101, 102].

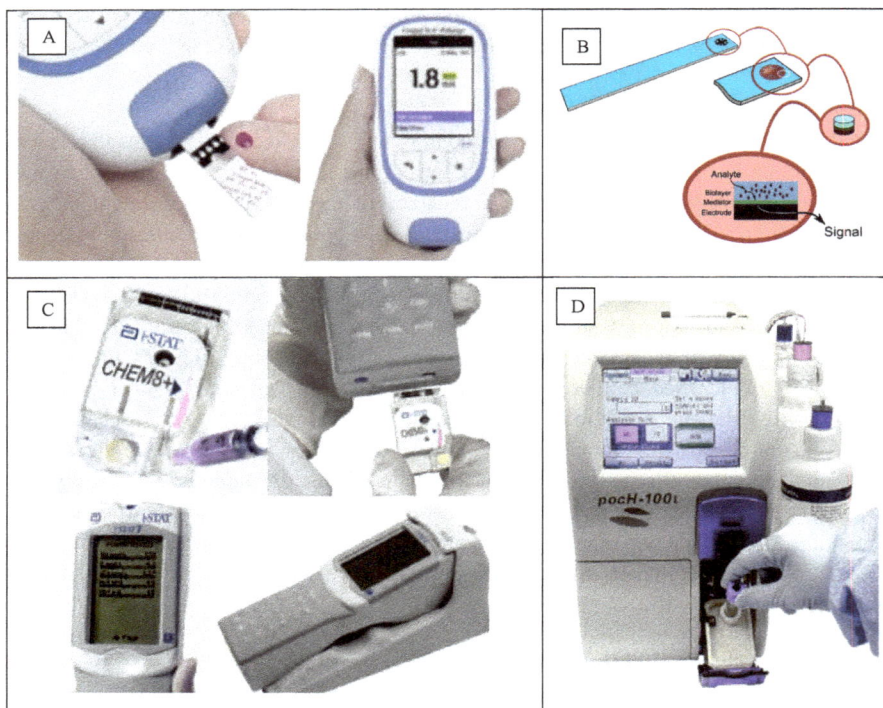

Figure 5. Electrochemical POCTs. *A) CoaguChek® (Roche) [100]. B) SMARTChip (Sarissa Biomedical Ltd.) [103]. C) i-STAT® BNP (Abbott) [111]. D) pocH-100i^{TM} (Sysmex).*

2.3.2 SMARTChip (Sarissa Biomedical)

A second electrochemical POCT is developed by Sarissa Biomedical, a spin-off company from Coventry (UK). SMARTChip [103] is a POCT device that monitor stroke biomarkers. It can be used for stroke diagnosis by a paramedic, which will allow faster identification of stroke victims at the point of injury, and facilitate rapid coordination of the clinical treatment pathway to maximize the chances of the best possible patient outcomes. The assay uses a hand-held reader and disposable biosensor. The SMARTChip can measure purines from just a drop of whole unprocessed blood and provide the results within minutes. Purines (e.g. adenosine, inosine and hypoxanthine) are neurochemicals that influence the function of the nervous system and can be used as stroke biomarkers as they are released after stroke onset. This SMARTChip POCT consists of a 3-layer structure, that converts a "dumb" microelectrode into a "smart" device capable of measuring specific analytes in real-time. The analyte sensitivity depends upon the nature of the enzymes included in the sensing layer. The upper sensing layer is termed 'biolayer' and contains the bioreporter (enzyme) and sample, with or without the target analyte. The layer underneath the 'biolayer' is termed 'mediator' layer, and the bottom layer is the electrode which transmits the signal. The current SMARTChip platforms available are used in the detection of the following analytes: adenosine triphosphate (ATP), adenosine, inosine, hypoxanthine, acetylcholine, choline, glutamate, glucose, lactate and D-serine. The user initially calibrates a SMARTChip by applying a drop of calibration fluid. Then, the fluid is removed and a finger-prick drop of blood from the patient is added. The reader is started and the purine level displayed as a numerical value on the device. The process from calibration to the final result takes approximately 3-5 minutes. SMARTChip was tested in one completed and two ongoing clinical studies: (1) a field deployable blood test for stroke, capable of detecting brain ischemia [104]; (2) validation of a purine biosensor in detecting acute cerebral ischemia: carotid endarterectomy model in SMARTChip [105]; (3) SMARTChip: a field deployable blood test for stroke, capable of detecting brain ischemia from the earliest stages of pathology [106].

2.3.3 i-STAT® (Abbott)

The i-STAT® is a portable, electrochemical clinical analyzer designed to be used at the patient's bedside for critical care tests for blood gases, electrolytes, metabolites, coagulation and cardiovascular biomarkers [107]. It was tested for use in stroke monitoring [93, 108, 109]. A clinical study, supervised by Abbott and conducted in an anonymous hospital [110], aimed to revitalize the ED's systems by including new emergent protocols that integrated bedside POCTs. As part of this new protocol, i-STAT® System CHEM8+ (basic metabolic panel), cTnI (troponin I), CG4+ (blood gas with lactate), and PT/INR were implemented in an effort to improve diagnostic efficiency and patient flow. Incorporating i-STAT CHEM8+, cTnI, CG4+, and PT/INR into nurse-driven emergent protocols made measurable advancements in the diagnostic-process efficiency. The ED staff was empowered to accelerate diagnosis, treatment and disposition of patients. The full parameters include: sodium, potassium, chloride, ionized calcium, BUN, glucose, creatinine, lactate, pH, PCO_2, PO_2, HCO_3, TCO_3, hematocrit, ACT celite, PT/INR, ACT kaolin, cTnI, CK-MB, BNP. Most of the test principles are different. For the cardiovascular biomarkers, cTnI, CK-MB, BNP, the test is determined amperometrically using a two-site ELISA method. Specific antibodies are located on an electrochemical sensor fabricated on a silicon chip. An antibody/alkaline phosphatase enzyme conjugate specific to a separate portion of the analyte molecule is also deposited in another location on the sensor silicon chip. The whole blood or plasma sample is brought into contact with the sensors, allowing the enzyme conjugate to dissolve into the sample. The analyte within the sample becomes labelled with alkaline phosphatase and is captured onto the surface of the electrochemical sensor during an incubation period of approximately seven minutes. The sample, as well as excess enzyme conjugate, is washed off the sensors. Within the wash fluid is a substrate for the alkaline phosphatase enzyme. The enzyme bound to the antibody/antigen/antibody sandwich cleaves the substrate releasing an electrochemically detectable product. The electrochemical (amperometric) sensor measures this enzyme product which is proportional to the concentration of analyte within the sample. Each test has a different measurement range, for example the i-STAT BNP test range is 15 to 5000 pg/mL. Results for BNP and cTnI are obtained within 10 minutes, while that for CK-MB are obtained within 5 minutes. Blood-gas, blood-chemistry, electrolytes and hematology results are obtained within only 2 minutes [111].

2.3.4 PocH-100i^{TM} (Sysmex)

Blood-count tests are also needed in stroke patient prognosis. A POCT specific for blood-count is the PocH-100i^{TM} hematology analyzer (Sysmex), which is a compact device designed specifically for a POCT environment. The technology is based on hydrodynamic focused impedance cell counting. The analyzer provides a full blood count and a 3-part differential leukocyte count [112]. This POCT is used for both pre-hospital stroke care [93] and in-hospital stroke care [82]. PocH-100i^{TM} is designed for laboratory testing of up to 25 samples per day and can detect red blood cells and platelets count. White blood cells (WBCs), red blood cells (RBCs) and platelets (PLTs) are counted using the direct current detection method with hydrodynamic focusing technology to minimize coincidence or recirculation. Hemoglobin analysis is conducted using a non-cyanide method. Hematocrit is directly determined based on the red cell count and volume detection of each individual RBC. The instrument requires only 15μL of EDTA whole blood for reporting of 17 clinical parameters including a 3-part WBC differential [113].

Organic Bioelectronics for Life Science and Healthcare   Materials Research Forum LLC
Materials Research Foundations **56** (2019) 263-314   doi: https://doi.org/10.21741/9781644900376-8

2.4 Optical POCTs Assays

2.4.1 Hemochron® (ITC/Accriva Diagnostics)

The optical POCT, Hemochron® (ITC/Accriva Diagnostics), is a whole blood micro-coagulation system, which was tested as an improved strategy for in-hospital stroke care [82, 114]. Being a micro-coagulation system, it offers point-of-care monitoring of: (1) ACT-LR: Low Range Activated Clotting Time, (2) ACT+: Activated Clotting Time Plus, (3) PT, (4) Citrate PT, (5) APTT and (6) Citrate APTT. The test technology is cuvette-based and uses finger-stick or venous samples with results received within minutes. The ACT+ assay accurately measures heparin (anticoagulant) effects from 1.0 to 6.0 u/mL of blood. Results are dependent on actual fibrin clot formation, and not thrombin generation. It uses a mixture of silica, kaolin, and phospholipids as an activator to create a rapid and highly sensitive alternative to existing ACT tests. The ACT-LR assay is sensitive to low doses of heparin (0.0 to 2.5 u/mL of blood) and it uses a celite activator due to its excellent heparin sensitivity. This POCT also reports INR, plasma-equivalent seconds and whole blood seconds, with an effective INR range from 0.8 to 10.0. As mentioned, the system measures whole blood clotting times using disposable single-use cuvettes. Each cuvette contains all of the reagents necessary for a specified test. First, the operator inserts a cuvette for the test into the instrument and then once the cuvette has warmed to 37 °C ±1.0 °C, the instrument will signal that a blood sample can be added to the cuvette to start the test. The instrument measures the required volume of blood and automatically moves it into the cuvette test channel, where it is mixed with reagents. The sample is then moved back and forth at a predetermined rate within the test channel and monitored for clot formation. The test channel is maintained at 37°C ±1.0°C during the test. The rate of movement of the sample is monitored by a series of LED optical detectors that are aligned with the test channel. When the blood clots, the flow of the blood sample within the test channel is impeded, reducing its rate of flow between the optical detectors. A reduction in flow below a predetermined value signals to the instrument that a clot has formed. An internal timer measures the elapsed time between the start of the test and the clot formation. During the test, the whole blood clotting time (in seconds) is displayed. APTT and APTT Citrate results are displayed as plasma equivalent (PE) values, PT and PT Citrate results are displayed as the International Normalized Ratios (INR) and PE values, and ACT+ and ACT-LR results are displayed as Celite ACT equivalent time [115].

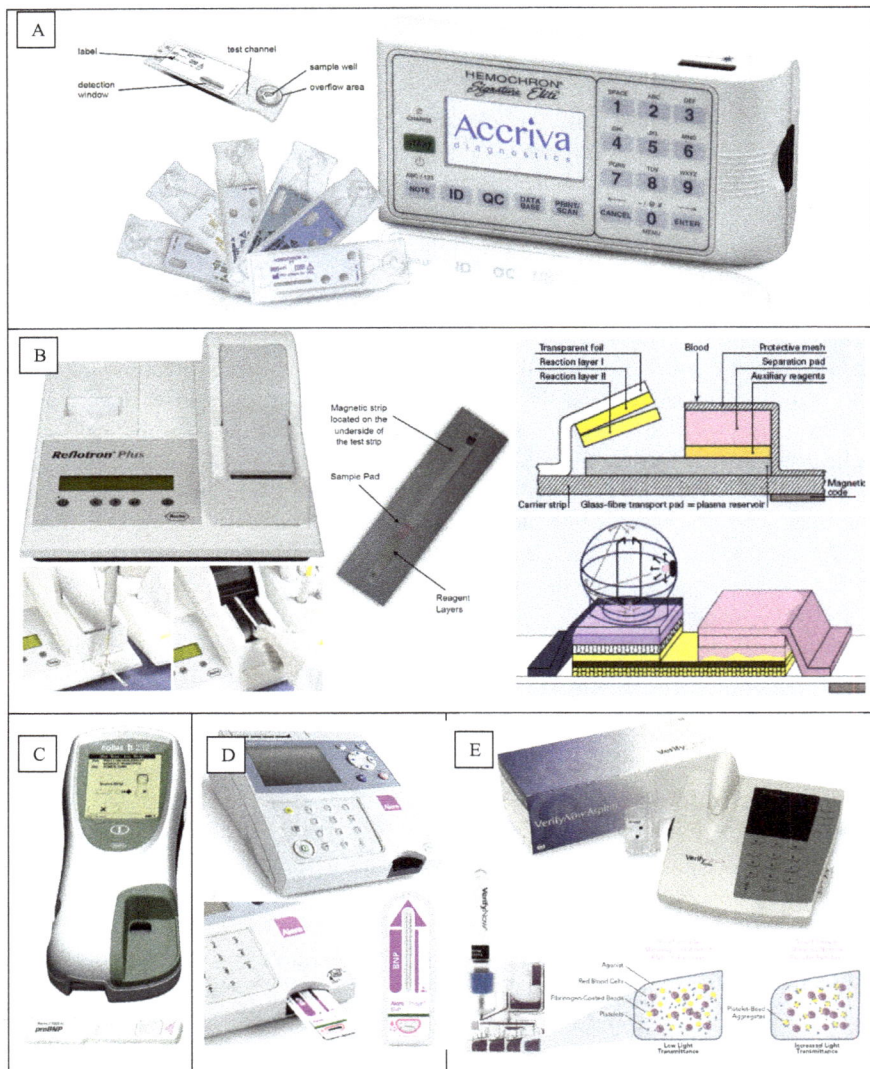

Figure 6. Optical POCTs. *A) Hemochron® (ITC/Accriva Diagnostics) [115]. B) Reflotron® plus analyzer (Roche, Cobas series) [117]. C) Cobas® h 232 (Roche) [147]. D) Triage® BNP Test (Alere) [148]. E) VerifyNow® (Accumetrics, Inc.) [174].*

2.4.2 Reflotron® plus analyzer (Roche)

Another optical POCT is the Reflotron® plus analyzer (Roche, Cobas series), which is also used for blood-chemistry measurements of c-glutamyltransferase, p-amylase, and glucose. It was tested as an improved strategy for in-hospital stroke care [82, 116]. This is a single-test clinical chemistry system which is able to measure whole blood, plasma or serum for: liver and pancreas enzymes, metabolites and blood lipids. The test gives on-site and reliable test results within 2-3 minutes, without sample or reagent preparation. It uses capillary or venous whole blood, plasma, serum, EDTA or heparin. The measuring principle is reflectance photometry, using LED as a light source (wavelengths 567 nm; 642 nm; 951 nm) and photodiodes as light detectors. A reflectance measurement is recorded based on the color change of the test strip. First, the operator applies 30μL of the sample material to the test strip, without any need for reagent preparation or calibration of the instrument. Then, the operator inserts the test strip into the measuring chamber. On the test strip, the sample passes through the separation pad which separates erythrocytes and other cellular components. Then, the pre-reaction layer helps to eliminate any interfering substances and/or helps the sample to be pre-incubated with special activators or auxiliary reagents. After, the plasma or serum that is obtained passes to the transport pad (glass fiber) and underneath the reaction zone. In this zone, the photometer starts the analyte determination reaction at a defined time by pushing the reaction zone onto the transport pad. The reaction layers absorb the quantity of plasma or serum required for the reaction from the transport pad by capillary action. The reagents that are required for the analyte determination reaction are present on carriers. Reagents that would normally be unstable together on a single carrier can be placed on separate carriers. The reaction starts when the reaction zone is pushed onto the filled plasma reservoir. The color intensity of the reaction product is measured at various times (kinetic measurement) or on completion of the reaction (end-point measurement). The depth of color is determined by the concentration of analyte in the sample being investigated. The instrument, a microprocessor-controlled reflectance photometer, takes full charge of the test and automatically measures reflectance values as well as performs calculations. It receives the test- and lot-specific information from the magnetic stripe affixed to the underside of each Reflotron® reagent strip. In this way, the microprocessor system is able to control and monitor all functions such as testing procedure, heating, automatic calibration and evaluation of reflectance values and calculation of results, specific to each test. At the heart of the photometer is the Ulbricht's sphere where the measurements take place. The full 17 parameters include (range): bilirubin (8.5-205μmol/L), cholesterol (2.59-12.9mmol/L), HDL (0.26-2.59mmol/L), creatinine (44.5-884μmol/L), glucose (0.56-33.3mmol/L), hemoglobin (5-20g/dL), potassium (2-12mmol/L), triglycerides (0.8-

6.86mmol/L), uric acid (120-1190μmol/L), urea (3.33-50mmol/L), alkaline phosphates (20-1250U/L), amylase (29-860U/L), pancreatic amylase (14-850U/L), creatine kinase (CK) (37°C: 24.4-1400IU/L, 30°C: 15.4-900IU/L, 25°C: 10-600IU/L), gamma glutamyl transpeptidase (γ-GT) (37°C: 5-3500U/L, 30°C: 3.85-2700U/L, 25°C: 2.8-2000U/L), aspartate aminotransferase (AST) (37°C: 5-500U/L, 30°C: 3.25-325U/L, 25°C: 2.25-225U/L) and alanine aminotransferase (ALT) (37°C: 5-2000U/L, 30°C: 3.8-1520U/L, 25°C: 2.66-1060U/L) [117].

2.4.3 Cobas® h 232 (Roche)

The Cobas h 232 point-of-care testing (POCT) system manufactured by Roche Diagnostics has been introduced as a near-patient instrument for measurement of a number of cardiac biomarkers including NT-proBNP. Recent studies demonstrated the usefulness of using the B-Type Natriuretic Peptide (BNP) POCT platform on suspected ischemic stroke patients in the ED in order to complete stroke prognosis [118-122]. BNP, a well-known biomarker for heart failure (HF), has now been identified as potentially useful for stroke patient prognostic and recovery management. BNP elevated serum levels in stroke patients show correlation with CEI stroke [90, 123-135], increased mortality [136-143] and indication of second stroke recurrence [144-146]. Cobas h 232 technology is based upon an immunological method and uses 150μL heparinized venous whole blood as sample material. The test strips (CARDIAC proBNP+) contain monoclonal and polyclonal antibodies against epitopes of the NT-proBNP molecule of which one is gold-labelled and the other biotinylated. The test strip detection zone for NT-proBNP has two lines: A signal line and a control line. The signal line indicates if NT-proBNP is present in the blood sample. The control line indicates that the test is valid. The antibodies form a sandwich complex with the analyte in the blood. Following removal of erythrocytes from the sample, plasma passes through the detection zone where the gold-labelled sandwich complexes accumulate and the positive signal is displayed as a reddish line (the signal line). Excess gold-labelled antibodies accumulate along the control line, signaling that the test is valid. An optical system converts the intensity of the signal line into a quantitative result which is shown on the display. Calibration is done automatically when a test strip is inserted. Each lot of the Roche CARDIAC proBNP+ test strip is calibrated against the Elecsys proBNP test. The measuring range is 7 – 1062pmol/L. The full 5 parameters includes: troponin T, NT-proBNP, D-dimer, myoglobin and CK-MB [147].

2.4.4 Triage® BNP Test (Alere)

Another optical POCT for BNP monitoring is the Alere Triage® BNP Test. It is a rapid fluorescence immunoassay used with Alere Triage® Meters for the quantitative measurement of BNP in EDTA anticoagulated whole blood or plasma specimens. The test is intended to be used as an aid in the diagnosis and assessment of severity of congestive heart failure. The test procedure involves the addition of several drops of an EDTA anticoagulated whole blood or plasma specimen to the sample port on the Test Device. After addition of the specimen, the whole blood cells are separated from the plasma using a filter contained in the Test Device. The specimen reacts with fluorescent antibody conjugates and flows through the Test Device by capillary action. Complexes of each fluorescent antibody conjugate are captured on discrete zones specific to the analyte. The Test Device is inserted into the Alere Triage® Meter (hereafter referred to as Meter), which is programmed to perform the BNP analysis after the specimen has reacted with the reagents within the Test Device. The concentration of BNP in the specimen is directly proportional to the fluorescence detected within a measurement zone on the Test Device. The results are displayed on the Meter screen in approximately 15 minutes from the addition of specimen. A larger amount of fluorescence detected by the Meter indicates a higher BNP concentration in the specimen. All results are stored in the Meter memory to display or print when needed. If connected, the Meter can transmit results to the lab or hospital information system. The Alere Triage® BNP Test contains all reagents necessary for the quantification of BNP in EDTA anticoagulated whole blood or plasma specimens. The Test Device contains: murine monoclonal antibodies and polyclonal antibodies against BNP, fluorescent dye and stabilizers. The test devices need to be stored in a refrigerator at 2-8°C, or stabilized for up to 14 days at room temperature. The lowest detectable concentration determined, with 95% confidence limit of the analytical sensitivity, was less than 5pg/mL [148].

2.4.4 Cornell University

Additional optical POCT directed against stroke biomarker measurement, was developed by researchers from Cornell University, State University of New York and the New York Presbyterian Hospital. This POCT is based on biomimicry-based approach to demonstrate a new diagnostic platform, based on enzymes tethered to nanoparticles (NPs), for the detection of neuron-specific enolase (NSE) [149]. Immobilization of pyruvate kinase (PK) and luciferase on silica NPs was used to achieve rapid and sensitive detection of NSE, a well-known and clinically relevant biomarker for stroke. Their approach couples production of ATP by PK with activity of firefly luciferase (Luc) to generate a highly rapid and sensitive bioluminescent readout for the detection of NSE. The researchers showed that their data matched well ($r = 0.815$) with the current gold standard for biomarker detection, ELISA. Moreover, they have a great advantage over ELISA, as they can achieve detection in 10 minutes as opposed to the several hours required for traditional ELISA. Although no single biomarker will likely provide a definitive diagnosis of any disease, the glycolytic enzyme, NSE, is released from damaged neurons and has been suggested to be valuable for the diagnosis of various brain injuries. NSE has been suggested to be useful in distinguishing stroke from mimics, an important first step in expediting the diagnostic process [150-152]. As an alternative to antibody capture, POCTs based on fluid phase enzymatic activities [153, 154] or semi-solid phase bioluminescence [155] are used for plasma NSE monitoring.

2.4.5 VerifyNow® (Accumetrics/Accriva Diagnostics)

A potentially useful POCT for stroke recovery is to screen for aspirin responsiveness after stroke [156]. Aspirin, the most commonly used antiplatelet agent, reduces the relative risk of major vascular events and vascular death by 20% after ischemic stroke [157]. However, the antiplatelet properties of aspirin are not uniform between individuals and recurrent events, which may be caused by 'aspirin non-responsiveness' [158-165]. There is evidence that aspirin non-responsive individuals may be at increased risk of ischemic vascular events [166, 167]. The response to aspirin should be monitored in post-hospital care of stroke patients for the prevention of second stroke recurrence [168, 169]. However, the platelet function tests for aspirin monitoring are time-consuming and difficult to follow as a routine practice. There are simpler platelet function POCTs, such as the PFA-100® Platelet Function Analyzer (Dade/Siemens) and the VerifyNow® Aspirin test (Accumetrics/Accriva Diagnostics) [170-172] available to screen stroke patients for aspirin non-responsiveness. Harrison et al compared the use of both POCTs in 100 patients with transient ischemic attack or stroke patients receiving daily low-dose aspirin treatment [156]. These POCTs offer the possibility of a rapid and reliable identification of aspirin non-responsive patients without the requirement of a specialized laboratory [173]. However, aspirin non-responsiveness is highly test-specific and large prospective studies should determine the prognostic value for each POCT used. The VerifyNow® aspirin test system is a qualitative turbidimetric based optical detection system which measures platelet induced aggregation as an increase in light transmittance. The system consists of a stand-alone instrument and disposable assay device with reagents based on microbead agglutination technology. The quality control system includes an electronic quality control, an assay device internal control, and two levels of external, wet quality controls. The instrument controls assay sequencing, temperature, reagent-sample mixing, determines the degree of platelet function, and displays the results and status information, as well as self-diagnostics. The assay device contains a lyophilized preparation of human fibrinogen coated beads, platelet agonist, a peptide, bovine serum albumin, buffer and stabilizer. The patient sample is citrated whole blood, which is automatically dispensed from the blood collection tube into the assay device, with no pre-treatment. Fibrinogen-coated microparticles are used in the assay device to bind activated platelet GP IIb/IIIa receptors. When the activated platelets are exposed to the fibrinogen-coated microparticles, aggregation occurs proportionally to the number of activated platelet receptors. To ensure consistent and uniform activation of the platelets, the agonist arachidonic acid is incorporated into the assay device. The results are reported in aspirin reaction units (ARU). Results ≤ 549 shows evidence of platelet dysfunction due to aspirin; > 550: no evidence of aspirin-induced platelet dysfunction [174]. The details for PFA-100® are presented under 3.4.4.

2.5   Other POCTs

2.5.4 PFA-100® (Dade/Siemens)

As previously mentioned, screening for aspirin non-responsiveness is useful for secondary stroke prevention. There are POCTs available to screen stroke patients for aspirin non-responsiveness, such as the PFA-100® system, which provides a tool for clinicians to use in the detection of platelet dysfunction. The PFA-100® system simulates, under high shear stress, the interaction of platelets with an injured blood vessel. These conditions allow the PFA-100® system to measure *in vitro* platelet function. The PFA-100® system measures the time required for platelets to adhere and aggregate and result in occlusion of an aperture under high shear conditions. The single use test cartridge unit consists of a capillary, a sample reservoir and a biologically active membrane with a central aperture coated with collagen and epinephrine. By applying constant vacuum, anticoagulated whole blood is aspirated from the sample reservoir through the capillary and the aperture under standardized rheological conditions that expose platelets to high shear stress. First, a trigger solution is dispensed to wet the membrane. Then, similar to an *in vivo* mechanism, platelets adhere and aggregate at the aperture, thereby gradually diminishing and finally arresting the blood flow. The instrument then determines the time from the start of the test until the platelet plug occludes the aperture and reports the time interval as the Closure Time (CT). The platelet plug formation in the PFA-100® is dependent on adequate platelet activity and adequate von willebrand factor status. Therefore, the closure time is an indicator of the platelet function in the analyzed whole blood sample [175].

2.5.2 Prediction Sciences LLC

Another POCT directed against a stroke biomarker was developed by Prediction Sciences LLC (California, USA). This POCT is directed for the measurement of proteomic marker cellular fibronectin (c-Fn) [176], which has been shown in recent studies to predict hemorrhagic transformation in tPA-treated patients with a sensitivity of 100%. As mentioned previously, administration of IV-tPA is limited to a critical time window of 3 hours and the amount of c-Fn in the blood of stroke patients at administration can identify if the patient is at high or low risk for a subsequent hemorrhage. This POCT platform is based on lateral flow technology detection of c-Fn, with the ability to obtain results within 10 minutes.

2.5.3 ReST™ (Valtari Bio™ Inc.)

Valtari Bio™ Inc., a company from West-Virginia (USA), is also developing a POCT for the detection of stroke related biomarkers. ReST™, is a rapid evaluation stroke triage test [177], which is aimed at improving the initial stroke versus no stroke determination in 10 minutes or less.   This approach is based on measurement of blood brain-specific biomarkers associated with immune responses, for better stroke identification. The degree and direction of the immune system activation, following stroke and brain injury, allow the accurate identification of acute stroke from non-stroke. In addition, they employ machine learning and pattern recognition tools in order to identify different immune response patterns in the peripheral blood following various types of neuro-related brain injury. Their method is optimized due to their use of pattern recognition and ratios of biomarkers, rather than the absolute measurement of specific biomarkers. The company conducted clinical studies on over 500 real-world patients and their preliminary clinical trial data suggests that the sensitivity and specificity for diagnosing stroke, using a pattern of expression of associated immune related biomarkers, are much higher than current clinical practice.

3.   Future Stroke POCTs

Novel stroke POCTs should show the use of stroke related biomarkers measurement. Stroke is associated with a variety of pathophysiological changes, which triggers different biochemical processes [56]. This results in a big variety of stroke related biomarkers, though their clinical practice values are yet to be fully determined (***Table 2***) [178]. A convenient way to categorize stroke biomarkers is by their origin, which includes the following groups: glial cells origin, neuronal cells origin, heart muscle cells (cardiomyocytes) origin, blood vessels cells (myocytes) origin, general inflammatory cytokines, cytoskeleton proteins, hemostatic proteins, lipids, metabolic proteins and others. Glial cells' origin stroke biomarkers include Protein S100-Beta (S100B) [57, 58, 179-183], Glial Fibrillary Acidic Protein (GFAP) [57, 179, 180, 184-189] and Myelin Basic Protein (MBP) [190-193], whereas that for Neuronal cells include, Neuron-Specific Enolase (NSE) [57, 58, 152, 179-181, 184, 194, 195], Ubiquitin Carboxyl-terminal Hydrolase L1 (UCH-L1) [57, 58, 196-199] and Creatine Kinase-BB (CK-BB) [200, 201]. On the other hand, Heart muscle cells (cardiomyocytes) include B-Type Natriuretic

Peptide (BNP) [90, 123-146, 202-204] and Blood vessels cells (myocytes) exhibit Matrix Metallo-Proteinase 9 (MMP-9) [205-209]. General inflammatory cytokines and proteins stroke biomarkers include, interleukin-6 (IL-6), interleukin-1b (IL-1b), tumor necrosis factor-α (TNF-α) [184, 210-217] and inflammatory protein Neutrophil Lymphocyte Ratios (NLR) [217-224], while Cytoskeleton proteins include, neurofilaments (NFs) [225-227], cleaved-tau (C-tau) [57, 58, 190, 228-231], microtubule-associated protein 2 (MAP2) [232-235] and alpha-II spectrin break-down products (SBDPs) [57, 236-239], Finally, Hemostatic stroke biomarkers include, D-dimer [26, 240-258], C-reactive protein (CRP) [26, 259-263], Fibrin monomer complex (FMC) [264], soluble fibrin (SF) [264], fibrinogen [264, 265], fibrin/fibrinogen degradation products (FDPs) [264] and von willebrand factor (vWF) [265], and those of Lipid origin include, Triglycerides [266-268], Low density lipoprotein (LDL)/High density lipoprotein (HDL) [266-268], heart fatty acid binding protein (H-FABP) [269, 270], free fatty acid (FFA) [271], ApoA [272, 273] and ApoE4 [267, 274], Whereas Metabolic proteins will have: lactate dehydrogenase (LD) [200, 275] and Albumin [276]. Another origin stroke biomarker is Decorin [277]. The use of a multi biomarker panel strategy, instead of the measurement of a single biomarker, is probably a more useful approach. However, it is still mostly research based and yet to be proven. While the results for certain biomarkers fulfill certain clinical requirements, there are currently no available biomarkers that can be recommended for immediate use in the clinical settings. Multi biomarker studies will make effective panels in clinical settings [184], with increased diagnostic accuracy [57]. The majority of the brain biomarkers tested so far for stroke also show association to other medical conditions. Hence, a multi-biomarker study would be especially significant [26, 207]. In addition, there is a need for a quantitative biomarker detection for a useful clinical value. The use of biomarker measurement as a diagnostic value also requires a specific cut-off value which will allow an improved decision-making process in clinical practice. In order to develop such a multiplex and quantitative POCT for stroke biomarker measurement panel, there is a need to engineer novel POCT platforms.

Table 2. Stroke Related Biomarkers [4].

Family	Biomarker	Expression Association	Stroke Application	Ref.
Glial cells origin	S100-Beta	Associated with blood–brain barrier (BBB) dysfunction	Diagnosis	[57, 58, 179-183]
	Glial Fibrillary Acidic Protein (GFAP)	Associated with the size of brain lesions, the neurological status and short-term functional outcome	Prognosis: outcome prediction	[57, 179, 180, 184-189]
	Myelin Basic Protein (MBP)	Associated with worsened outcome	Prognosis: outcome prediction	[190-193]
Neuronal cells origin	Neuron-Specific Enolase (NSE)	Associated with neurological outcomes and infarct volume	Prognosis: outcome prediction	[57, 58, 152, 179-181, 184, 194, 195]
	Ubiquitin Carboxyl-terminal Hydrolase L1 (UCH-L1)	Associated with the extent of the neuronal injury	Prognosis: outcome prediction	[57, 58, 196-199]
	Creatinine Kinase-BB (CKBB)	Associated with the extent and severity of the brain damage and the recovery potential	Prognosis and recovery	[200, 201]
Heart muscle cells (cardio-myocytes) origin	B-Type Natriuretic Peptide (BNP)	Associated with CEI ischemic stroke, increased mortality and second stroke indication	Prognosis: subtype classification and recovery: second stroke prevention	[90, 123-146, 202-204]
Blood vessels cells (myocytes) origin	Matrix Metallo-Proteinase 9 (MMP-9)	Associated with blood–brain barrier (BBB) dysfunction and second stroke indication	Diagnosis and recovery: second stroke prevention	[205-209]
General inflammatory cytokines and proteins	Interleukin-6 (IL-6), interleukin-1b (IL-1b), tumor necrosis factor-α (TNF-α)	Associated with CEI ischemic stroke subtype	Prognosis: subtype classification	[184, 210-217]
	Neutrophil Lymphocyte Ratios (NLR)	Associated with increased mortality and LAA ischemic stroke subtype	Prognosis: subtype classification and recovery	[217-224]
Cytoskeleton proteins	Neurofilaments (NF)	Associated with disconnected axons	Prognosis	[225-227]
	Cleaved-tau (C-tau)	Associated with neuronal degeneration and disease progression	Prognosis	[57, 58, 190, 228-231]
	Microtubule-associated protein 2 (MAP2)	Associated with dendritic injury	Prognosis	[232-235]
	Alpha-II spectrin break-down products (SBDPs)	Associated with apoptosis and necrotic neuronal death	Prognosis	[57, 236-239]
Hemostatic proteins	D-dimer (DD)	Associated with CEI ischemic stroke subtype and recurrence strokes	Prognosis: subtype classification	[26, 240-258]
	C-reactive protein (CRP)	Associated with AIS diagnosis, stroke severity and LAA ischemic stroke subtype	Prognosis: subtype classification	[26, 259-263]

	Fibrin monomer complex (FMC)	Associated with stroke early recognition and CEI ischemic stroke subtype	Diagnosis and prognosis: subtype classification	[264]
	Soluble fibrin (SF)	Associated with stroke early recognition and CEI ischemic stroke subtype	Diagnosis and prognosis: subtype classification	[264]
	Fibrinogen	Associated with stroke early recognition	Diagnosis	[264, 265]
	Fibrin/fibrinogen degradation products (FDPs)	Associated with stroke early recognition	Diagnosis	[264]
	Von willebrand factor (vWF)	Associated with stroke early recognition and LAC ischemic stroke subtype	Diagnosis and prognosis: subtype classification	[265]
Lipids origin	Triglycerides	Associated with LAA ischemic stroke subtype	Prognosis: subtype classification	[266-268]
	Low density lipoprotein (LDL)/High density lipoprotein (HDL)	Associated with LAA ischemic stroke subtype	Prognosis: subtype classification	[266-268]
	Heart fatty acid binding protein (H-FABP)	Associated with early diagnosis of stroke	Diagnosis	[269, 270]
	Free fatty acid (FFA)	Associated with CEI ischemic stroke subtype	Prognosis: subtype classification	[271]
	ApoA	Associated with LAA ischemic stroke subtype, and stroke severity	Prognosis: subtype classification and outcome prediction	[272, 273]
	ApoE4	Associated with ischemic stroke diagnosis (vs. hemorrhagic stroke) and with LAC and LAA ischemic stroke subtypes	Diagnosis and prognosis: subtype classification	[267, 274]
Metabolic proteins	lactate dehydrogenase (LD)	Associated with the extent and severity of the brain damage and the recovery potential	Prognosis and recovery	[200, 275]
	Albumin	Associated with CEI ischemic stroke subtype	Prognosis: subtype classification	[276]
Others	Decorin	Associated with LAA ischemic stroke subtype	Prognosis: subtype classification	[277]

Acknowledgments

This publication is supported by the National Research Foundation (NRF) of Singapore under the Campus for Research Excellence and Technological Enterprise (CREATE) and the Singapore-MIT Alliance for Research and Technology (SMART), The Institute for Sport Research (ISR) and the Singapore International Graduate Award (SINGA). The chapter is based on the following publication: "Point-of-Care-Testing in Acute Stroke Management: An Unmet Need Ripe for Technological Harvest" (Biosensors 2018). The authors thank Yeo Loo Pin for correcting the manuscript.

Conflicts of Interest: The authors declare no conflict of interest.

References

[1]   Strong, K., C. Mathers, and R. Bonita, *Preventing stroke: saving lives around the world.* Lancet Neurol, 2007. **6**(2): p. 182-7. https://doi.org/10.1016/S1474-4422(07)70031-5

[2]   Goldstein, L.B., et al., Primary prevention of ischemic stroke: a guideline from the American Heart Association/American Stroke Association Stroke Council: cosponsored by the Atherosclerotic Peripheral Vascular Disease Interdisciplinary Working Group; Cardiovascular Nursing Council; Clinical Cardiology Council; Nutrition, Physical Activity, and Metabolism Council; and the Quality of Care and Outcomes Research Interdisciplinary Working Group: the American Academy of Neurology affirms the value of this guideline. Stroke, 2006. **37**(6): p. 1583-633. https://doi.org/10.1161/01.STR.0000223048.70103.F1

[3]   Wellesley MA USA, B.J., project analyst, Stroke Diagnostics and Therapeutics: Global Markets, BCC Research. 2015. **HLC180A.**

[4]   Harpaz, D., et al., Point-of-Care-Testing in Acute Stroke Management: An Unmet Need Ripe for Technological Harvest. Biosensors, 2017. **7**(3): p. 30. https://doi.org/10.3390/bios7030030

[5]   American Heart & Stroke Association, I., 2016.

[6]   Govindarajan, P., et al., Comparative evaluation of stroke triage algorithms for emergency medical dispatchers (MeDS): prospective cohort study protocol. BMC Neurol, 2011. **11**: p. 14. https://doi.org/10.1186/1471-2377-11-14

[7]   Watkins, C.L., et al., Training emergency services' dispatchers to recognise stroke: an interrupted time-series analysis. BMC Health Serv Res, 2013. **13**: p. 318. https://doi.org/10.1186/1472-6963-13-318

[8]   Deng, Y.Z., et al., IV tissue plasminogen activator use in acute stroke: experience from a statewide registry. Neurology, 2006. **66**(3): p. 306-12. https://doi.org/10.1212/01.wnl.0000196478.77152.fc

[9]   Wojner-Alexandrov, A.W., et al., Houston paramedic and emergency stroke treatment and outcomes study (HoPSTO). Stroke, 2005. **36**(7): p. 1512-8. https://doi.org/10.1161/01.STR.0000170700.45340.39

[10]   Zhai, S., et al., The Cost-Effectiveness of a Stroke Unit in Providing Enhanced Patient Outcomes in an Australian Teaching Hospital. J Stroke Cerebrovasc Dis, 2017. https://doi.org/10.1016/j.jstrokecerebrovasdis.2017.05.025

Organic Bioelectronics for Life Science and Healthcare   Materials Research Forum LLC
Materials Research Foundations **56** (2019) 263-314   doi: https://doi.org/10.21741/9781644900376-8

[11]   Miller, E.C., C. Blum, and S.K. Rostanski, *Developing a Stroke Center.* Stroke, 2017. **48**(7): p. e155-e156. https://doi.org/10.1161/STROKEAHA.117.017745

[12]   E., F.-L., Stroke prevention and care in France: report for the Minister of Health and Sport. 2009.

[13]   Weber, J.E., et al., Prehospital thrombolysis in acute stroke: results of the PHANTOM-S pilot study. Neurology, 2013. **80**(2): p. 163-8. https://doi.org/10.1212/WNL.0b013e31827b90e5

[14]   Handschu, R., et al., Telemedicine in emergency evaluation of acute stroke: interrater agreement in remote video examination with a novel multimedia system. Stroke, 2003. **34**(12): p. 2842-6. https://doi.org/10.1161/01.STR.0000102043.70312.E9

[15]   Demaerschalk, B.M., et al., Smartphone teleradiology application is successfully incorporated into a telestroke network environment. Stroke, 2012. **43**(11): p. 3098-101. https://doi.org/10.1161/STROKEAHA.112.669325

[16]   Audebert, H.J., et al., Prehospital stroke care: new prospects for treatment and clinical research. Neurology, 2013. **81**(5): p. 501-8. https://doi.org/10.1212/WNL.0b013e31829e0fdd

[17]   Katz, B.S., et al., Estimated Impact of Emergency Medical Service Triage of Stroke Patients on Comprehensive Stroke Centers: An Urban Population-Based Study. Stroke, 2017. https://doi.org/10.1161/STROKEAHA.116.015971

[18]   Waqar Faiz, K., et al., *Prehospital path in acute stroke.* Tidsskr Nor Laegeforen, 2017. **137**(11): p. 798-802. https://doi.org/10.4045/tidsskr.16.0512

[19]   Brott, T., et al., Measurements of acute cerebral infarction: a clinical examination scale. Stroke, 1989. **20**(7): p. 864-70. https://doi.org/10.1161/01.STR.20.7.864

[20]   Kidwell, C.S., et al., Comparison of MRI and CT for detection of acute intracerebral hemorrhage. Jama, 2004. **292**(15): p. 1823-30. https://doi.org/10.1001/jama.292.15.1823

[21]   Assis, Z.A., B.K. Menon, and M. Goyal, Imaging department organization in a stroke center and workflow processes in acute stroke. Eur J Radiol, 2017. https://doi.org/10.1016/j.ejrad.2017.06.014

[22]   Ferro, J.M., A.R. Massaro, and J.L. Mas, *Aetiological diagnosis of ischaemic stroke in young adults.* Lancet Neurol, 2010. **9**(11): p. 1085-96. https://doi.org/10.1016/S1474-4422(10)70251-9

[23]   Singh, T.P., J.R. Weinstein, and S.P. Murphy, *Stroke: Basic and Clinical.* Adv Neurobiol, 2017. **15**: p. 281-293. https://doi.org/10.1007/978-3-319-57193-5_10

[24] Patel, R.A.G. and C.J. White, *Stroke Treatment and Prevention.* Prog Cardiovasc Dis, 2017. **59**(6): p. 525-526. https://doi.org/10.1016/j.pcad.2017.05.006

[25] Tsivgoulis, G., O. Kargiotis, and A.V. Alexandrov, *Intravenous thrombolysis for acute ischemic stroke: a bridge between two centuries.* Expert Rev Neurother, 2017. **17**(8): p. 819-837. https://doi.org/10.1080/14737175.2017.1347039

[26] Glickman, S.W., et al., Discriminative capacity of biomarkers for acute stroke in the emergency department. J Emerg Med, 2011. **41**(3): p. 333-9. https://doi.org/10.1016/j.jemermed.2010.02.025

[27] Group, T.N.I.o.N.D.a.S.r.-P.S.S., *Tissue plasminogen activator for acute ischemic stroke.* N Engl J Med, 1995. **333**(24): p. 1581-7. https://doi.org/10.1056/NEJM199512143332401

[28] Adams, H.P., Jr., et al., Guidelines for the early management of adults with ischemic stroke: a guideline from the American Heart Association/American Stroke Association Stroke Council, Clinical Cardiology Council, Cardiovascular Radiology and Intervention Council, and the Atherosclerotic Peripheral Vascular Disease and Quality of Care Outcomes in Research Interdisciplinary Working Groups: The American Academy of Neurology affirms the value of this guideline as an educational tool for neurologists. Circulation, 2007. **115**(20): p. e478-534.

[29] Radu, R.A., et al., *Etiologic classification of ischemic stroke: Where do we stand?* Clin Neurol Neurosurg, 2017. **159**: p. 93-106. https://doi.org/10.1016/j.clineuro.2017.05.019

[30] Amarenco, P., et al., *Classification of stroke subtypes.* Cerebrovasc Dis, 2009. **27**(5): p. 493-501. https://doi.org/10.1159/000210432

[31] Ingelheim, B., 2015.

[32] Lee, L.J., et al., Impact on stroke subtype diagnosis of early diffusion-weighted magnetic resonance imaging and magnetic resonance angiography. Stroke, 2000. **31**(5): p. 1081-9. https://doi.org/10.1161/01.STR.31.5.1081

[33] Adams, H.P., Jr., et al., Classification of subtype of acute ischemic stroke. Definitions for use in a multicenter clinical trial. TOAST. Trial of Org 10172 in Acute Stroke Treatment. Stroke, 1993. **24**(1): p. 35-41. https://doi.org/10.1161/01.STR.24.1.35

[34] Landau, W.M. and A. Nassief, *Editorial comment--time to burn the TOAST.* Stroke, 2005. **36**(4): p. 902-4. https://doi.org/10.1161/str.36.4.902

[35] Amarenco, P., Patent foramen ovale and the risk of stroke: smoking gun guilty by association? Heart, 2005. **91**(4): p. 441-3. https://doi.org/10.1136/hrt.2004.052241

Organic Bioelectronics for Life Science and Healthcare   Materials Research Forum LLC
Materials Research Foundations **56** (2019) 263-314   doi: https://doi.org/10.21741/9781644900376-8

[36]   Gokcal, E., E. Niftaliyev, and T. Asil, Etiological classification of ischemic stroke in young patients: a comparative study of TOAST, CCS, and ASCO. Acta Neurol Belg, 2017. https://doi.org/10.1007/s13760-017-0813-8

[37]   Sacco, R.L., et al., *Infarcts of undetermined cause: the NINCDS Stroke Data Bank.* Ann Neurol, 1989. **25**(4): p. 382-90. https://doi.org/10.1002/ana.410250410

[38]   Bamford, J., et al., Classification and natural history of clinically identifiable subtypes of cerebral infarction. Lancet, 1991. **337**(8756): p. 1521-6. https://doi.org/10.1016/0140-6736(91)93206-O

[39]   Lindley, R.I., et al., Interobserver reliability of a clinical classification of acute cerebral infarction. Stroke, 1993. **24**(12): p. 1801-4. https://doi.org/10.1161/01.STR.24.12.1801

[40]   Thomas S. Maldonado, M.D., and Thomas S. Riles, M.D., F.A.C.S, *Stroke and Transient Ischemic Attack - 1 1 STROKE AND TRANSIENT ISCHEMIC ATTACK.* © WebMD, Inc. All rights reserved. , 2007. **6 VASCULAR SYSTEM 1**(ACS Surgery: Principles and Practice).

[41]   Go, A.S., et al., Heart disease and stroke statistics--2013 update: a report from the American Heart Association. Circulation, 2013. **127**(1): p. e6-e245.

[42]   Teo, K. and J. Slark, A systematic review of studies investigating the care of stroke survivors in long-term care facilities. Disabil Rehabil, 2015: p. 1-9.

[43]   Chollet, F. and J.F. Albucher, *Strategies to augment recovery after stroke.* Curr Treat Options Neurol, 2012. **14**(6): p. 531-40. https://doi.org/10.1007/s11940-012-0196-3

[44]   Lees, K.R., et al., Time to treatment with intravenous alteplase and outcome in stroke: an updated pooled analysis of ECASS, ATLANTIS, NINDS, and EPITHET trials. Lancet, 2010. **375**(9727): p. 1695-703. https://doi.org/10.1016/S0140-6736(10)60491-6

[45]   Siegel, J., et al., *Update on Neurocritical Care of Stroke.* Curr Cardiol Rep, 2017. **19**(8): p. 67. https://doi.org/10.1007/s11886-017-0881-7

[46]   Khaku, A.D., S., *Stroke*, in *StatPearls*. 2017, StatPearls Publishing LLC.: Treasure Island (FL).

[47]   Nolte, C.H. and H.J. Audebert, *[Management of acute ischemic stroke].* Dtsch Med Wochenschr, 2015. **140**(21): p. 1583-6. https://doi.org/10.1055/s-0041-106309

[48]   Hacke, W., et al., Thrombolysis with alteplase 3 to 4.5 hours after acute ischemic stroke. N Engl J Med, 2008. **359**(13): p. 1317-29. https://doi.org/10.1056/NEJMoa0804656

Organic Bioelectronics for Life Science and Healthcare   Materials Research Forum LLC
Materials Research Foundations **56** (2019) 263-314   doi: https://doi.org/10.21741/9781644900376-8

[49]   Sandercock, P., et al., The benefits and harms of intravenous thrombolysis with recombinant tissue plasminogen activator within 6 h of acute ischaemic stroke (the third international stroke trial [IST-3]): a randomised controlled trial. Lancet, 2012. **379**(9834): p. 2352-63. https://doi.org/10.1016/S0140-6736(12)60768-5

[50]   Leys, D., et al., Facilities available in European hospitals treating stroke patients. Stroke, 2007. **38**(11): p. 2985-91. https://doi.org/10.1161/STROKEAHA.107.487967

[51]   Evenson, K.R., W.D. Rosamond, and D.L. Morris, *Prehospital and in-hospital delays in acute stroke care.* Neuroepidemiology, 2001. **20**(2): p. 65-76. https://doi.org/10.1159/000054763

[52]   Tai, Y.J. and B. Yan, Minimising time to treatment: targeted strategies to minimise time to thrombolysis for acute ischaemic stroke. Intern Med J, 2013. **43**(11): p. 1176-82. https://doi.org/10.1111/imj.12204

[53]   Bustamante, A., et al., Ischemic stroke outcome: A review of the influence of post-stroke complications within the different scenarios of stroke care. Eur J Intern Med, 2015. https://doi.org/10.1016/j.ejim.2015.11.030

[54]   McMullan, J.T., et al., Time-critical neurological emergencies: the unfulfilled role for point-of-care testing. Int J Emerg Med, 2010. **3**(2): p. 127-31. https://doi.org/10.1007/s12245-010-0177-9

[55]   Rooney, K.D. and U.M. Schilling, Point-of-care testing in the overcrowded emergency department – can it make a difference? Critical Care, 2014. **18**(6): p. 692. https://doi.org/10.1186/s13054-014-0692-9

[56]   Ng, G.J., et al., Stroke biomarkers in clinical practice: A critical appraisal. Neurochem Int, 2017. https://doi.org/10.1016/j.neuint.2017.01.005

[57]   Yokobori, S., et al., Biomarkers for the clinical differential diagnosis in traumatic brain injury--a systematic review. CNS Neurosci Ther, 2013. **19**(8): p. 556-65. https://doi.org/10.1111/cns.12127

[58]   Cata, J.P., B. Abdelmalak, and E. Farag, *Neurological biomarkers in the perioperative period.* Br J Anaesth, 2011. **107**(6): p. 844-58. https://doi.org/10.1093/bja/aer338

[59]   AP., D., Biomarkers: coming of age for environmental health and risk assessment. Environ Sci Technol, 1997. **31: 1837-48**. https://doi.org/10.1021/es960920a

[60]   Cummins, B.M., F.S. Ligler, and G.M. Walker, *Point-of-care diagnostics for niche applications.* Biotechnol Adv, 2016. **34**(3): p. 161-76. https://doi.org/10.1016/j.biotechadv.2016.01.005

[61]   Vasan, A.S., et al., *Point-of-care biosensor system.* Front Biosci (Schol Ed), 2013. **5**: p. 39-71. https://doi.org/10.2741/S357

[62]　Price, C.P., *Point of care testing.* BMJ : British Medical Journal, 2001. **322**(7297): p. 1285-1288. https://doi.org/10.1136/bmj.322.7297.1285

[63]　St John, A. and C.P. Price, *Existing and Emerging Technologies for Point-of-Care Testing.* The Clinical Biochemist Reviews, 2014. **35**(3): p. 155-167.

[64]　Yoo, E.H. and S.Y. Lee, *Glucose biosensors: an overview of use in clinical practice.* Sensors (Basel), 2010. **10**(5): p. 4558-76. https://doi.org/10.3390/s100504558

[65]　Shah, D. and D. Maghsoudlou, *Enzyme-linked immunosorbent assay (ELISA): the basics.* British Journal of Hospital Medicine, 2016. **77**(7): p. C98-C101. https://doi.org/10.12968/hmed.2016.77.7.C98

[66]　Drain, P.K., et al., *Diagnostic point-of-care tests in resource-limited settings.* Lancet Infect Dis, 2014. **14**(3): p. 239-49. https://doi.org/10.1016/S1473-3099(13)70250-0

[67]　Kumar, S., et al., Microfluidic-integrated biosensors: prospects for point-of-care diagnostics. Biotechnol J, 2013. **8**(11): p. 1267-79. https://doi.org/10.1002/biot.201200386

[68]　Eltzov, E., A. Cohen, and R.S. Marks, Bioluminescent liquid light guide pad biosensor for indoor air toxicity monitoring. Anal Chem, 2015. **87**(7): p. 3655-61. https://doi.org/10.1021/ac5038208

[69]　Eltzov, E., et al., Lateral Flow Immunoassays – from Paper Strip to Smartphone Technology. Electroanalysis, 2015. **27**(9): p. 2116-2130. https://doi.org/10.1002/elan.201500237

[70]　Eltzov, E. and R.S. Marks, *Colorimetric stack pad immunoassay for bacterial identification.* Biosens Bioelectron, 2016. **87**: p. 572-578. https://doi.org/10.1016/j.bios.2016.08.044

[71]　Eltzov, E. and R.S. Marks, Miniaturized Flow Stacked Immunoassay for Detecting Escherichia coli in a Single Step. Anal Chem, 2016. **88**(12): p. 6441-9. https://doi.org/10.1021/acs.analchem.6b01034

[72]　Algaar, F., et al., Fiber-optic immunosensor for detection of Crimean-Congo hemorrhagic fever IgG antibodies in patients. Anal Chem, 2015. **87**(16): p. 8394-8. https://doi.org/10.1021/acs.analchem.5b01728

[73]　Eltzov, E., S. Cosnier, and R.S. Marks, Biosensors based on combined optical and electrochemical transduction for molecular diagnostics. Expert Rev Mol Diagn, 2011. **11**(5): p. 533-46. https://doi.org/10.1586/erm.11.38

[74]　Eltzov, E. and R.S. Marks, *Fiber-optic based cell sensors.* Adv Biochem Eng Biotechnol, 2010. **117**: p. 131-54. https://doi.org/10.1007/10_2009_6

[75]   Thevenot, D.R., et al., *Electrochemical biosensors: recommended definitions and classification.* Biosensors & Bioelectronics, 2001. **16**(1-2): p. 121-131. https://doi.org/10.1016/S0956-5663(01)00115-4

[76]   Eltzov, E., S. Cosnier, and R.S. Marks, *Biosensors based on combined optical and electrochemical transduction for molecular diagnostics.* Expert Review of Molecular Diagnostics, 2011. **11**(5): p. 533-546. https://doi.org/10.1586/erm.11.38

[77]   Eltzov, E. and R.S. Marks, *Whole-cell aquatic biosensors.* Analytical and Bioanalytical Chemistry, 2011. **400**(4): p. 895-913. https://doi.org/10.1007/s00216-010-4084-y

[78]   Collings, A.F. and F. Caruso, *Biosensors: recent advances.* Reports on Progress in Physics, 1997. **60**(11): p. 1397-1445. https://doi.org/10.1088/0034-4885/60/11/005

[79]   Eltzov, E., A. Kushmaro, and R.S. Marks, *Biosensors and related techniques for endocrine disruptors*, in *Endocrine disrupting chemicals in food*, I. Snow, Editor. 2009, Woodhead Publishing: Cambridge, UK. https://doi.org/10.1201/9781439829158.ch8

[80]   Marks, R.S., et al., *Handbook of biosensors and biochips.* 2007: John Wiley & Sons, Ltd.

[81]   Chin, C.D., V. Linder, and S.K. Sia, *Commercialization of microfluidic point-of-care diagnostic devices.* Lab on a Chip, 2012. **12**(12): p. 2118-2134. https://doi.org/10.1039/c2lc21204h

[82]   Walter, S., et al., Point-of-care laboratory halves door-to-therapy-decision time in acute stroke. Ann Neurol, 2011. **69**. https://doi.org/10.1002/ana.22355

[83]   Karlinski, M., M. Gluszkiewicz, and A. Czlonkowska, The accuracy of prehospital diagnosis of acute cerebrovascular accidents: an observational study. Arch Med Sci, 2015. **11**(3): p. 530-5. https://doi.org/10.5114/aoms.2015.52355

[84]   Brandler, E.S., et al., Prehospital stroke scales in urban environments: a systematic review. Neurology, 2014. **82**(24): p. 2241-9. https://doi.org/10.1212/WNL.0000000000000523

[85]   Drescher, M.J., et al., Point-of-Care Testing for Coagulation Studies in an Emergency Department Stroke Protocol: A Time-Saving Innovation. Annals of Emergency Medicine, 2007. **50**(3): p. S25. https://doi.org/10.1016/j.annemergmed.2007.06.109

[86]   Rizos, T., et al., Point-of-Care International Normalized Ratio Testing Accelerates Thrombolysis in Patients With Acute Ischemic Stroke Using Oral Anticoagulants. Stroke, 2009. **40**(11): p. 3547-3551. https://doi.org/10.1161/STROKEAHA.109.562769

[87]   Fassbender, K., et al., Mobile stroke units for prehospital thrombolysis, triage, and beyond: benefits and challenges. Lancet Neurol, 2017. **16**(3): p. 227-237. https://doi.org/10.1016/S1474-4422(17)30008-X

[88]   Walter, S., et al., Diagnosis and treatment of patients with stroke in a mobile stroke unit versus in hospital: a randomised controlled trial. Lancet Neurol, 2012. **11**(5): p. 397-404. https://doi.org/10.1016/S1474-4422(12)70057-1

[89]   Fassbender, K., et al., *"Mobile stroke unit" for hyperacute stroke treatment.* Stroke, 2003. **34**(6): p. e44. https://doi.org/10.1161/01.STR.0000075573.22885.3B

[90]   Parker, S.A., et al., Establishing the first mobile stroke unit in the United States. Stroke, 2015. **46**(5): p. 1384-91. https://doi.org/10.1161/STROKEAHA.114.007993

[91]   Ebinger, M., et al., PHANTOM-S: the prehospital acute neurological therapy and optimization of medical care in stroke patients - study. Int J Stroke, 2012. **7**(4): p. 348-53. https://doi.org/10.1111/j.1747-4949.2011.00756.x

[92]   Wendt, M., et al., Improved prehospital triage of patients with stroke in a specialized stroke ambulance: results of the pre-hospital acute neurological therapy and optimization of medical care in stroke study. Stroke, 2015. **46**(3): p. 740-5. https://doi.org/10.1161/STROKEAHA.114.008159

[93]   Gomes, J.A., et al., Prehospital Reversal of Warfarin-Related Coagulopathy in Intracerebral Hemorrhage in a Mobile Stroke Treatment Unit. Stroke, 2015. **46**(5): p. e118-e120. https://doi.org/10.1161/STROKEAHA.115.008483

[94]   Samsung/Neurologica, *CereTom®.* http://www.neurologica.com/ceretom, 2018.

[95]   Siemens, *SOMATOM Scope* https://www.healthcare.siemens.com/computed-tomography/single-source-ct/somatom-scope, 2018.

[96]   Kraft, P., et al., Feasibility and diagnostic accuracy of point-of-care handheld echocardiography in acute ischemic stroke patients – a pilot study. BMC Neurology, 2017. **17**(1): p. 159. https://doi.org/10.1186/s12883-017-0937-8

[97]   Healthcare, G., *Vivid q®.* http://www3.gehealthcare.com.sg/en-gb/products/categories/ultrasound/vivid/vivid_q, 2018.

[98]   Nusa, D., et al., Assessment of point-of-care measurement of international normalised ratio using the CoaguChek XS Plus system in the setting of acute ischaemic stroke. Intern Med J, 2013. **43**(11): p. 1205-9. https://doi.org/10.1111/imj.12255

[99]   Green, T.L., et al., *Reliability of point-of-care testing of INR in acute stroke.* Can J Neurol Sci, 2008. **35**(3): p. 348-51. https://doi.org/10.1017/S0317167100008945

[100] Roche, *CoaguChek®* http://www.coaguchek.com/coaguchek_patient/landing, 2017.

[101] Thorne, K., H. McNaughton, and M. Weatherall, An audit of coagulation screening in patients presenting to the emergency department for potential stroke thrombolysis. Intern Med J, 2017. **47**(2): p. 189-193. https://doi.org/10.1111/imj.13323

[102] Zenlander, R., et al., *Point-of-care versus central laboratory testing of INR in acute stroke.* Acta Neurol Scand, 2018. **137**(2): p. 252-255. https://doi.org/10.1111/ane.12860

[103] Nick Dale, T.P., Diagnostics by Sarissa Biomedical, *SMARTChip.* School of Life Sciences, Gibbet Hill Campus, The University of Warwick, Coventry, CV4 7AL, 2016. **diagnostics@sarissa-biomedical.com.**

[104] Trust, U.H.C.a.W.N., *SMARTCap Stroke Study: A Field Deployable Blood Test for Stroke (SMARTCAP).* ClinicalTrials.gov Identifier: NCT02308605. https://clinicaltrials.gov/ct2/show/NCT02308605, 2015 December

[105] sponsor: University Hospitals Coventry and Warwickshire NHS Trust, c.U.H.o.N.M.N.T., *Validation of a Purine Biosensor in Detecting Acute Cerebral Ischaemia: Carotid Endarterectomy Model in SMARTChip (CEMS).* ClinicalTrials.gov Identifier: NCT02545166. https://clinicaltrials.gov/ct2/show/NCT02545166, 2017, July 6.

[106] Trust, U.H.C.a.W.N., *The SMARTChip Stroke Study.* ClinicalTrials.gov Identifier: NCT02795481. https://clinicaltrials.gov/ct2/show/NCT02795481, 2016 July.

[107] Inc., A.P.o.C., *i-STAT®.* https://www.pointofcare.abbott/int/en/home, 2017.

[108] Drescher, M.J., et al., Point-of-care testing for coagulation studies in a stroke protocol: a time-saving innovation. Am J Emerg Med, 2011. **29**(1): p. 82-5. https://doi.org/10.1016/j.ajem.2009.09.020

[109] Nanduri, S., et al., An analysis of discrepancy between point-of-care and central laboratory international normalized ratio testing in ED patients with cerebrovascular disease. Am J Emerg Med, 2012. **30**(9): p. 2025-9. https://doi.org/10.1016/j.ajem.2012.04.003

[110] Inc., A.P.o.C., 2016.

[111] Abbott, *i-STAT®.* https://www.pointofcare.abbott/int/en/offerings/istat/istat-handheld, 2017.

[112] Briggs, C., et al., Performance evaluation of a new compact hematology analyzer, the Sysmex pocH-100i. Lab Hematol, 2003. **9**(4): p. 225-33.

[113] Sysmex, *PocH-100iTM* http://www.sysmex-ap.com/products/diagnostics/hematology/poch-100i/, 2018.

[114] Diagnostics, I.A., *Hemochron®*. Accriva Diagnostics, representing ITC and Accumetrics. All rights reserved. Accriva is ISO 13485 certified. , © 2015 http://www.accriva.com/products#.

[115] Diagnostics, I.A., *Hemochron®* http://www.accriva.com/products/hemochron-signature-elite-whole-blood-microcoagulation-system, 2015.

[116] Ltd, R.D.I., *Reflotron® plus analyzer* http://www.cobas.com/home/product/point-of-care-testing/reflotron-plus-sprint-system.html, © 2017 .

[117] Roche, *Reflotron® plus analyzer.* http://www.cobas.com/home/product/point-of-care-testing/reflotron-plus-sprint-system.html, 2017.

[118] Wu, Z., et al., Experiences and the use of BNP POCT platform on suspected ischemic stroke patients in the emergency department setting. Clin Neurol Neurosurg, 2014. **123**: p. 199-200. https://doi.org/10.1016/j.clineuro.2014.04.033

[119] He, M., et al., A new algorithm of suspected stroke patient management with brain natriuretic peptide/N-terminal pro-brain natriuretic peptide point of care testing platform in the emergency department. Ann Indian Acad Neurol, 2017. **20**(1): p. 81-82. https://doi.org/10.4103/0972-2327.194316

[120] Wu, Z., et al., Validation of the use of B-type natriuretic peptide point-of-care test platform in preliminary recognition of cardioembolic stroke patients in the ED. Am J Emerg Med, 2015. **33**(4): p. 521-6. https://doi.org/10.1016/j.ajem.2015.01.013

[121] Nayer, J., P. Aggarwal, and S. Galwankar, Utility of point-of-care testing of natriuretic peptides (brain natriuretic peptide and n-terminal pro-brain natriuretic peptide) in the emergency department. International Journal of Critical Illness and Injury Science, 2014. **4**(3): p. 209-215. https://doi.org/10.4103/2229-5151.141406

[122] Gils, C., et al., NT-proBNP on Cobas h 232 in point-of-care testing: Performance in the primary health care versus in the hospital laboratory. Scandinavian Journal of Clinical and Laboratory Investigation, 2015. **75**(7): p. 602-609. https://doi.org/10.3109/00365513.2015.1066846

[123] Llombart, V., et al., B-type natriuretic peptides help in cardioembolic stroke diagnosis: pooled data meta-analysis. Stroke, 2015. **46**(5): p. 1187-95. https://doi.org/10.1161/STROKEAHA.114.008311

[124] Kawase, S., et al., Plasma Brain Natriuretic Peptide is a Marker of Prognostic Functional Outcome in Non-Cardioembolic Infarction. J Stroke Cerebrovasc Dis, 2015. **24**(10): p. 2285-90. https://doi.org/10.1016/j.jstrokecerebrovasdis.2015.06.006

[125] Clerico, A., et al., State of the art of immunoassay methods for B-type natriuretic peptides: An update. Crit Rev Clin Lab Sci, 2015. **52**(2): p. 56-69. https://doi.org/10.3109/10408363.2014.987720

[126] Chaudhuri, J.R., et al., Association of plasma brain natriuretic peptide levels in acute ischemic stroke subtypes and outcome. J Stroke Cerebrovasc Dis, 2015. **24**(2): p. 485-91. https://doi.org/10.1016/j.jstrokecerebrovasdis.2014.09.025

[127] Yang, H.L., et al., Predicting cardioembolic stroke with the B-type natriuretic peptide test: a systematic review and meta-analysis. J Stroke Cerebrovasc Dis, 2014. **23**(7): p. 1882-9. https://doi.org/10.1016/j.jstrokecerebrovasdis.2014.02.014

[128] Qihong, G., et al., Experiences and the use of BNP POCT platform on suspected stroke patients by a Chinese emergency department. Ann Indian Acad Neurol, 2014. **17**(2): p. 243-4. https://doi.org/10.4103/0972-2327.132670

[129] Maruyama, K., et al., *Brain natriuretic peptide in acute ischemic stroke.* J Stroke Cerebrovasc Dis, 2014. **23**(5): p. 967-72. https://doi.org/10.1016/j.jstrokecerebrovasdis.2013.08.003

[130] Kara, K., et al., B-type natriuretic peptide predicts stroke of presumable cardioembolic origin in addition to coronary artery calcification. Eur J Neurol, 2014. **21**(6): p. 914-21. https://doi.org/10.1111/ene.12411

[131] Balion, C., et al., *B-type natriuretic peptide-guided therapy: a systematic review.* Heart Fail Rev, 2014. **19**(4): p. 553-64. https://doi.org/10.1007/s10741-014-9451-x

[132] Sakai, K., et al., Brain natriuretic peptide as a predictor of cardioembolism in acute ischemic stroke patients: brain natriuretic peptide stroke prospective study. Eur Neurol, 2013. **69**(4): p. 246-51. https://doi.org/10.1159/000342887

[133] Hajsadeghi, S., et al., The diagnostic value of N-terminal pro-brain natriuretic peptide in differentiating cardioembolic ischemic stroke. J Stroke Cerebrovasc Dis, 2013. **22**(4): p. 554-60. https://doi.org/10.1016/j.jstrokecerebrovasdis.2013.01.012

[134] Cojocaru, I.M., et al., Could pro-BNP, uric acid, bilirubin, albumin and transferrin be used in making the distinction between stroke subtypes? Rom J Intern Med, 2013. **51**(3-4): p. 188-95.

[135] Shibazaki, K., et al., Plasma brain natriuretic peptide can be a biological marker to distinguish cardioembolic stroke from other stroke types in acute ischemic stroke. Intern Med, 2009. **48**(5): p. 259-64. https://doi.org/10.2169/internalmedicine.48.1475

[136] Garcia-Berrocoso, T., et al., B-type natriuretic peptides and mortality after stroke: a systematic review and meta-analysis. Neurology, 2013. **81**(23): p. 1976-85. https://doi.org/10.1212/01.wnl.0000436937.32410.32

[137] Jickling, G.C. and C. Foerch, *Predicting stroke mortality: BNP could it be?* Neurology, 2013. **81**(23): p. 1970-1. https://doi.org/10.1212/01.wnl.0000436949.75473.79

[138] Shibazaki, K., et al., Brain natriuretic peptide on admission as a biological marker of long-term mortality in ischemic stroke survivors. Eur Neurol, 2013. **70**(3-4): p. 218-24. https://doi.org/10.1159/000351777

[139] Chen, X., et al., The prognostic value of combined NT-pro-BNP levels and NIHSS scores in patients with acute ischemic stroke. Intern Med, 2012. **51**(20): p. 2887-92. https://doi.org/10.2169/internalmedicine.51.8027

[140] Montaner, J., et al., Brain natriuretic peptide is associated with worsening and mortality in acute stroke patients but adds no prognostic value to clinical predictors of outcome. Cerebrovasc Dis, 2012. **34**(3): p. 240-5. https://doi.org/10.1159/000341858

[141] Mäkikallio, A.M., et al., *Natriuretic Peptides and Mortality After Stroke.* Stroke, 2005. **36**(5): p. 1016-1020. https://doi.org/10.1161/01.STR.0000162751.54349.ae

[142] Shibazaki, K., et al., Plasma Brain Natriuretic Peptide as an Independent Predictor of In-Hospital Mortality after Acute Ischemic Stroke. Internal Medicine, 2009. **48**(18): p. 1601-1606. https://doi.org/10.2169/internalmedicine.48.2166

[143] Jensen, J.K., et al., Usefulness of Natriuretic Peptide Testing for Long-Term Risk Assessment Following Acute Ischemic Stroke. American Journal of Cardiology, 2009. **104**(2): p. 287-291. https://doi.org/10.1016/j.amjcard.2009.03.029

[144] Shibazaki, K., et al., Brain natriuretic peptide level on admission predicts recurrent stroke after discharge in stroke survivors with atrial fibrillation. Clin Neurol Neurosurg, 2014. **127**: p. 25-9. https://doi.org/10.1016/j.clineuro.2014.09.028

[145] Shibazaki, K., et al., Plasma brain natriuretic Peptide as a predictive marker of early recurrent stroke in cardioembolic stroke patients. J Stroke Cerebrovasc Dis, 2014. **23**(10): p. 2635-40. https://doi.org/10.1016/j.jstrokecerebrovasdis.2014.06.003

[146] Mortezabeigi, H.R., et al., *ABCD2 score and BNP level in patients with TIA and cerebral stroke.* Pak J Biol Sci, 2013. **16**(21): p. 1393-7. https://doi.org/10.3923/pjbs.2013.1393.1397

[147] Roche, *Cobas® h 232* http://www.cobas.com/home/product/point-of-care-testing/cobas-h-232.html, 2017.

[148] Alere, *Triage® BNP Test* https://www.alere.com/kr/en/product-details/triage-bnp-test.html, 2018.

[149] Cohen, R., et al., Use of Tethered Enzymes as a Platform Technology for Rapid Analyte Detection. PLoS ONE, 2015. **10**(11): p. e0142326. https://doi.org/10.1371/journal.pone.0142326

[150] Dash, P.K., et al., Biomarkers for the diagnosis, prognosis, and evaluation of treatment efficacy for traumatic brain injury. Neurotherapeutics, 2010. **7**(1): p. 100-14. https://doi.org/10.1016/j.nurt.2009.10.019

[151] Ahmad, O., J. Wardlaw, and W.N. Whiteley, Correlation of levels of neuronal and glial markers with radiological measures of infarct volume in ischaemic stroke: a systematic review. Cerebrovasc Dis, 2012. **33**(1): p. 47-54. https://doi.org/10.1159/000332810

[152] Wunderlich, M.T., et al., Neuron-specific enolase and tau protein as neurobiochemical markers of neuronal damage are related to early clinical course and long-term outcome in acute ischemic stroke. Clin Neurol Neurosurg, 2006. **108**(6): p. 558-63. https://doi.org/10.1016/j.clineuro.2005.12.006

[153] Wevers, R.A., A.A. Jacobs, and O.R. Hommes, A bioluminescent assay for enolase (EC 4.2.1.11) activity in human serum and cerebrospinal fluid. Clin Chim Acta, 1983. **135**(2): p. 159-68. https://doi.org/10.1016/0009-8981(83)90131-6

[154] Viallard, J.L., M.R. Murthy, and B. Dastugue, *An ultramicro bioluminescence assay of enolase: application to human cerebrospinal fluid.* Neurochem Res, 1985. **10**(12): p. 1555-66. https://doi.org/10.1007/BF00988598

[155] Wevers, R.A., A.W. Theunisse, and G. Rijksen, An immunobioluminescence assay for gamma-gamma enolase activity in human serum and cerebrospinal fluid. Clin Chim Acta, 1988. **178**(2): p. 141-50. https://doi.org/10.1016/0009-8981(88)90220-3

[156] Harrison, P., et al., *Screening for Aspirin Responsiveness After Transient Ischemic Attack and Stroke.* Comparison of 2 Point-of-Care Platelet Function Tests With Optical Aggregometry, 2005. **36**(5): p. 1001-1005. https://doi.org/10.1161/01.STR.0000162719.11058.bd

[157] Collaborative meta-analysis of randomised trials of antiplatelet therapy for prevention of death, myocardial infarction, and stroke in high risk patients. Bmj, 2002. **324**(7329): p. 71-86. https://doi.org/10.1136/bmj.324.7329.71

[158] Altman, R., et al., The antithrombotic profile of aspirin. Aspirin resistance, or simply failure? Thromb J, 2004. **2**(1): p. 1.

[159] Gum, P.A., et al., Profile and prevalence of aspirin resistance in patients with cardiovascular disease. Am J Cardiol, 2001. **88**(3): p. 230-5. https://doi.org/10.1016/S0002-9149(01)01631-9

[160] Helgason, C.M., et al., *Aspirin response and failure in cerebral infarction.* Stroke, 1993. **24**(3): p. 345-50. https://doi.org/10.1161/01.STR.24.3.345

[161] Howard, P.A., *Aspirin resistance.* Ann Pharmacother, 2002. **36**(10): p. 1620-4. https://doi.org/10.1345/aph.1C013

[162] McKee, S.A., D.C. Sane, and E.N. Deliargyris, Aspirin resistance in cardiovascular disease: a review of prevalence, mechanisms, and clinical significance. Thromb Haemost, 2002. **88**(5): p. 711-5. https://doi.org/10.1055/s-0037-1613290

[163] Patrono, C., Aspirin resistance: definition, mechanisms and clinical read-outs. J Thromb Haemost, 2003. **1**(8): p. 1710-3. https://doi.org/10.1046/j.1538-7836.2003.00284.x

[164] Hankey, G.J. and J.W. Eikelboom, *Aspirin resistance.* Bmj, 2004. **328**(7438): p. 477-9. https://doi.org/10.1136/bmj.328.7438.477

[165] Eikelboom, J.W. and G.J. Hankey, Failure of aspirin to prevent atherothrombosis: potential mechanisms and implications for clinical practice. Am J Cardiovasc Drugs, 2004. **4**(1): p. 57-67. https://doi.org/10.2165/00129784-200404010-00006

[166] Gum, P.A., et al., A prospective, blinded determination of the natural history of aspirin resistance among stable patients with cardiovascular disease. J Am Coll Cardiol, 2003. **41**(6): p. 961-5. https://doi.org/10.1016/S0735-1097(02)03014-0

[167] Eikelboom, J.W. and G.J. Hankey, *Aspirin resistance: a new independent predictor of vascular events?* J Am Coll Cardiol, 2003. **41**(6): p. 966-8. https://doi.org/10.1016/S0735-1097(02)03013-9

[168] Baigent, C., et al., Aspirin in the primary and secondary prevention of vascular disease: collaborative meta-analysis of individual participant data from randomised trials. Lancet, 2009. **373**(9678): p. 1849-60. https://doi.org/10.1016/S0140-6736(09)60503-1

[169] Rothwell, P.M., et al., Effects of aspirin on risk and severity of early recurrent stroke after transient ischaemic attack and ischaemic stroke: time-course analysis of randomised trials. Lancet, 2016. **388**(10042): p. 365-75. https://doi.org/10.1016/S0140-6736(16)30468-8

[170] Harrison, P., *Progress in the assessment of platelet function.* Br J Haematol, 2000. **111**(3): p. 733-44. https://doi.org/10.1046/j.1365-2141.2000.02269.x

[171] Rand, M.L., R. Leung, and M.A. Packham, *Platelet function assays.* Transfus Apher Sci, 2003. **28**(3): p. 307-17. https://doi.org/10.1016/S1473-0502(03)00050-8

[172] Pearson, C., et al., Utility of point of care assessment of platelet reactivity (using the PFA-100(R)) to aid in diagnosis of stroke. Am J Emerg Med, 2017. **35**(5): p. 802.e1-802.e5. https://doi.org/10.1016/j.ajem.2016.11.036

[173] Lordkipanidzé, M., et al., A comparison of six major platelet function tests to determine the prevalence of aspirin resistance in patients with stable coronary artery

disease. European Heart Journal, 2007. **28**(14): p. 1702-1708.
https://doi.org/10.1093/eurheartj/ehm226

[174] Diagnostics, A.A., *VerifyNow®.* http://www.accriva.com/products/verifynow-system-platelet-reactivity-test, 2015.

[175] Dade/Siemens, *PFA-100®*
https://usa.healthcare.siemens.com/hemostasis/systems/pfa-100/technical-specifications, 2018.

[176] SBIR*STTR, A.s.S.F.b.S., *Development of Stroke Point of Care Immunoassay for Cellular Fibronectin.* 2009(Agency Tracking Number: NS057921, Contract: 2R44NS057921-02A1).

[177] Bio, V., *ReST™, a rapid evaluation stroke triage test.* Eight Medical Center Drive | Morgantown, WV 26506 | 304-825-3131 2014. http://valtaribio.com/executive-overview/.

[178] Monbailliu, T., J. Goossens, and S. Hachimi-Idrissi, *Blood protein biomarkers as diagnostic tool for ischemic stroke: a systematic review.* Biomark Med, 2017. **11**(6): p. 503-512. https://doi.org/10.2217/bmm-2016-0232

[179] Herrmann, M. and H. Ehrenreich, Brain derived proteins as markers of acute stroke: their relation to pathophysiology, outcome prediction and neuroprotective drug monitoring. Restor Neurol Neurosci, 2003. **21**(3-4): p. 177-90.

[180] Herrmann, M., et al., Release of glial tissue-specific proteins after acute stroke: A comparative analysis of serum concentrations of protein S-100B and glial fibrillary acidic protein. Stroke, 2000. **31**(11): p. 2670-7.
https://doi.org/10.1161/01.STR.31.11.2670

[181] Wunderlich, M.T., et al., Early neurobehavioral outcome after stroke is related to release of neurobiochemical markers of brain damage. Stroke, 1999. **30**(6): p. 1190-5.
https://doi.org/10.1161/01.STR.30.6.1190

[182] Kapural, M., et al., Serum S-100beta as a possible marker of blood-brain barrier disruption. Brain Res, 2002. **940**(1-2): p. 102-4. https://doi.org/10.1016/S0006-8993(02)02586-6

[183] Gazzolo, D., et al., *Neuromarkers and unconventional biological fluids.* J Matern Fetal Neonatal Med, 2010. **23 Suppl 3**: p. 66-9.
https://doi.org/10.3109/14767058.2010.507960

[184] Mir, I.N. and L.F. Chalak, *Serum biomarkers to evaluate the integrity of the neurovascular unit.* Early Hum Dev, 2014. **90**(10): p. 707-11.
https://doi.org/10.1016/j.earlhumdev.2014.06.010

Materials Research Forum LLC
doi: https://doi.org/10.21741/9781644900376-8

[185] Brouns, R., et al., Neurobiochemical markers of brain damage in cerebrospinal fluid of acute ischemic stroke patients. Clin Chem, 2010. **56**(3): p. 451-8. https://doi.org/10.1373/clinchem.2009.134122

[186] Lian, T., et al., Identification of Site-Specific Stroke Biomarker Candidates by Laser Capture Microdissection and Labeled Reference Peptide. Int J Mol Sci, 2015. **16**(6): p. 13427-41. https://doi.org/10.3390/ijms160613427

[187] Pelinka, L.E., et al., GFAP versus S100B in serum after traumatic brain injury: relationship to brain damage and outcome. J Neurotrauma, 2004. **21**(11): p. 1553-61. https://doi.org/10.1089/neu.2004.21.1553

[188] Dvorak, F., et al., Characterisation of the diagnostic window of serum glial fibrillary acidic protein for the differentiation of intracerebral haemorrhage and ischaemic stroke. Cerebrovasc Dis, 2009. **27**(1): p. 37-41. https://doi.org/10.1159/000172632

[189] Yang, Z. and K.K. Wang, Glial fibrillary acidic protein: from intermediate filament assembly and gliosis to neurobiomarker. Trends Neurosci, 2015. **38**(6): p. 364-74. https://doi.org/10.1016/j.tins.2015.04.003

[190] Teunissen, C.E., C. Dijkstra, and C. Polman, *Biological markers in CSF and blood for axonal degeneration in multiple sclerosis.* Lancet Neurol, 2005. **4**(1): p. 32-41. https://doi.org/10.1016/S1474-4422(04)00964-0

[191] Shibata, D., et al., Myelin basic protein autoantibodies, white matter disease and stroke outcome. J Neuroimmunol, 2012. **252**(1-2): p. 106-12. https://doi.org/10.1016/j.jneuroim.2012.08.006

[192] Zierath, D., et al., *Promiscuity of autoimmune responses to MBP after stroke.* J Neuroimmunol, 2015. **285**: p. 101-5. https://doi.org/10.1016/j.jneuroim.2015.05.024

[193] Becker, K.J., et al., Antibodies to myelin basic protein are associated with cognitive decline after stroke. J Neuroimmunol, 2016. **295-296**: p. 9-11. https://doi.org/10.1016/j.jneuroim.2016.04.001

[194] Basile, A.M., et al., S-100 protein and neuron-specific enolase as markers of subclinical cerebral damage after cardiac surgery: preliminary observation of a 6-month follow-up study. Eur Neurol, 2001. **45**(3): p. 151-9. https://doi.org/10.1159/000052114

[195] Karkela, J., E. Bock, and S. Kaukinen, CSF and serum brain-specific creatine kinase isoenzyme (CK-BB), neuron-specific enolase (NSE) and neural cell adhesion molecule (NCAM) as prognostic markers for hypoxic brain injury after cardiac arrest in man. J Neurol Sci, 1993. **116**(1): p. 100-9. https://doi.org/10.1016/0022-510X(93)90095-G

Organic Bioelectronics for Life Science and Healthcare
Materials Research Foundations 56 (2019) 263-314

Materials Research Forum LLC
doi: https://doi.org/10.21741/9781644900376-8

[196] Liu, M.C., et al., Ubiquitin C-terminal hydrolase-L1 as a biomarker for ischemic and traumatic brain injury in rats. Eur J Neurosci, 2010. **31**(4): p. 722-32. https://doi.org/10.1111/j.1460-9568.2010.07097.x

[197] Papa, L., et al., Ubiquitin C-terminal hydrolase is a novel biomarker in humans for severe traumatic brain injury. Crit Care Med, 2010. **38**(1): p. 138-44. https://doi.org/10.1097/CCM.0b013e3181b788ab

[198] Ren, C., et al., Different expression of ubiquitin C-terminal hydrolase-L1 and alphaII-spectrin in ischemic and hemorrhagic stroke: Potential biomarkers in diagnosis. Brain Res, 2013. **1540**: p. 84-91. https://doi.org/10.1016/j.brainres.2013.09.051

[199] Tongaonkar, P., et al., Evidence for an interaction between ubiquitin-conjugating enzymes and the 26S proteasome. Mol Cell Biol, 2000. **20**(13): p. 4691-8. https://doi.org/10.1128/MCB.20.13.4691-4698.2000

[200] Vaagenes, P., et al., Enzyme level changes in the cerebrospinal fluid of patients with acute stroke. Arch Neurol, 1986. **43**(4): p. 357-62. https://doi.org/10.1001/archneur.1986.00520040043017

[201] Ingebrigtsen, T. and B. Romner, Biochemical serum markers for brain damage: a short review with emphasis on clinical utility in mild head injury. Restor Neurol Neurosci, 2003. **21**(3-4): p. 171-6.

[202] Atisha, D., et al., A prospective study in search of an optimal B-natriuretic peptide level to screen patients for cardiac dysfunction. Am Heart J, 2004. **148**(3): p. 518-23. https://doi.org/10.1016/j.ahj.2004.03.014

[203] Ioannou, A., et al., Biomarkers associated with stroke risk in atrial fibrillation. Curr Med Chem, 2017.

[204] Bustamante, A., et al., Blood Biomarkers for the Early Diagnosis of Stroke: The Stroke-Chip Study. Stroke, 2017. https://doi.org/10.1161/STROKEAHA.117.017076

[205] Castellanos, M., et al., Plasma metalloproteinase-9 concentration predicts hemorrhagic transformation in acute ischemic stroke. Stroke, 2003. **34**(1): p. 40-6. https://doi.org/10.1161/01.STR.0000046764.57344.31

[206] Sood, R., et al., Increased apparent diffusion coefficients on MRI linked with matrix metalloproteinases and edema in white matter after bilateral carotid artery occlusion in rats. J Cereb Blood Flow Metab, 2009. **29**(2): p. 308-16. https://doi.org/10.1038/jcbfm.2008.121

[207] Laskowitz, D.T., et al., Clinical usefulness of a biomarker-based diagnostic test for acute stroke: the Biomarker Rapid Assessment in Ischemic Injury (BRAIN) study. Stroke, 2009. **40**(1): p. 77-85. https://doi.org/10.1161/STROKEAHA.108.516377

[208] Zhang, X.Q., et al., Exploring the optimal operation time for patients with hypertensive intracerebral hemorrhage: tracking the expression and progress of cell apoptosis of prehematomal brain tissues. Chin Med J (Engl), 2010. **123**(10): p. 1246-50.

[209] Taurino, M., et al., Metalloproteinase expression in carotid plaque and its correlation with plasma levels before and after carotid endarterectomy. Vasc Endovascular Surg, 2007. **41**(6): p. 516-21. https://doi.org/10.1177/1538574407307405

[210] Tayal, V. and B.S. Kalra, *Cytokines and anti-cytokines as therapeutics--an update.* Eur J Pharmacol, 2008. **579**(1-3): p. 1-12. https://doi.org/10.1016/j.ejphar.2007.10.049

[211] Rostene, W., et al., Neurochemokines: a menage a trois providing new insights on the functions of chemokines in the central nervous system. J Neurochem, 2011. **118**(5): p. 680-94. https://doi.org/10.1111/j.1471-4159.2011.07371.x

[212] Tuttolomondo, A., et al., *Inflammation in ischemic stroke subtypes.* Curr Pharm Des, 2012. **18**(28): p. 4289-310. https://doi.org/10.2174/138161212802481200

[213] Licata, G., et al., Immuno-inflammatory activation in acute cardio-embolic strokes in comparison with other subtypes of ischaemic stroke. Thromb Haemost, 2009. **101**(5): p. 929-37. https://doi.org/10.1160/TH08-06-0375

[214] Abbott, N.J., L. Ronnback, and E. Hansson, *Astrocyte-endothelial interactions at the blood-brain barrier.* Nat Rev Neurosci, 2006. **7**(1): p. 41-53. https://doi.org/10.1038/nrn1824

[215] Bailey, S.L., et al., *Innate and adaptive immune responses of the central nervous system.* Crit Rev Immunol, 2006. **26**(2): p. 149-88. https://doi.org/10.1615/CritRevImmunol.v26.i2.40

[216] Vela, J.M., et al., Interleukin-1 regulates proliferation and differentiation of oligodendrocyte progenitor cells. Mol Cell Neurosci, 2002. **20**(3): p. 489-502. https://doi.org/10.1006/mcne.2002.1127

[217] Rodriguez-Yanez, M. and J. Castillo, *Role of inflammatory markers in brain ischemia.* Curr Opin Neurol, 2008. **21**(3): p. 353-7. https://doi.org/10.1097/WCO.0b013e3282ffafbf

[218] Gokhan, S., et al., Neutrophil lymphocyte ratios in stroke subtypes and transient ischemic attack. Eur Rev Med Pharmacol Sci, 2013. **17**(5): p. 653-7.

[219] Huang, G., et al., Significance of white blood cell count and its subtypes in patients with acute coronary syndrome. Eur J Clin Invest, 2009. **39**(5): p. 348-58. https://doi.org/10.1111/j.1365-2362.2009.02107.x

[220] Papa, A., et al., Predictive value of elevated neutrophil-lymphocyte ratio on cardiac mortality in patients with stable coronary artery disease. Clin Chim Acta, 2008. **395**(1-2): p. 27-31. https://doi.org/10.1016/j.cca.2008.04.019

[221] Cook, E.J., et al., Post-operative neutrophil-lymphocyte ratio predicts complications following colorectal surgery. Int J Surg, 2007. **5**(1): p. 27-30. https://doi.org/10.1016/j.ijsu.2006.05.013

[222] Karabinos, I., et al., Neutrophil count on admission predicts major in-hospital events in patients with a non-ST-segment elevation acute coronary syndrome. Clin Cardiol, 2009. **32**(10): p. 561-8. https://doi.org/10.1002/clc.20624

[223] Buck, B.H., et al., Early neutrophilia is associated with volume of ischemic tissue in acute stroke. Stroke, 2008. **39**(2): p. 355-60. https://doi.org/10.1161/STROKEAHA.107.490128

[224] Elkind, M.S., et al., Relative elevation in baseline leukocyte count predicts first cerebral infarction. Neurology, 2005. **64**(12): p. 2121-5. https://doi.org/10.1212/01.WNL.0000165989.12122.49

[225] Petzold, A., Neurofilament phosphoforms: surrogate markers for axonal injury, degeneration and loss. J Neurol Sci, 2005. **233**(1-2): p. 183-98. https://doi.org/10.1016/j.jns.2005.03.015

[226] Chen, X.H., et al., Evolution of neurofilament subtype accumulation in axons following diffuse brain injury in the pig. J Neuropathol Exp Neurol, 1999. **58**(6): p. 588-96. https://doi.org/10.1097/00005072-199906000-00003

[227] Jafari, S.S., et al., *Axonal cytoskeletal changes after non-disruptive axonal injury.* J Neurocytol, 1997. **26**(4): p. 207-21. https://doi.org/10.1023/A:1018588114648

[228] Ost, M., et al., Initial CSF total tau correlates with 1-year outcome in patients with traumatic brain injury. Neurology, 2006. **67**(9): p. 1600-4. https://doi.org/10.1212/01.wnl.0000242732.06714.0f

[229] Folkerts, M.M., et al., Disruption of MAP-2 immunostaining in rat hippocampus after traumatic brain injury. J Neurotrauma, 1998. **15**(5): p. 349-63. https://doi.org/10.1089/neu.1998.15.349

[230] Zemlan, F.P., et al., C-tau biomarker of neuronal damage in severe brain injured patients: association with elevated intracranial pressure and clinical outcome. Brain Res, 2002. **947**(1): p. 131-9. https://doi.org/10.1016/S0006-8993(02)02920-7

[231] Zemlan, F.P., J.J. Mulchahey, and G.A. Gudelsky, Quantification and localization of kainic acid-induced neurotoxicity employing a new biomarker of cell death: cleaved microtubule-associated protein-tau (C-tau). Neuroscience, 2003. **121**(2): p. 399-409. https://doi.org/10.1016/S0306-4522(03)00459-7

Organic Bioelectronics for Life Science and Healthcare   Materials Research Forum LLC
Materials Research Foundations 56 (2019) 263-314   doi: https://doi.org/10.21741/9781644900376-8

[232] Olmsted, J.B., *Microtubule-associated proteins.* Annu Rev Cell Biol, 1986. **2**: p. 421-57. https://doi.org/10.1146/annurev.cb.02.110186.002225

[233] Matus, A., Neurofilament protein phosphorylation--where, when and why. Trends Neurosci, 1988. **11**(7): p. 291-2. https://doi.org/10.1016/0166-2236(88)90086-0

[234] Garner, C.C., R.P. Tucker, and A. Matus, Selective localization of messenger RNA for cytoskeletal protein MAP2 in dendrites. Nature, 1988. **336**(6200): p. 674-7. https://doi.org/10.1038/336674a0

[235] Kobeissy, F.H., et al., Neuroproteomics and systems biology-based discovery of protein biomarkers for traumatic brain injury and clinical validation. Proteomics Clin Appl, 2008. **2**(10-11): p. 1467-83. https://doi.org/10.1002/prca.200800011

[236] Buki, A., et al., The role of calpain-mediated spectrin proteolysis in traumatically induced axonal injury. J Neuropathol Exp Neurol, 1999. **58**(4): p. 365-75. https://doi.org/10.1097/00005072-199904000-00007

[237] Reeves, T.M., et al., Proteolysis of submembrane cytoskeletal proteins ankyrin-G and alphaII-spectrin following diffuse brain injury: a role in white matter vulnerability at Nodes of Ranvier. Brain Pathol, 2010. **20**(6): p. 1055-68. https://doi.org/10.1111/j.1750-3639.2010.00412.x

[238] Wang, K.K., et al., Simultaneous degradation of alphaII- and betaII-spectrin by caspase 3 (CPP32) in apoptotic cells. J Biol Chem, 1998. **273**(35): p. 22490-7. https://doi.org/10.1074/jbc.273.35.22490

[239] Cox, C.D., et al., Dicyclomine, an M1 muscarinic antagonist, reduces biomarker levels, but not neuronal degeneration, in fluid percussion brain injury. J Neurotrauma, 2008. **25**(11): p. 1355-65. https://doi.org/10.1089/neu.2008.0671

[240] Adam, S.S., N.S. Key, and C.S. Greenberg, *D-dimer antigen: current concepts and future prospects.* Blood, 2009. **113**(13): p. 2878-87. https://doi.org/10.1182/blood-2008-06-165845

[241] Mai, H., et al., Clinical presentation and imaging characteristics of occult lung cancer associated ischemic stroke. J Clin Neurosci, 2015. **22**(2): p. 296-302. https://doi.org/10.1016/j.jocn.2014.05.039

[242] Kim, K. and J.H. Lee, Risk factors and biomarkers of ischemic stroke in cancer patients. J Stroke, 2014. **16**(2): p. 91-6. https://doi.org/10.5853/jos.2014.16.2.91

[243] Liu, L.B., et al., The role of hs-CRP, D-dimer and fibrinogen in differentiating etiological subtypes of ischemic stroke. PLoS One, 2015. **10**(2): p. e0118301. https://doi.org/10.1371/journal.pone.0118301

Organic Bioelectronics for Life Science and Healthcare   Materials Research Forum LLC
Materials Research Foundations **56** (2019) 263-314   doi: https://doi.org/10.21741/9781644900376-8

[244] Zi, W.J. and J. Shuai, Plasma D-dimer levels are associated with stroke subtypes and infarction volume in patients with acute ischemic stroke. PLoS One, 2014. **9**(1): p. e86465. https://doi.org/10.1371/journal.pone.0086465

[245] Zecca, B., et al., A bioclinical pattern for the early diagnosis of cardioembolic stroke. Emerg Med Int, 2014. **2014**: p. 242171. https://doi.org/10.1155/2014/242171

[246] Yuan, W. and Z.H. Shi, The relationship between plasma D-dimer levels and outcome of Chinese acute ischemic stroke patients in different stroke subtypes. J Neural Transm (Vienna), 2014. **121**(4): p. 409-13. https://doi.org/10.1007/s00702-013-1113-y

[247] Wiseman, S., et al., Blood markers of coagulation, fibrinolysis, endothelial dysfunction and inflammation in lacunar stroke versus non-lacunar stroke and non-stroke: systematic review and meta-analysis. Cerebrovasc Dis, 2014. **37**(1): p. 64-75. https://doi.org/10.1159/000356789

[248] Okazaki, T., et al., The ratio of D-dimer to brain natriuretic peptide may help to differentiate between cerebral infarction with and without acute aortic dissection. J Neurol Sci, 2014. **340**(1-2): p. 133-8. https://doi.org/10.1016/j.jns.2014.03.011

[249] Isenegger, J., et al., *D-dimers predict stroke subtype when assessed early.* Cerebrovasc Dis, 2010. **29**(1): p. 82-6. https://doi.org/10.1159/000256652

[250] Ilhan, D., et al., Evaluation of platelet activation, coagulation, and fibrinolytic activation in patients with symptomatic lacunar stroke. Neurologist, 2010. **16**(3): p. 188-91. https://doi.org/10.1097/NRL.0b013e318198d8bc

[251] Haapaniemi, E. and T. Tatlisumak, *Is D-dimer helpful in evaluating stroke patients? A systematic review.* Acta Neurol Scand, 2009. **119**(3): p. 141-50. https://doi.org/10.1111/j.1600-0404.2008.01081.x

[252] Montaner, J., et al., Etiologic diagnosis of ischemic stroke subtypes with plasma biomarkers. Stroke, 2008. **39**(8): p. 2280-7. https://doi.org/10.1161/STROKEAHA.107.505354

[253] Dougu, N., et al., Differential diagnosis of cerebral infarction using an algorithm combining atrial fibrillation and D-dimer level. Eur J Neurol, 2008. **15**(3): p. 295-300. https://doi.org/10.1111/j.1468-1331.2008.02063.x

[254] Squizzato, A., et al., D-dimer is not a long-term prognostic marker following acute cerebral ischemia. Blood Coagul Fibrinolysis, 2006. **17**(4): p. 303-6. https://doi.org/10.1097/01.mbc.0000224850.57872.d0

[255] Squizzato, A. and W. Ageno, D-dimer testing in ischemic stroke and cerebral sinus and venous thrombosis. Semin Vasc Med, 2005. **5**(4): p. 379-86. https://doi.org/10.1055/s-2005-922484

[256] Koch, H.J., et al., The relationship between plasma D-dimer concentrations and acute ischemic stroke subtypes. J Stroke Cerebrovasc Dis, 2005. **14**(2): p. 75-9. https://doi.org/10.1016/j.jstrokecerebrovasdis.2004.12.002

[257] Ageno, W., et al., Plasma measurement of D-dimer levels for the early diagnosis of ischemic stroke subtypes. Arch Intern Med, 2002. **162**(22): p. 2589-93. https://doi.org/10.1001/archinte.162.22.2589

[258] Takano, K., T. Yamaguchi, and K. Uchida, *Markers of a hypercoagulable state following acute ischemic stroke.* Stroke, 1992. **23**(2): p. 194-8. https://doi.org/10.1161/01.STR.23.2.194

[259] Elkind, M.S., et al., High-sensitivity C-reactive protein, lipoprotein-associated phospholipase A2, and outcome after ischemic stroke. Arch Intern Med, 2006. **166**(19): p. 2073-80. https://doi.org/10.1001/archinte.166.19.2073

[260] Muir, K.W., et al., *C-reactive protein and outcome after ischemic stroke.* Stroke, 1999. **30**(5): p. 981-5. https://doi.org/10.1161/01.STR.30.5.981

[261] Audebert, H.J., et al., Systemic inflammatory response depends on initial stroke severity but is attenuated by successful thrombolysis. Stroke, 2004. **35**(9): p. 2128-33. https://doi.org/10.1161/01.STR.0000137607.61697.77

[262] Suwanwela, N.C., A. Chutinet, and K. Phanthumchinda, Inflammatory markers and conventional atherosclerotic risk factors in acute ischemic stroke: comparative study between vascular disease subtypes. J Med Assoc Thai, 2006. **89**(12): p. 2021-7.

[263] Masotti, L., et al., Prognostic role of C-reactive protein in very old patients with acute ischaemic stroke. J Intern Med, 2005. **258**(2): p. 145-52. https://doi.org/10.1111/j.1365-2796.2005.01514.x

[264] Hirano, K., et al., Study of hemostatic biomarkers in acute ischemic stroke by clinical subtype. J Stroke Cerebrovasc Dis, 2012. **21**(5): p. 404-10. https://doi.org/10.1016/j.jstrokecerebrovasdis.2011.08.013

[265] Iskra, T., et al., [Hemostatic markers of endothelial injury in ischaemic stroke caused by large or small vessel disease]. Pol Merkur Lekarski, 2006. **21**(125): p. 429-33.

[266] Bang, O.Y., et al., Association of serum lipid indices with large artery atherosclerotic stroke. Neurology, 2008. **70**(11): p. 841-7. https://doi.org/10.1212/01.wnl.0000294323.48661.a9

[267] Slowik, A., et al., LDL phenotype B and other lipid abnormalities in patients with large vessel disease and small vessel disease. J Neurol Sci, 2003. **214**(1-2): p. 11-6. https://doi.org/10.1016/S0022-510X(03)00166-7

[268] Iskra, T., et al., *[LDL phenotype A and B in ischemic stroke].* Przegl Lek, 2002. **59**(1): p. 7-10.

[269] Zimmermann-Ivol, C.G., et al., Fatty acid binding protein as a serum marker for the early diagnosis of stroke: a pilot study. Mol Cell Proteomics, 2004. **3**(1): p. 66-72. https://doi.org/10.1074/mcp.M300066-MCP200

[270] Watson, M.A. and M.G. Scott, *Clinical utility of biochemical analysis of cerebrospinal fluid.* Clin Chem, 1995. **41**(3): p. 343-60.

[271] Sun, G.J., et al., Cerebrospinal Fluid Free Fatty Acid Levels Are Associated with Stroke Subtypes and Severity in Chinese Patients with Acute Ischemic Stroke. World Neurosurg, 2015. **84**(5): p. 1299-304. https://doi.org/10.1016/j.wneu.2015.06.006

[272] Zambrelli, E., et al., Apo(a) size in ischemic stroke: relation with subtype and severity on hospital admission. Neurology, 2005. **64**(8): p. 1366-70. https://doi.org/10.1212/01.WNL.0000158282.83369.1D

[273] Iskra, T., et al., [Lipoprotein (a) in stroke patients with large and small vessel disease]. Przegl Lek, 2002. **59**(11): p. 877-80.

[274] Saidi, S., et al., Association of apolipoprotein E gene polymorphism with ischemic stroke involving large-vessel disease and its relation to serum lipid levels. J Stroke Cerebrovasc Dis, 2007. **16**(4): p. 160-6. https://doi.org/10.1016/j.jstrokecerebrovasdis.2007.03.001

[275] Lampl, Y., et al., Cerebrospinal fluid lactate dehydrogenase levels in early stroke and transient ischemic attacks. Stroke, 1990. **21**(6): p. 854-7. https://doi.org/10.1161/01.STR.21.6.854

[276] Alvarez-Perez, F.J., M. Castelo-Branco, and J. Alvarez-Sabin, *Albumin level and stroke. Potential association between lower albumin level and cardioembolic aetiology.* Int J Neurosci, 2011. **121**(1): p. 25-32. https://doi.org/10.3109/00207454.2010.523134

[277] Xu, Y.Z., et al., Dynamic reduction of plasma decorin following ischemic stroke: a pilot study. Neurochem Res, 2012. **37**(9): p. 1843-8. https://doi.org/10.1007/s11064-012-0787-0

[278] Department, C.C.E.I.E.E., *Mobile Stroke Unit.* Copyright © All rights reserved, 2015.

Conclusions and Outlook

The race between optical detection principles (based on, e.g., fluorescence, surface plasmons, optical waveguides, etc.) and electrical/ electrochemical/ electronic concepts in bio-molecular diagnostics is not decided yet but goes on. Both scientific communities continue to offer application- relevant solutions for fast, multiplexed, simple and cheap detection of peptides, proteins, oligonucleotides, PCR amplicons, small molecules like disease markers, food toxins, flavors, odorants, etc. Most likely, the competition will never see a single winner that meets all needs because the different practical formats and boundary conditions for applications, as well as, market requirements may ask for specific and unique solutions that could be better achieved in one case by optics and in another situation by electronics.

Generally speaking and focussing on label-free sensing, the field of optical devices and systems for market applications in biosensing at the moment seems to be in the lead. Certain commercial instruments have set a standard in biosensing that acts as a benchmark for any other technology approaching the market.

However, the use and performance of these (mostly) surface-plasmon wave-based instruments show weaknesses and limitations for two particular scenarios: (i) for any surface-affinity reaction (between a receptor immobilized on the transducer surface and the analyte binding from the aqueous solution) that leads to only a minute change in the (optical) interfacial architecture the resulting change in the dispersion- relation (the energy-momentum relation) of the surface plasmon wave may be too small to be detected. This could be the result of a very dilute surface coverage of the bound analyte or of analyte molecules of interest that are simply too small to induce such a modification of the functional surface layer and, hence, of the dispersion curve. (ii) Any bioreaction of interest that is associated with an electronic response, e.g., a change in the redox-state of a protein, a ligand-gated ion channel opening, a whole action potential induced by a stimulus, etc., cannot be monitored directly by optical techniques.

For both scenarios electronic (or electrical/ electrochemical) detection principles offer for certain analytes promising alternative solutions. E.g., if the small molecular marker that needs to be detected carries an electrical charge the binding to a transducer surface-immobilized receptor leads to a significant accumulation of interfacial charges that can be monitored by electrochemical or electronic methods. The analyte- (ligand-) gated opening and closing of membrane- integral ion channels can be read by impedance spectroscopy or by a transistor device that translates the change in membrane resistance into a change of the gate voltage controlling the source-drain current.

Given the fact that electrochemical/ electronic sensing started with the pioneering work on electrochemical biosensors by Leland C. Clark Jr., the "father of biosensors", already in the fifties and sixties and by Piet Bergveld, the inventor of the ion-sensitive field effect transistor (IS-FET) in the 1970s, it is somewhat surprising that the field has not developed more dynamically and has not reached - in practical applications - a level competitive to optical transduction principles.

Recently, however, a new momentum in this research could be observed: progress in organic electronics and its promises for cheap, printable, flexible, disposable, and wearable electronic devices has spurred a new interest in using these modules also for biosensing applications. This trend is summarized in the first Section of this book with descriptions of how an organic FET with a classical back-gate can be modified by a functional bio interlayer so as to act as a biosensor; with a chapter on a liquid-gated OFETs, an architecture in which the gate electrode is functionalized by the biorecognition element that induces a change in the gate voltage upon the binding of the analyte; and a special transistor device architecture based on a floating gate configuration with a specific bio-sensing area next to the control gate.

These more classical approaches are complemented in Section 2 by introducing electronic devices based on novel carbon materials; in particular, applying graphene as the semiconducting material that offers new and exciting opportunities also for biosensing. And finally, the last chapters presented in Section 3 give some details of a very economic fabrication protocol based on inkjet printing and a chapter on specific applications in a point-of-care scenario.

With this selection of chapters, we hope to give a good overview of the current status on organic bioelectronics for life science and healthcare - but what's next?

What we see are three major trends for research and development and challenges for the scientific and engineering communities:

(i) From a fundamental point of view the current understanding of the performance parameters of electronic sensing devices and their coupling to the molecular processes of the bioaffinity reactions is far from satisfactory and not acceptable. The understanding of essential elementary steps of the interaction of molecules and charges at the sensor surface and how they modify the electronic characteristics of the electronic device is still missing and many questions remain unanswered: E.g., which are the dominant factors that need to be optimized for a sensitive biosensing platform? Is it the change in the capacity of the interfacial layer that is modified upon analyte binding? Is it the charge (re-) distribution that counts? Or the modification of dipole layers? Or is the main influence of the ligand binding seen as a change in the carrier mobility in the channel of the transistor? A lot more research is needed before we have the answers that will allow for a better translation of the basics into useful applications.

(ii) As mentioned above, the main field of applications where electronic sensing may outrival optical techniques is the huge area of small molecule sensing. This has been and still is a notoriously difficult task for optical sensors. Obviously, if the analyte carries charges like, e.g., miRNA electronic sensing is the method of choice. But even in cases where the analyte is of low molar mass, has no charge nor a dipole, and is too small to be labelled with a chromophore, a label-free optical detection concept is extremely tedious. However, with a suitable receptor one might still have a good detection signal provided the binding of the analyte induces a change in the electronic configuration of the receptor that can couple to the electrical circuitry of the sensing device, e.g., modifies the source-drain current of a transistor. A very promising example in this respect is the monitoring of odorant molecules. These are typically small molecules, but if recognized by surface-immobilized odorant binding proteins induce a change in the charge- and dipole-architecture of this receptor that modifies the current through a FET device. Similarly, promising results have been reported for the monitoring of food toxins, like mycotoxins, that are also very small units, difficult to detect by their mere change of the (optical) mass of the sensor surface architecture upon binding. But binding of these toxins to antibodies immobilized on a transistor surface gave rise to very sensitive electrical signals.

(iii) And finally, future needs for sensors able to monitor simultaneously a whole multitude of analytes will favor electronic concepts as multiplexing can be more easily achieved by an array of electronic sensors, with each sensor element being highly integrated in a chip and easily read by algorithms that apply methods of artificial intelligence. This way, one will be able to deal with the huge amounts of data that will be produced once the internet of things will ask for and will eventually have sensors everywhere. The historic example of the development of microelectronics has given evidence of what device integration is able to achieve in this context.

317

Keyword Index

About the Editors

Dr. Akio Yasuda is currently a consultant for medical, healthcare, materials science, global open innovation in industry-academia collaboration.

He had been leading materials-based research at Sony worldwide for more than 35 years in the field of Medical Electronics, Organic bioelectronics, Nano-Bio fusion domain, Nanotechnology, Molecular Engineering, Spectroscopy (IR, Raman), Functional Organic Molecules, Electrochemistry, Display Materials and Technologies (Liquid Crystals, Electrochromics, OLEDs), Batteries. He has published more than 130 peer reviewed scientific papers in the international journals and has more than 200 patents.

He established and operated as a director Materials Science Laboratories of Sony Europe GmbH in Germany, Life Science Laboratory of Sony Corporation in Japan and Medical Electronics Laboratory of Sony Asia Pacific in Singapore.

He is currently also a Guest Professor at Tokyo Medical and Dental University (2009～), Guest Professor at The University of Natural Resources and Life Sciences Vienna, Austria (2010～), Guest Professor at Cyber University (online university, Japan), Senior Industrial Advisor at Austrian Institute of Technology in Austria (2009～), Advisory Board Member of the Center for Biomimetic Sensor and Science, Nanyang Technological University in Singapore(2011～), Executive Board Member of the Society of the Future Medicine (2009～).

Dr. Wolfgang Knoll earned a PhD degree in Biophysics from the University of Konstanz in 1976. From 1991-1999 he was the laboratory director for Exotic Nanomaterials in Wako, Japan, at the Institute of Physical and Chemical Research (RIKEN). From 1993 to 2008, he was furthermore Director of the Materials Science Department at the Max Planck Institute for Polymer Research in Mainz, Germany. Since April 1, 2008 he is the Scientific Managing Director of the AIT Austrian Institute of Technology. Since 2010 he is a Regular Member of the Austrian Academy of Sciences, received in 2012 an Honorary Doctorate from the University of Twente, the Netherlands, and became a member of the Academia Europaea in 2017.

www.ingramcontent.com/pod-product-compliance
Lightning Source LLC
Chambersburg PA
CBHW071325210326
41597CB00015B/1352